Springer Tracts in Advanced Robotics

Volume 44

Editors: Bruno Siciliano · Oussama Khatib · Frans Groen

Herman Bruyninckx, Libor Přeučil,
Miroslav Kulich (Eds.)

European Robotics Symposium 2008

 Springer

Professor Bruno Siciliano, Dipartimento di Informatica e Sistemistica, Università di Napoli Federico II, Via Claudio 21, 80125 Napoli, Italy, E-mail: siciliano@unina.it

Professor Oussama Khatib, Robotics Laboratory, Department of Computer Science, Stanford University, Stanford, CA 94305-9010, USA, E-mail: khatib@cs.stanford.edu

Professor Frans Groen, Department of Computer Science, Universiteit van Amsterdam, Kruislaan 403, 1098 SJ Amsterdam, The Netherlands, E-mail: groen@science.uva.nl

Editors

Dr. ir. Herman Bruyninckx
Katholieke Universiteit Leuven
Department of Mechanical Engineering
Division of Production Engineering
Machine Design & Automation (PMA)
Celestijnenlaan 300B
B-3001 Leuven (Heverlee)
Belgium
Email: Herman.Bruyninckx@mech.kuleuven.be

Dr. Miroslav Kulich
Department of Cybernetics
Faculty of Electrical Engineering
Czech Technical University
Technická 2
166 27 Prague 6
Czech Republic
E-mail: kulich@labe.felk.cvut.cz

Dr. Libor Přeučil
Department of Cybernetics
Faculty of Electrical Engineering
Czech Technical University
Technická 2
166 27 Prague 6
Czech Republic
E-mail: preucil@labe.felk.cvut.cz

ISBN 978-3-540-78315-2   e-ISBN 978-3-540-78317-6

DOI 10.1007/978-3-540-78317-6

Springer Tracts in Advanced Robotics   ISSN 1610-7438

Library of Congress Control Number: 2008921244

©2008 Springer-Verlag Berlin Heidelberg

Typesetting by the authors and Scientific Publishing Services Pvt. Ltd.

Printed in acid-free paper

5 4 3 2 1 0

springer.com

STAR (Springer Tracts in Advanced Robotics) has been promoted **ROBOTICS** under the auspices of EURON (European Robotics Research Network)

Foreword

At the dawn of the new millennium, robotics is undergoing a major transformation in scope and dimension. From a largely dominant industrial focus, robotics is rapidly expanding into the challenges of unstructured environments. Interacting with, assisting, serving, and exploring with humans, the emerging robots will increasingly touch people and their lives.

The goal of the *Springer Tracts in Advanced Robotics (STAR)* series is to bring, in a timely fashion, the latest advances and developments in robotics on the basis of their significance and quality. It is our hope that the wider dissemination of research developments will stimulate more exchanges and collaborations among the research community and contribute to further advancement of this rapidly growing field.

The *European Robotics Symposium* (EUROS) was launched in 2006 as an international scientific single-track event promoted by EURON, the European Robotics Network linking most of the European research teams since its inception in 2000. Since then, EUROS has found its parental home under STAR, together with the other thematic symposia devoted to excellence in robotics research: FSR, ISER, ISRR, WAFR.

The Second edition has been prepared by Herman Bruyninckx, the new EURON Coordinator, together with Miroslav Kulich and Libor Přeučil, the local organizers of the event. The four key areas of the symposium are: cognition, autonomy, adaptivity, and robustness. The contents of the thirty-five–chapter volume represent a cross-section of forefront robotics research, with contributions on: vision and navigation, localization and mapping, cooperation, coordination and multirobot systems, human–robot interaction, and robotics applications. Thanks to the conscientious cooperation of the contributors, to the committed work of the international programme committee, and last but not least to the devotion of the editors, this volume has been prepared in a record time to be available at the symposium.

From its warm social program to its excellent technical program, EUROS culminates with this unique reference about robotics research in Europe and beyond. A fine confirmation in the STAR series!

Naples, Italy   Bruno Siciliano
January 2008   STAR Editor

Preface

The European Robotics Research Network (EURON) together with the Czech Technical University in Prague is proud to present the second European Robotics Symposium 2008 (EUROS 2008). The event aims to bring participants into the symbolic and attractive environment of Prague, Czech Republic, the birthplace of the word "robot" and bids to be very successful successor of the 1st EUROS 2006 in Palermo, founding a new robotics symposia track in Europe.

EUROS 2008 continues to provide a single-track and high quality scientific forum and presents leading edge research on robotics across the latest topics in fundamental research and applications. This years' symposium key areas cover topics as: **COGNITION, AUTONOMY, ADAPTIVITY** and **ROBUSTNESS** in robotics with main targeting on the issues related namely to: Vision and Navigation, Localization and Mapping, Cooperation, Coordination and Multirobot Systems, Human-Robot Interaction, and Robotics Applications.

The initial and core objective for organizing EUROS track of symposia is to crossbreed latest research and application results in order to improve mutual interlinking of the European robotics community in the field of academia with industrial partners. Moreover, this years' added value of the EUROS 2008 stands in implementation of a "2in1 concept" representing both the top-quality scientific event as well as promotion of extended features for young researchers and PhD students. The young scientists are offered to come in touch with European leading research labs co-workers and their results. This is aimed to create a space to establish "Ph.D. to industry" meetings with present robotics companies and research labs representatives via informal presentation of own research, to launch fruitful discussions, and even to found future co-operations and other joint R&D activities.

EUROS track of symposia has always been organized in terms to ensure the best possible quality European event, paying a special attention to stay a single-track event and to set high demands on the accepted papers. All contributions submitted to this symposium were reviewed by at least two independent reviewers and selection of 20 best papers for oral presentation and 15 poster presentations from 95 papers submitted can be found hereunder.

Therefore, the organizing committee would like to thank all the authors, reviewers and many others for their contributions and excellent work spent in order to make the EUROS 2008 possible.

EUROS 2008   Herman Bruyninckx
                                                                  Libor Přeučil
                                                                 Miroslav Kulich

European Robotics Symposium 2008
-
EUROS 2008

Editors

Herman Bruyninckx Katholieke Universiteit Leuven, Belgium
Miroslav Kulich Czech Technical University in Prague, Czech Republic
Libor Přeučil Czech Technical University in Prague, Czech Republic

General Chair

Herman Bruyninckx Katholieke Universiteit Leuven, Belgium

Local Chair

Libor Přeučil Czech Technical University in Prague, Czech Republic

Program Committee

Antonio Bicchi University of Pisa, Italy
Hendrik van Brussel Katholieke Universiteit Leuven, Belgium
Martin Buss Technische Universitaet Muenchen, Germany
Alícia Casals Technical University of Catalonia, Spain
Raja Chatila Laboratoire d'Analyse et d'Architecture des Systemes, France
Henrik I Christensen Georgia Institute of Technology, USA
Paolo Dario Scuola Superiore Sant' Anna, Italy
Jorge Dias Universidade de Coimbra, Portugal
Rüdiger Dillmann Universität Karlsruhe, Germany

John Hallam University of Southern Denmark, Denmark
Gerd Hirzinger German Aerospace Center, Germany
Oussama Khatib Stanford University, USA
Suk Han Lee Sungkyunkwan University, Korea
Jadran Lenarcic Jožef Stefan Institute, Slovenia
Amiram Moshaiov Tel-Aviv University, Israel
Ulrich Nehmzow University of Essex, United Kingdom
Angel Pasqual del Pobil Jaume I University, Spain

Hubert Roth University of Siegen, Germany
Klaus Schilling University of Würzburg, Germany

Bruno Siciliano The University of Naples "Federico II", Italy
Roland Siegwart ETH Zurich, Switzerland

Organizing Committee

Jan Faigl Czech Technical University in Prague, Czech Republic
Zuzana Hochmeisterová Czech Technical University in Prague, Czech Republic
Jan Chudoba Czech Technical University in Prague, Czech Republic
Karel Košnar Czech Technical University in Prague, Czech Republic
Tomáš Krajník Czech Technical University in Prague, Czech Republic

Hana Krautwurmová Czech Technical University in Prague, Czech Republic
Miroslav Kulich Czech Technical University in Prague, Czech Republic
Petr Štěpán Czech Technical University in Prague, Czech Republic
Milena Zeithamlová Action M Agency, Czech Republic

Co-organizers

 European Robotics Research Network

 Czech Technical University in Prague

 Gerstner Laboratory

Contents

List of Contributors

Anibal T. de Almeida, Institute for Systems and Robotics - University of Coimbra, 3030-290 Coimbra, Portugal
anibal@isr.uc.pt

Luís Almeida, R&D Division, IntRoSys, S.A., Campus FCT/ UNL Edifício Uninova. Quinta da Torre, 2829-516 Monte da Caparica,Portugal

Kerem Altun, Department of Electrical and Electronics Engineering, Bilkent University, TR-06800 Bilkent, Ankara, Turkey
kaltun@ee.bilkent.edu.tr

Michel Banâtre, INRIA Rennes - ACES Team, IRISA, Campus Universitaire de Beaulieu, F-35042 Rennes cedex, France
banatre@irisa.fr

José Barata, UNINOVA, New University of Lisbon, Quinta da Torre, 2825-114 Monte de Caparica,Portugal
jab@uninova.pt

Billur Barshan, Department of Electrical and Electronics Engineering, Bilkent University, TR-06800 Bilkent, Ankara, Turkey
billur@ee.bilkent.edu.tr

Victor M. Becerra, School of Systems Engineering, University of Reading, Whiteknights, RG6 6AY Reading, UK
V.M.Becerra@reading.ac.uk

Giorgio Belloni, Polo Scientifico Didattico di Terni, University of Perugia, Via G. Duranti, 93-06125 Perugia, Italy
belloni@diei.unipg.it

Yvan Bourquin, Cyberbotics Ltd., PSE C - EPFL, 1015 Lausanne, Switzerland
Yvan.Bourquin@cyberbotics.com

Marvin K. Bugeja, Department of Systems and Control Engineering, University of Malta, MSD 2080 Msida, Malta
mkbuge@eng.um.edu.mt

Hans-Dieter Burkhard, Institut für Informatik, LFG Künstliche Intelligenz, Humboldt-Universität zu Berlin, Unter den Linden 6, 10099 Berlin, Germany
hdb@informatik.hu-berlin.de

Carlos Cândido, R&D Division, IntRoSys, S.A., Campus FCT/ UNL Edifício Uninova. Quinta da Torre, 2829-516 Monte da Caparica, Portugal

Luiz Castro, Industrial Electronics' Department, University of Minho, Campus de Azurem, 4800-058 Guimarães, Portugal
luizcastro@gmail.com

François Charpillet, MAIA team, LORIA, Campus Scientifique, BP 239, 54506 Vandœuvre-lès-Nancy Cedex, France
francois.charpillet@loria.fr

Sonya Coleman, School of Computing and Intelligent Systems, University of Ulster, Magee campus, BT48 7JL Londonderry, Northern Ireland
SA.Coleman@ulster.ac.uk

Luís Correia, LabMAg, University of Lisbon, Edifício C6, Campo Grande, 1749-016 Lisbon, Portugal
Luis.Correia@di.fc.ul.pt

Paul Couderc, INRIA Rennes - ACES Team, IRISA, Campus Universitaire de Beaulieu, F-35042 Rennes cedex, France
pcouderc@irisa.fr

Vivien Delsart, Inria Rhône-Alpes, LIG-CNRS, Grenoble Universities, Inovallée, 655 avenue de l'Europe, Montbonnot, 38 334 Saint Ismier Cedex, France
vivien.delsart@inrialpes.fr

Rüdiger Dillmann, University of Karlsruhe, Institute of Computer Science and Engineering, Kaiserstr. 12, 76128 Karlsruhe, Germany
dillmann@ira.uka.de

Can Ulas Dogruer, Mechanical Engineering Department, Hacettepe University, 06800 Ankara, Turkey
dogruer@hacettepe.edu.tr

Melik Dolen, Mechanical Engineering Department, Middle East Technical University, 06532 Ankara, Turkey
dolen@metu.edu.tr

Zoe Doulgeri, Aristotle University of Thessaloniki, Department of Electrical and Computer Engineering, 54124 Thessloniki, Greece
doulgeri@eng.auth.gr

Julia Downes, School of Systems Engineering, University of Reading, Whiteknights, RG6 6AY Reading, UK
J.Downes@reading.ac.uk

Boris Durán, Italian Institute of Technology - University of Genova, Via Morego 30, 16145 Genova, Italy
boris@unige.it

Simon G. Fabri, Department of Systems and Control Engineering, University of Malta, MSD 2080 Msida, Malta
sgfabr@eng.um.edu.mt

Michele Feroli, Siralab Robotics srl, Terni, Via G. Duranti, 93-06125 Perugia, Italy
michele.feroli@siralab.com

Antonio Ficola, Department of Electronic and Information Engineering, University of Perugia, Via G. Duranti, 93-06125 Perugia, Italy
ficola@diei.unipg.it

Sigurd A. Fjerdingen, SINTEF ICT, NO-7465 Trondheim, Norway
sigurd.fjerdingen@sintef.no

Thierry Fraichard, Inria Rhône-Alpes, LIG-CNRS, Grenoble Universities, Inovallée, 655 avenue de l'Europe, Montbonnot, 38 334 Saint Ismier Cedex, France
thierry.fraichard@inria.fr

Simon Le Gloannec, MAD team, GREYC UMR 6072, Campus Côte de Nacre, bd Maréchal Juin, BP 5186, 14032 Caen Cedex, France
slegloan@info.unicaen.fr

Daniel Göhring, Institut für Informatik, LFG Künstliche Intelligenz, Humboldt-Universität zu Berlin, Unter den Linden 6, 10099 Berlin, Germany
goehring@informatik.hu-berlin.de

Grzegorz Granosik, Technical University of Lodz, Stefanowskiego 18/22, 90-924 Lodz, Poland
granosikp.lodz.pl

Mark W. Hammond, School of Systems Engineering, University of Reading, Whiteknights, RG6 6AY Reading, UK
M.W.Hammond@reading.ac.uk

Auke Ijspeert, School of Computer and Communication Sciences, Department of Computer Science, EPFL, Swiss Federal Institute of Technology, Campus de Azurem, 4800-058 Guimarães, Portugal
auke.ijspeert@epfl.ch

Rainer Jäkel, University of Karlsruhe, Institute of Computer Science and Engineering, Kaiserstr. 12, 76128 Karlsruhe, Germany
jaekel@ira.uka.de

Robin Jaulmes, Délégation Générale pour l'Armement (French Defence Procurement Agency), Technical Expertise Centre, 16 bis avenue Prieur de la Cote d'Or, 94114 ARCUEIL CEDEX
robin.jaulmes@dga.defense.gouv.fr

Ekaitz Jauregi, Robotics and Autonomous Systems Group, University of Basque Country, Paseo de Manuel Lardizábal, 1, 20018 Donostia/San Sebastián, Spain
ekaitzji@hotmail.com

Kristof Jebens, Institute of Parallel and Distributed Systems, University of Stuttgart, Universitätsstr. 38, D-70569 Stuttgart, Germany
jebenskf@ipvs.uni-stuttgart.de

Ondřej Jež, Dept. of Control and Instrumentation, Brno University of Technology, Kolejni 4, 612 00 Brno Czech Republic
ondrej.jez@phd.feec.vutbr.cz

Edward Jezierski, Technical University of Lodz, Stefanowskiego 18/22, 90-924 Lodz, Poland
edward.jezierski@p.lodz.pl

Marcin Kaczmarski, Technical University of Lodz, Stefanowskiego 18/22, 90-924 Lodz, Poland
marcin.kaczmarski@p.lodz.pl

Yiannis Karayiannidis, Aristotle University of Thessaloniki, Department of Electrical and Computer Engineering, 54124 Thessloniki, Greece
yiankar@auth.gr

Ulrich Kaufman, University of Ulm, Department of Neural Information Processing, 89069 Ulm, Germany
ulrich.kaufmann@uni-ulm.de

Serge Kernbach, Institute of Parallel and Distributed Systems, University of Stuttgart, Universitätsstr. 38, D-70569 Stuttgart, Germany
korniesi@ipvs.uni-stuttgart.de

Oussama Khatib, Computer Science Department, Stanford University, Stanford, 94305 California, USA
ok@cs.stanford.edu

Bálint Kiss, Dept. of Control Engineering and Information Technology Budapest University of Technology and Economics, XI. Magyar tudósok körútja 2., H-1117 Budapest, Hungary
bkiss@iit.bme.hu

A. Bugra Koku, Mechanical Engineering Department, Middle East Technical University, 06532 Ankara, Turkey
kbugra@metu.edu.tr

Lukas König, Institute of Parallel and Distributed Systems, University of Stuttgart, Universitätsstr. 38, D-70569 Stuttgart, Germany
lukas-koenig@gmx.net

Karel Košnar, The Gerstner Laboratory for Intelligent Decision Making and Control, Department of Cybernetics, Faculty of Electrical Engineering, Czech Technical University in Prague, Technicka 2, 166 27, Prague 6, Czech Republic
kosnar@labe.felk.cvut.cz

Tomáš Krajník, The Gerstner Laboratory for Intelligent Decision Making and Control, Department of Cybernetics, Faculty of Electrical Engineering, Czech Technical University in Prague, Technicka 2, 166 27, Prague 6, Czech Republic
tkrajnik@labe.felk.cvut.cz

Yasuo Kuniyoshi, University of Tokyo, Laboratory for Intelligent Systems and Informatics, Eng. Bldg. 2. 7-3-1 Hongo, Bunkyo-ku, Tokyo, Japan
kuniyosh@isi.imi.i.u-tokyo.ac.jp

Erik Kyrkjebø, SINTEF ICT, NO-7465 Trondheim, Norway
erik.kyrkjebo@sintef.no

Florent Lamiraux, University of Toulouse, av. du Colonel Roche, 31077 Toulouse, France
florent@laas.fr

Elena Lazkano, Robotics and Autonomous Systems Group, University of Basque Country, Paseo de Manuel Lardizábal, 1, 20018 Donostia/San Sebastián, Spain
e.lazkano@ehu.es

Paul Levi, Institute of Parallel and Distributed Systems, University of Stuttgart, Universitätsstr. 38, D-70569 Stuttgart, Germany
levipl@ipvs.uni-stuttgart.de

Martin Lösch, University of Karlsruhe, Institute of Computer Science and Engineering, Kaiserstr. 12, 76128 Karlsruhe, Germany
loesch@ira.uka.de

Abed C. Malti, University of Toulouse, av. du Colonel Roche, 31077 Toulouse, France
amalti@laas.fr

Lino Marques, Institute for Systems and Robotics - University of Coimbra, 3030-290 Coimbra, Portugal
lino@isr.uc.pt

Simon Marshall, School of
Pharmacy, University of Reading,
Whiteknights, RG6 6AP Reading, UK
S.Marshall@reading.ac.uk

Damien Martin-Guillerez, INRIA
Rennes - ACES Team, IRISA, Campus
Universitaire de Beaulieu, F-35042
Rennes cedex, France
dmartin@irisa.fr

José María Martínez-Otzeta,
Robotics and Autonomous Systems
Group, University of Basque Country,
Paseo de Manuel Lardizábal, 1, 20018
Donostia/San Sebastián, Spain
ccbmaotj@si.ehu.es

John. R. Mathiassen, SINTEF
ICT, NO-7465 Trondheim, Norway
john.r.mathiassen@sintef.no

Heinrich Mellmann, Institut für
Informatik, LFG Künstliche Intelli-
genz, Humboldt-Universität zu Berlin,
Unter den Linden 6, 10099 Berlin,
Germany
mellmann@informatik.hu-berlin.de

Olivier Michel, Cyberbotics Ltd.,
PSE C - EPFL, 1015 Lausanne,
Switzerland
Olivier.Michel@cyberbotics.com

Eric Moliné, Délégation Générale
pour l'Armement (French Defence
Procurement Agency), Technical
Expertise Centre, 16 bis avenue Prieur
de la Cote d'Or, 94114 ARCUEIL
CEDEX
eric.moline@dga.defense.gouv.fr

Abdel Illah Mouaddib, MAD
team, GREYC UMR 6072, Campus
Côte de Nacre, bd Maréchal Juin, BP
5186, 14032 Caen Cedex, France
mouaddib@info.unicaen.fr

Slawomir J. Nasuto, School of
Systems Engineering, University of
Reading, Whiteknights, RG6 6AY
Reading, UK
S.J.Nasuto@reading.ac.uk

Viet Nguyen, Autonomous Systems
Laboratory, Swiss Federal Insti-
tute of Technology (ETH Zurich),
Rämistrasse 101, CH-8092 Zürich,
Switzerland
viet.nguyen@mavt.ethz.ch

Daniel J. Norcott, School of
Systems Engineering, University of
Reading, Whiteknights, RG6 6AY
Reading, UK
D.J.Norcott@reading.ac.uk

Goro Obinata, EcoTopia Science Institute, Nagoya University, Furo-cho,
Chikusa-ku, 464-8603 Nagoya, Japan
obinata@mech.nagoya-u.ac.jp

Miguel Oliveira, Industrial Electronics' Department, University of
Minho, Campus de Azurem, 4800-058
Guimarães, Portugal
miglobito@gmail.com

Stefano Pagnottelli, Department
of Electronic and Information Engineering, University of Perugia. Via G.
Duranti, 93-06125 Perugia, Italy
pagnottelli@diei.unipg.it

Guenther Palm, University of Ulm,
Department of Neural Information
Processing, 89069 Ulm, Germany

Jae-Heung Park, Computer Science
Department, Stanford University,
Stanford, 94305 California, USA
park73@cs.stanford.edu

Libor Přeučil, The Gerstner Laboratory for Intelligent Decision Making and Control, Department of Cybernetics, Faculty of Electrical Engineering, Czech Technical University in Prague, Technicka 2, 166 27, Prague 6, Czech Republic
preucil@labe.felk.cvut.cz

Fabien Rohrer, Swiss Federal Institute of Technology in Lausanne (EPFL), EPFL-STI-I2S-LIS, Station 11, 1015 Lausanne, Switzerland
Fabien.Rohrer@epfl.ch

Giulio Sandini, Italian Institute of Technology - University of Genova, Via Morego 30, 16145 Genova, Italy
giulio.sandini@iit.it

Pedro Santana, LabMAg, University of Lisbon, Edifício C6, Campo Grande, 1749-016 Lisbon, Portugal
pfs@uninova.pt

Cristina P. Santos, Industrial Electronics' Department, University of Minho, Campus de Azurem, 4800-058 Guimarães, Portugal
cristina@dei.uminho.pt

Paulo Santos, R&D Division, IntRoSys, S.A., Campus FCT/ UNL Edifício Uninova. Quinta da Torre, 2829-516 Monte da Caparica, Portugal

Sven R. Schmidt-Rohr, University of Karlsruhe, Institute of Computer Science and Engineering, Kaiserstr. 12, 76128 Karlsruhe, Germany
srsr@ira.uka.de

Henrik Schumann-Olsen , SINTEF ICT, NO-7465 Trondheim, Norway
henrik.s.olsen@sintef.no

Bryan Scotney, School of Computing and Information Engineering, University of Ulster, Coleraine campus, Cromore Road, Coleraine, BT52 1SA Co. Londonderry, Northern Ireland
BW.Scotney@ulster.ac.uk

Luis Sentis, Computer Science Department, Stanford University, Stanford, 94305 California, USA
lsentis@cs.stanford.edu

João Sequeira, Instituto Superior Técnico / Institute for Systems and Robotic, Torre Norte - 7° Piso, Av.Rovisco Pais, 1, 1049-001 Lisboa, Portugal
jseq@isr.ist.utl.pt

Roland Siegwart, Autonomous Systems Laboratory, Swiss Federal Institute of Technology (ETH Zurich), Rämistrasse 101, CH-8092 Zürich, Switzerland
roland.siegwart@mavt.ethz.ch

Basi Sierra, Robotics and Autonomous Systems Group, University of Basque Country, Paseo de Manuel Lardizábal, 1, 20018 Donostia/San Sebastián, Spain
ccpsiarb@si.ehu.es

Ufuk Y. Sisli, Istanbul Technical University, Department of Electrical Engineering, Robotics Lab., 34409 Istanbul, Turkey
sisli@itu.edu.tr

Bayu A. Slamet, Intelligent Systems Laboratorium Amsterdam, Universiteit van Amsterdam (UvA), Kruislaan 403, 1098 SJ Amsterdam, The Netherlands

Shanmugalingam Suganthan, School of Computing and Intelligent Systems, University of Ulster, Magee campus, BT48 7JL Londonderry, Northern Ireland
S.Suganthan@ulster.ac.uk

Emese Szádeczky-Kardoss, Dept. of Control Engineering and Information Technology Budapest University of Technology and Economics, XI. Magyar tudósok körútja 2., H-1117 Budapest, Hungary
szadeczky@iit.bme.hu

Michel Taïx, University of Toulouse, av. du Colonel Roche, 31077 Toulouse, France
taix@laas.fr

Mahmoud Tavakoli, Institute for Systems and Robotics - University of Coimbra, 3030-290 Coimbra, Portugal
mahmoodtavakoli@gmail.com

Hakan Temeltas, Istanbul Technical University, Department of Electrical Engineering, Robotics Lab., 34409 Istanbul, Turkey
temeltas@elk.itu.edu.tr

Paolo Valigi, Polo Scientifico Didattico di Terni, University of Perugia, Via G. Duranti, 93-06125 Perugia, Italy
valigi@diei.unipg.it

Arnoud Visser, Intelligent Systems Laboratorium Amsterdam, Universiteit van Amsterdam (UvA), Kruislaan 403, 1098 SJ Amsterdam, The Netherlands
arnoud@science.uva.nl

Philipp Vorst, Department of Computer Science, University of Tübingen, Sand 1, D-72076 Tübingen, Germany
philipp.vorst@uni-tuebingen.de

Kevin Warwick, School of Systems Engineering, University of Reading, Whiteknights, RG6 6AY Reading, UK
K.Warwick@reading.ac.uk

Norinao Watanabe, Graduate School of Engineering, Nagoya University, Furo-cho, Chikusa-ku, 464-8603 Nagoya, Japan
n.watanabe@dynamics.mech.nagoya-u.ac.jp

Benjamin J.Whalley, School of Pharmacy, University of Reading, Whiteknights, RG6 6AP Reading, UK
B.J.Whalley@reading.ac.uk

Dimitris Xydas, School of Systems Engineering, University of Reading, Whiteknights, RG6 6AY Reading, UK
D.Xydas@reading.ac.uk

Andreas Zell, Department of Computer Science, University of Tübingen, Sand 1, D-72076 Tübingen, Germany
andreas.zell@uni-tuebingen.de

Adaptive Multiple Resources Consumption Control for an Autonomous Rover

Simon Le Gloannec[1], Abdel Illah Mouaddib[1], and François Charpillet[2]

[1] MAD team, GREYC UMR 6072, Campus Côte de Nacre, bd Maréchal Juin,
BP 5186, 14032 Caen Cedex, France
{slegloan,mouaddib}@info.unicaen.fr
[2] MAIA team, LORIA, Campus Scientifique, BP 239,
54506 Vandœuvre-lès-Nancy Cedex, France
francois.charpillet@loria.fr

Summary. Resources consumption control is crucial in the autonomous rover context. Most of the time, the resources consumption is probabilistic. During execution time, the rover has to adapt its resources consumption, in order to keep more resources for important tasks or avoid to fail. Progressive processing is a model that describes tasks that can be performed in several ways. Therefore, it allows the agent to adapt and to control its resources consumption during the mission. The resource control is obtained via a Markov decision process that we solve off-line. In its original form, Progressive processing could only control one resource : time. We present an extension for multiple resources. Our main contribution is a new state space representation for multiple consumable resources, which is compact and has suitable access time.

Introduction

Resources consumption control is crucial in the autonomous rover context. For example, a rover must know when to go back to a docking station before having no energy left. Resources are multiple, it can be batteries, memory capacity, or for example a water tank for a fire fighter robot. The questions for the agent are : where, when, and the resources quantity it has to spend for a particular task. We want this agent to be adaptive : it has to choose the resources quantity to consume in each task by taking into account the tasks priority. For these reasons, we use resource-bounded reasoning based on progressive processing [8] to model and to control the mission. This model of reasoning describes tasks that can be performed progressively, with several options at each step. The resources consumption is probabilistic. Progressive processing provides an optimal meta-level resources consumption control via a Markov decision process. In its original form, Progressive processing could only control one resource : time [10]. We present an extension for multiple resources. While we are facing to a combinatory explosion problem, we propose an aggregate state space representation.

This paper is divided into four main sections. In the next section, we present some related works. In the second section, we present Progressive Processing. Firstly, we explain how to model the agent mission, and especially the progressive processing multiple resources extension. Secondly we show how to represent the corresponding MDP. Finally, we study the complexity of the control calculation. In the third section, we

H. Bruyninckx et al. (Eds.): European Robotics Symposium 2008, STAR 44, pp. 1–11, 2008.
springerlink.com   © Springer-Verlag Berlin Heidelberg 2008

propose an algorithm that uses the smart multiple resources state space representation and solve the MDP. The algorithm principle is illustrated. In the final section, we compare time performance for the MDP solving with and without the aggregate state space representation.

1   Related Work on Resource-Bounded Reasoning

Some recent researches deals with handling multiple resources for an autonomous agent. [7] propose to deals with multiple continuous resources. They avoid to enumerate the state space by using an hybrid AO* algorithm [6]. The heuristic is based on the previous work developed in [5]. In a specific military context, [2] are interesting in building a planner that must choose between multiple tasks that consumes multiple resources.

In our context, we want the agent to be adaptive, and control resources that can be either time, battery level, or any kind of consumable resource like a water tank. Each resource quantity is decreasing during the mission. In this paper, we assume the resources quantity to be discrete. In [1], we first tried to represent the corresponding MDP state space in a smart way. Although the state space was represented compactly, the access to the state data was time consuming. Indeed, the access time to a state value is crucial while the agent is solving the MDP.

Progressive Processing is a way to model and to control a set of tasks that can be performed progressively. It also allows an agent to perform its tasks in many several ways. It is suitable for agents that have to be adaptive, especially in a time consumption context. This processing has been used in some applications such as railway train control [9] and an information retrieval engine [3]. There is also an autonomous rover application proposition [4]. This rover mission is depicted in Figure 1. Before we start discussing the problem of handling multiple resources, let us first introduce resource-bounded reasoning called progressive processing and its suitability to the control of the operations of robots.

2   Progressive Processing

The problem formalism will be presented hierarchically: the mission is divided into Progressive pRocessing Units (PRU), which are splitted up into levels.

2.1   Formalism

A mission is a set of tasks, and each task is an acyclic graph vertice (see Figure 1). To facilitate access to the formalism, we assume that this acyclic graph is an ordered sequence. This involves no loss of generality. You can see this sequence as a particular path in the graph (for example A, B, E, F). There are **P** tasks in the sequence. Each task is modelled by a progressive processing unit. A progressive processing unit (PRU) is structured hierarchically. Each PRU_p, $p \in \{1, \ldots, P\}$ is a level ordered finite sequence $[L_{p,1}, \ldots, L_{p,L}]$. An agent can process a level only if the preceding level is finished. The process can be interrupted after each level. It means that the agent can stop the PRU execution, but it receives no reward for it. Other situations have been proposed when the agent can have a reward at each level. However, in our case, only the end of the accomplishment of the task provides a reward.

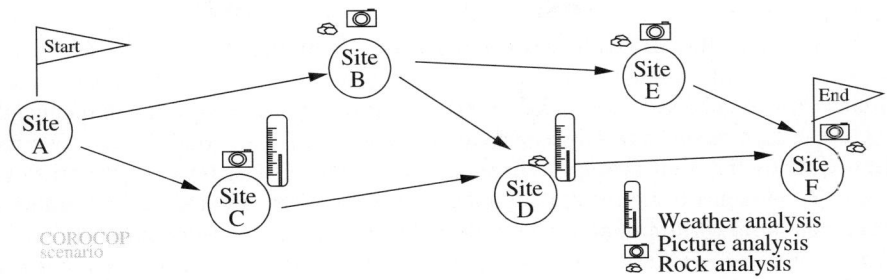

Fig. 1. A mission

Each level $L_{p,\ell}$ contains one or more modules $[m_{p,\ell,1}, \ldots, m_{p,\ell,M}]$. A module is a specific way to execute a level. The agent can only execute one module per level. The execution of a module $m_{p,\ell,m}$ produces a quality $Q_{p,\ell,m}$ and consumes some resources. A progressive processing unit definition is illustrated on Figure 2 for a picture task. The execution is performed from bottom to top.

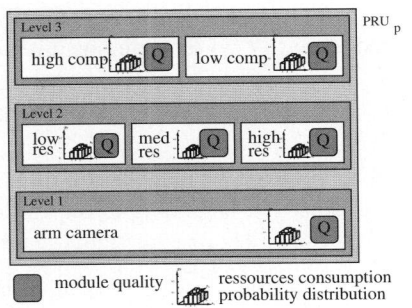

Fig. 2. PRU

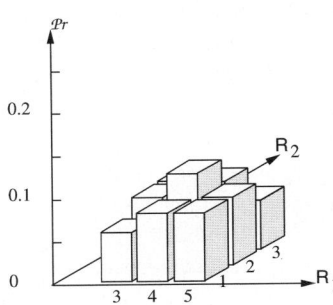

Fig. 3. Resources consumption probability distribution

The quality $Q_{p,\ell,m} \in \mathbb{R}^+$ is a criterion to measure the module execution impact. There is no immediate reward after each level processing. The agent receives the sum of all the $Q_{p,\ell,m}$ only when the last level is performed.

The resource consumption in a module $m_{p,\ell,m}$ is probabilistic. We denote as $\mathcal{P}r_{p,\ell,m}$ the probability distribution of resources consumption depicted in Figure 3. We make no particular assumption on this joint probability distribution. However, we assume the resources to be discrete. It is required that the supports of all the $\mathcal{P}r_{p,\ell,m}$ are discrete and bounded (or finite).

Resources are denoted as : $\bar{r} = (r_1, \ldots, r_\omega, \ldots, r_\Omega)$. There are exactly Ω different resources. Each $r_\omega \in \mathbb{R}^+$ is a remaining resource quantity.

2.2   Mission Control

The problem of control we address in this paper consists of a robot exploring an area where it has some sites to visit and perform some exploration tasks. The problem is that the robot cannot know in advance its resource consumption, and then, it has to develop a policy how to behave at each site according to its local state. The decision depends only on the current state and consequently the modelisation can be Markovian. We use then a Markov decision process to control the mission. The agent is supposed to be rational. It must maximise the mathematical expected value. It computes a policy that correspond to this criterion before executing the mission. The on-line mission control process consists in following this policy. In the next section, we present the Markov decision process model and the control policy calculation. At each site the agent must make a decision to **stay for continuing** the exploration or to **move to another site**.

Modelling the Mission as an Markov decision process

Formally, a Markov decision process is tuple $\{\mathscr{S}, \mathscr{A}, \mathscr{P}r, \mathscr{R}\}$ where :

- $\mathscr{S}$ is a finite set of states,
- $\mathscr{A}$ is a finite set of actions,
- $\mathscr{P}r$ is a transition model mapping $\mathscr{S} \times \mathscr{A} \times \mathscr{S} \rightarrow [0,1]$,
- $\mathscr{R} : \mathscr{S} \rightarrow \mathbb{R}$ is a reward function.

Given a particular rational criterion, algorithms for solving MDP can return a policy π, that maps from $\mathscr{S}$ to $\mathscr{A}$, and a real-valued function $V : \mathscr{S} \rightarrow \mathbb{R}$. Here, the criterion is to maximise the expected reward sum.

States

In the progressive processing model, a state is a tuple $\langle \bar{r}, \mathbf{Q}, p, \ell \rangle$. $\bar{r}$ indicates the amount of remaining resources. $\mathbf{Q}$ is the cumulated quality since the beginning of the current PRU. p and ℓ indicates the last executed level $L_{p,\ell}$ done. The failure state, denoted as $s_{failure}$ is reachedwhen at least one resource is negative. Indeed, each r_ω is a real positive number. The reward is assigned to the agent only when the last level of the PRU has been succesfully executed. Then, it is necessary to store the cumulated quality $\mathbf{Q}$ in the state description. We have introduced level 0 in order to represent the situation where the agent is at the beginning of the PRU execution.

Actions

There are two kinds of actions in the progressive processing model. An agent can execute only one module in the next level or move to the next PRU. These two actions are depicted in Figure 4. When the agent reaches the last level in a PRU_p, it directly moves to the next PRU_{p+1}.

Actually, $\mathbf{E_m}$ is an improvement of the the current PRU whereas $\mathbf{M}$ is an interruption. The agent can progressively improve the PRU execution, it can also stop it at any moment. Formally, $\mathscr{A} = \{\mathbf{E_m}, \mathbf{M}\}$.

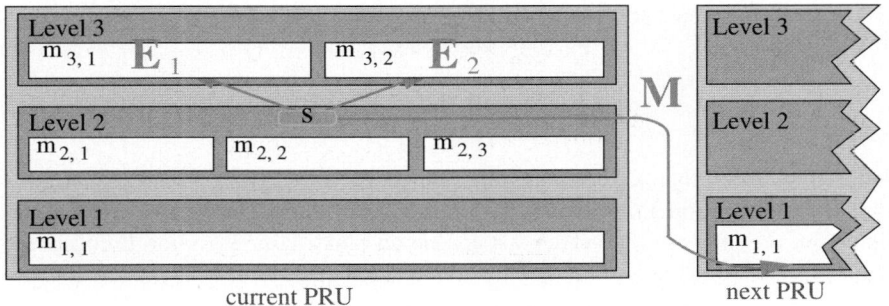

Fig. 4. Actions

Transitions

M is a deterministic action whereas $\mathbf{E}_m$ is not. Indeed, the module execution consumes resources and this consumption is probabilistic. After executing a given module, the agent always cumulates a fixed quality. the uncertainty is related to the resources consumption probability distribution depicted in Figure3. Thus,

$$\mathscr{P}r(\langle \bar{\mathbf{r}}, \mathbf{Q}, \mathbf{p}, \ell \rangle, \mathbf{M}, \langle \bar{\mathbf{r}}, 0, \mathbf{p}+1, 0 \rangle) = 1 \tag{1}$$

$$\mathscr{P}r(\langle \bar{\mathbf{r}}, \mathbf{p}, \mathbf{Q}, \ell \rangle, \mathbf{E}_m, \langle \bar{\mathbf{r}} - \Delta \bar{\mathbf{r}}, \mathbf{Q} + \mathbf{Q}_{\mathbf{p}, \ell, \mathbf{m}}, \mathbf{p}, \ell+1 \rangle) = \mathscr{P}r(\Delta \bar{\mathbf{r}} | \mathbf{m}_{\mathbf{p}, \ell, \mathbf{m}}) \tag{2}$$

Reward

A reward is given to the agent as soon as it finishes a PRU. This reward corresponds to the cumulated quality through the modules path. If the agent leaves a PRU without finished it, it receives no reward. This makes a sense for exploration task where the robot has no reward if the task is not completely finished.

$$\mathscr{R}(\langle \mathbf{r}, \mathbf{Q}, \mathbf{p}, \mathbf{L_p} \rangle) = \mathbf{Q} \tag{3}$$

$$\mathscr{R}(\langle \mathbf{r}, \mathbf{Q}, \mathbf{p}, \ell < \mathbf{L_p} \rangle) = 0 \tag{4}$$

where $\mathbf{L_p}$ is the number of levels in PRU_p.

Value Function

The control policy $\pi : \mathscr{S} \to \mathscr{A}$ depends on a value function that is calculated thanks to the Bellman equation 5. We assume that $V(\mathbf{s}_{\texttt{failure}}) = 0$.

$$V(\langle \bar{\mathbf{r}}, \mathbf{Q}, \mathbf{p}, \ell \rangle) = \begin{cases} 0 & \text{if } \exists \omega, r_\omega < 0 \quad \text{(failure)} \\ \mathscr{R}(\langle \bar{\mathbf{r}}, \mathbf{Q}, \mathbf{p}, \mathbf{L_p} \rangle) + \max(V_\mathbf{M}, V_\mathbf{E}) & \text{otherwise} \end{cases} \tag{5}$$

$$V_\mathbf{M}(\langle \bar{\mathbf{r}}, \mathbf{Q}, \mathbf{p}, \ell \rangle) = \begin{cases} 0 & \text{if } \mathbf{p} = \mathbf{P} \\ V(\langle \bar{\mathbf{r}}, \mathbf{0}, \mathbf{p}+1, 0 \rangle) & \text{otherwise} \end{cases} \tag{6}$$

$$V_{\mathbf{E}}(\langle \overline{\mathbf{r}}, \mathbf{Q}, \mathrm{p}, \ell \rangle) = \begin{cases} 0 & \text{if } \ell = \mathrm{L_p} \\ \max_{\mathbf{E_m}} \sum_{\Delta \overline{\mathbf{r}}} \mathscr{P}r(\Delta \overline{\mathbf{r}} | \mathrm{m}_{\mathrm{p},\ell,\mathrm{m}}).V(\langle \overline{\mathbf{r}}', \mathbf{Q}', \mathrm{p}, \ell+1 \rangle) \\ \quad \text{where} \quad \mathbf{Q}' = \mathbf{Q} + \mathbf{Q}_{\mathrm{p},\ell,\mathrm{m}} \\ \quad \text{and} \quad \overline{\mathbf{r}}' = \overline{\mathbf{r}} - \Delta \overline{\mathbf{r}} \end{cases} \tag{7}$$

Equation 6 shows that action $\mathbf{M}$ is deterministic. It consumes no resources. Equation 6 shows that action $\mathbf{E}$ consumes resources. We consider that the failure value is 0. Hence, we do not write the values that corresponds to a failure. Action $\mathbf{E}$ consumes some resources. The agent will execute the module that gives it the best expected value.

2.3   Complexity Study

Since we are dealing with an MDP, the size of the state space $\mathscr{S}$ really matter. Furthermore we are facing to an acyclic graph in which the resources are decreasing. Hence, the complexity of the algorithm that we use to solve the control problem is proportional to #$\mathbf{s} \in \mathscr{S}$. Figure 5 shows a state space example for three PRU composed of two levels with only one module.

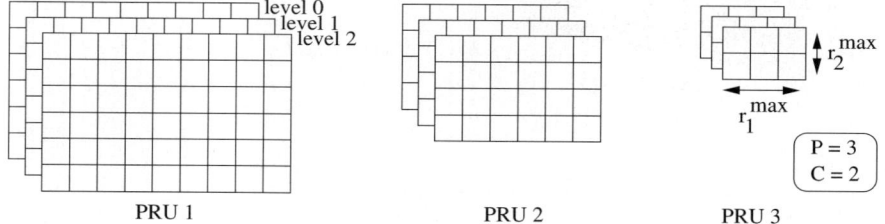

Fig. 5. State space enumeration

A state is a 4-uplet $\langle \overline{\mathbf{r}}, \mathbf{Q}, \mathrm{p}, \ell \rangle$. The state space size depends on this for attributes. For all the PRU in the mission, the state space size is:

$$\#\mathbf{s} \in \mathscr{S} \leq \sum_{\mathrm{p}=1}^{\mathbf{P}} \left(\left(\prod_{\omega=1}^{\Omega} \sum_{\mathrm{p}'=\mathrm{p}}^{\mathbf{P}} \mathrm{r}_{\omega \mathrm{p}'}^{\max} \right) \times \left(1 + \sum_{\ell=1}^{\mathrm{L_p}} \prod_{\ell'=1}^{\ell} \#\mathrm{m}_{\mathrm{p},\ell',} \right) \right), \tag{8}$$

where $\mathbf{P}$ is the number of PRU in the mission, Ω the number of different resources, r_ω is a resource. The right part of this equation represent the number of different accumulated quality $\mathbf{Q}$. To make this equation more understandable, we can consider that all PRU are identical and all resources r_ω are defined on the same interval.

$$\#\mathbf{s} \in \mathscr{S} \leq \left(\sum_{\mathrm{p}=1}^{\mathbf{P}} \mathrm{p}^\Omega \right) . \left(\mathrm{r}_{\mathbf{P}}^{\max} \right)^\Omega . \mathrm{C},$$

where C represents the average number of modules in a PRU. The value function evaluation algorithm is linear in the state space size. Our objective is to reduce this state space size, in order to address real exploration missions.

3 Algorithm

To calculate the value function, we can process backward chain among the whole state space (from the end to the beginning). Because there is no cycle in the MDP, each state is evaluated only once. First, we generate the states for the last $\mathsf{PRU_P}$. We begin with the states in the sub spaces $\mathscr{S}_{\mathbf{Q},\mathbf{P},\mathrm{L}}$ (for all $\mathbf{Q}$), then we evaluate the sub spaces $\mathscr{S}_{\mathbf{Q},\mathbf{P},\mathrm{L}-1}$ until the level 0. Each $\mathscr{S}_{\mathbf{Q},\mathbf{P},\ell}$ has a size of $\mathbf{r}_{\mathbf{P}}^{\max}$. We continue in the same way to obtain a value for each possible state. Each evaluation step generates a value function The problem is that the state space remains exponentially large. That's why we propose an other state space representation, based on some properties of the value function.

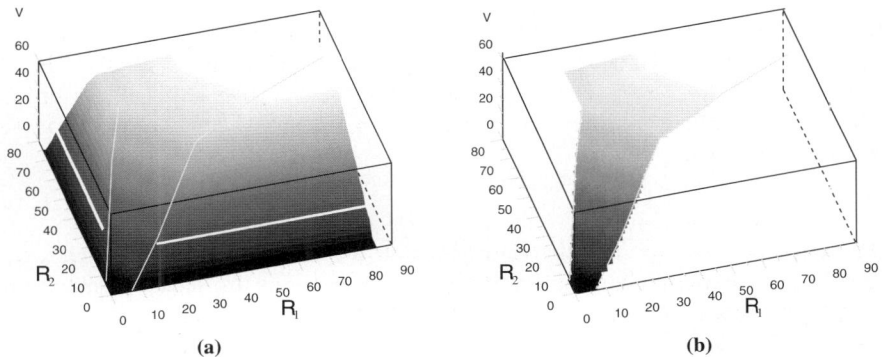

Fig. 6. Value function on a subspace $\mathscr{S}_{\mathbf{Q},\mathrm{p},\ell}$

3.1 Value Function Properties

On a given subspace : $\mathscr{S}_{\mathbf{Q},\mathrm{p},\ell}$, the value function V is increasing with the remaining resources. $\forall \omega, \mathbf{r}_{\omega}^{a} < \mathbf{r}_{\omega}^{b} \Rightarrow V(\langle \overline{\mathbf{r}}^{a}, \mathbf{Q}, \mathrm{p}, \ell \rangle) < V(\langle \overline{\mathbf{r}}^{b}, \mathbf{Q}, \mathrm{p}, \ell \rangle)$. Thus, we propose to exploit this property to avoid some state evaluation. On a subspace : $\mathscr{S}_{\mathbf{Q},\mathrm{p},\ell}$, the value function V is stable for some states (see the white lines in Figure 6.a). We call these white lines *floors*. In this Figure, the agent must consume two different resources $\mathbf{r}_1$ and $\mathbf{r}_2$ to execute each module. If there is not enough resource $\mathbf{r}_1$, even if the agent has an infinite among of $\mathbf{r}_2$, it will not be able to perform a module that consumes many $\mathbf{r}_1$ resources.

We formalise this as follows ($\mathbf{Q}, \mathrm{p}$ and ℓ are fixed):

$$\forall \mathbf{r}_1 \in \mathbb{R}, \exists \mathbf{r}_2^{\max} \in \mathbb{R}, \forall \mathbf{r}_2 \in \mathbb{R}, \quad V(\langle\langle(\mathbf{r}_1, \mathbf{r}_2)\rangle\rangle) \leq V(\langle\langle(\mathbf{r}_1, \mathbf{r}_2^{\max})\rangle\rangle),$$
$$\forall \mathbf{r}_2 \in \mathbb{R}, \exists \mathbf{r}_1^{\max} \in \mathbb{R}, \forall \mathbf{r}_1 \in \mathbb{R}, \quad V(\langle\langle(\mathbf{r}_1, \mathbf{r}_2)\rangle\rangle) \leq V(\langle\langle(\mathbf{r}_1^{\max}, \mathbf{r}_2)\rangle\rangle)$$

Therefore, we only have to calculate the states values which are in the "middle" of the value function (see Figure 6.b).

3.2   Bounded Space Data

Basically, when the agent has to take 2 different resources into account, each $\mathscr{S}_{Q,p,\ell}$ is a rectangle. Typically, if we want to evaluate each state, we have to path through the grid line per line or column per column. Instead of using this basic sweep, we propose to follow the sweep shown on Figure 7.**b**. The advantage of this particular path is that we follow the gradient of the value function.

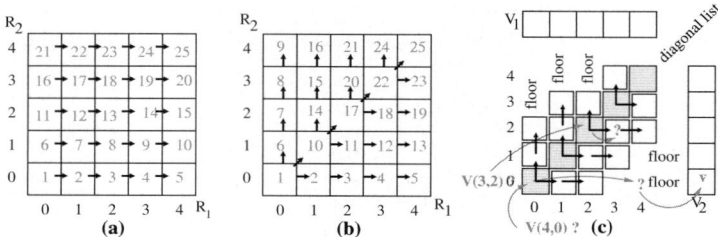

Fig. 7. Diagonal path trough a 2-dimensional state space

Furthermore, we know that the value function is bounded. When we reach a state which value is a max value, we know that all the states in this line have the same value. We neither calculate their value nor create them. We save both memory and computation time. To really take avantage of this method, we have first to calculate the bounds (see Figure 8.a). In the case of two resources, two value bounding functions $V_1 = V(r_1, +\infty)$ and $V_2 = V(+\infty, r_2)$ are calculated. After that, we calculate $V(r_1, r_2)$ (see Figure 8.b and Algorithm 1). The floors are now aggregated in the max value functions V_1 and V_2.

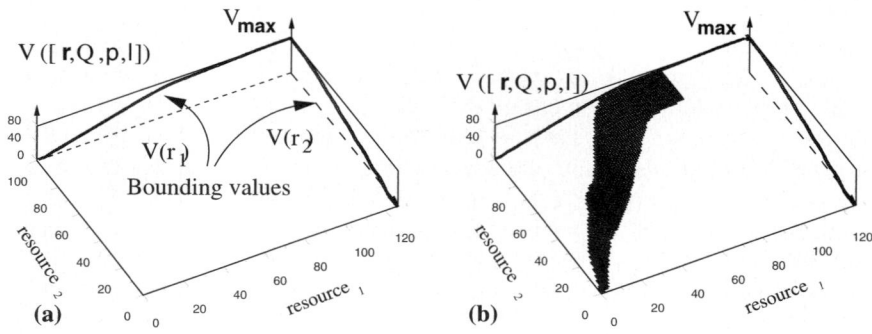

Fig. 8. Fast value function calculation

Where **evaluate** is the value function given in Equation 5. The global backward valuation algorithm consists in evaluating the last level in the last **PRU**, and so on until the level 0 of the first **PRU** is evaluated. Thus, at each evaluation step, we have to access

---

Algorithm 1. A calculation step of V, on a subspace $\mathcal{S}_{\mathbf{Q},\mathrm{p},\ell}$

---

1 $i_{diag} = 0$; $j_{diag} = 0$;

2 **while** $i_{diag} \leq r_1^{max}$ *et* $j_{diag} \leq r_2^{max}$ **do**

3   **for** i *in* $[i_{diag}, r_1^{max}]$ **do**

4      $value =$ **evaluate**$\left([(i, j_{diag}), \mathbf{Q}, \mathrm{p}, \ell]\right)$

5   **if** $value == V([(r_1^{max}, j_{diag}), \mathbf{Q}, \mathrm{p}, \ell])$ **then break**

    end

6   **for** j *in* $[j_{diag}, r_2^{max}]$ **do**

7      $value =$ **evaluate**$\left([(i_{diag}, j), \mathbf{Q}, \mathrm{p}, \ell]\right)$

8   **if** $value == V([(i_{diag}, r_2^{max}), \mathbf{Q}, \mathrm{p}, \ell])$ **then break**

    end

9    $i_{diag} ++$; $j_{diag} ++$;

  end

---

to values in the previous step. It is crucial to have a quick access to theses values. The data structure is divided in many sub spaces $\mathcal{S}_{\mathbf{Q},\mathbf{P},\ell}$ (see Figure 5). Each sub space (in a 2-resources example) is composed of a *diagonal list* that contains the values for the states where $r_1 = r_2$ (see Figure 7.c). Each cell of this diagonal list is connected to two states lists : the one where $r_1 > r_2$ (the row) and the other one where $r_1 < r_2$ (the column).

The getValue function works as follow : for a given point (r_1, r_2) (for example $(4, 0)$, we look for the minimum (here) between r_1 and r_2. Then, at this index (0) in the diagonal list, we look for the list that correspond to the maximum between r_1 and r_2. If there is a point at the given index in this list, we find the value (example with $(2, 3)$), else this value is stored in the corresponding bounding value function (example with $(4, 0)$).

This technique also works for more than $\Omega > 2$ resources. We just have to compute Ω bounding value functions. For each Ω bounding value functions, we have to calculate $\Omega - 1$ functions on a $\Omega - 2$ dimensional state space. The calculation of the bounding values functions does not provide any extra calculation, because we should calculate them anyway if we didn't aggregate the state space.

Theorem 3.1. *The policy obtained with the backward-chaining based on Algorithm 1 is **optimal**.*

Proof. The state space aggregation technique is exact, and non-uniform. Then, there is no approximation in the value function. As the previous technique provides an optimal policy, this policy is also optimal.

4   Experiment and Results

We have made some experiments with different PRU. We compare the time that the agent needs to calculate the policy with and without the aggregation. In an example with 2 resources, we can now calculate an optimal policy for a mission of 10 PRU in 10 seconds while it takes 2 minutes without aggregation technique. The number of

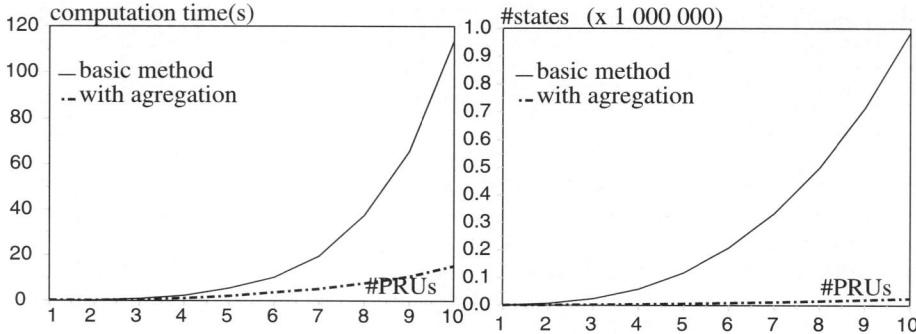

Fig. 9. Computation time and memory saved

generated states is divided per 100 000 for this experiment. The global complexity for the value function calculation remains in #PRU^3, but there is less states to explore. Now, the agent will be able to calculate an optimal policy for larger missions..

In the worst case, our algorithm takes as much time as the basic one. This worst case is a mission where all the modules are : $(\mathbb{R}, \mathbb{R})[0, 1] \rightarrow \mathbb{R}$, $m_a : (1, 0), 1 \rightarrow 1$, $m_b : (1, 1), 1 \rightarrow 2$ and $m_c : (0, 1), 1 \rightarrow 1$. In this particular case, there is no floor in the value function.

Conclusion and Discussion

We can now model and manage multiple resources consumption for an autonomous robot using the progressive processing. The progressive processing task are adaptive and interruptible. With this model of reasoning we can calculate an optimal policy for the resources consumption. We have succeeded in limiting the size of the state space with an aggregating technique. Furthermore, we have reduced the computation time by creating a suitable state space representation.

We have already implement some experiments on a real robot (Koala) with only one resource : time. In the rover context, the first level in each PRU is compulsory. Indeed, it is not possible to interrupt the rover during the movement between two sites. The **M** action can not represent a movement between two sites because it does not consume any resource. **M** is just a way to interrupt the current PRU execution. In the rover model, the movement is then included in the first level which must absolutely be performed. The resulting behaviour is interesting because of the optimal time consumption control. We intend also to implement the multiple resources extension on a real robot in a near future.

References

1. Le Gloannec, S., Mouaddib, A.I., Charpillet, F.: Meta-level control under uncertainty for handling multiple consumable resources of robots. In: proceedings of (IROS) (2005)
2. Aberdeen, D., Thiébaux, S., Zhang, L.: Decision-theoretic military operations planning. In: proceedings of ICAPS, Whistler, British Columbia, Canada (2004)

3. Arnt, A., Zilberstein, S., Allan, J., Mouaddib, A.I.: Dynamic composition of information retrieval techniques. Journal of Intelligent Information Systems 23(1), 67–97 (2004)
4. Cardon, S., Mouaddib, A.I., Zilberstein, S., Washington, R.: Adaptive control of acyclic progressive processing task structures. In: proceedings of IJCAI, Seattle, Washington, USA, pp. 701–706 (2001)
5. Feng, Z., Dearden, R., Meuleau, N., Washington, R.: Dynamic programming for structured continuous markov decision problems. In: proceedings of UAI 2004, pp. 154–161 (2004)
6. Hansen, E.A., Zilberstein, S.: LAO*: A heuristic search algorithm that finds solutions with loops. Artif. Intell. 129(1-2), 35–62 (2001)
7. Mausam, B.E., Brafman, R., Meuleau, N., Hansen, E.A.: Planning with continuous resources in stochastic domains. In: Mausam, E., Benazera, R., Brafman, N. (eds.) proceedings of IJCAI, Edinburgh, Scotland, pp. 12–44 (2005)
8. Mouaddib, A.I., Zilberstein, S.: Optimal scheduling for dynamic progressive processing. In: Proceedings of ECAI, pp. 499–503 (1998)
9. Mouaddib, A.-I., Zilberstein, S.: Knowledge-based anytime computation. In: proceedings of the 14th International Joint Conference on Artificial Intelligence (IJCAI), pp. 775–783 (1995)
10. Zilberstein, S., Washington, R., Berstein, D., Mouaddib, A.I.: Decision-theoretic control of planetary rovers. In: Beetz, M., Hertzberg, J., Ghallab, M., Pollack, M.E. (eds.) Dagstuhl Seminar 2001. LNCS (LNAI), vol. 2466, pp. 270–289. Springer, Heidelberg (2002)

Adaptive Snake Robot Locomotion: A Benchmarking Facility for Experiments

S.A. Fjerdingen, J.R. Mathiassen, H. Schumann-Olsen, and E. Kyrkjebø

INTEF ICT, N-7465 Trondheim, Norway
{sigurd.fjerdingen,john.r.mathiassen,henrik.s.olsen,
erik.kyrkjebo}@sintef.no

Summary. A benchmarking facility for snake robot locomotion is presented, including the design of a snake-like robot extended with a sensor setup combining three-dimensional vision and an array of force sensors to register friction and impulse forces. A surrounding, modular environment consisting of a reconfigurable obstacle course and a ceiling mounted camera system is also presented. This enables research into adaptive obstacle-based and non-obstacle-based movement patterns for robotic snakes. Experimental results show possibilities for detailed data analysis of snake robot locomotion. Thus, the facility may be a common reference on which to experiment and evaluate future ideas.

Keywords: Snake robot, autonomous mobile robot, benchmarking facility.

1 Introduction

Snake-like robots have proven to be flexible and adaptable in environments considered difficult for more conventional robots ([9],[16],[20],[21]). The ability to make use of its own body as both propulsive force and operation tool, combined with a slim appearance, makes for a worthy candidate for operations ranging from search-and-rescue in difficult environments to subsea operations, pipe inspection, service robots, and even fire fighting ([11]). Snake robot locomotion studies have generally been conducted as single experiments showing the ability of robotic snakes to locomote theirselves in specific and, in general, easy terrain. Thus, no conclusion regarding the quality of locomotion is easily made. This paper presents a novel benchmarking facility for investigating the quality of snake robot locomotion in terms of path planning, velocity over ground, target acquisition and adaptivity to the environment. The benchmarking facility consists of a robotic snake, connected sensor setup and research environment giving reproducible and reconfigurable experiments. Thus, the facility provides a complete setup for research into adaptivity and autonomy for robotic snake locomotion.

Research into snake robots and snake robot locomotion can be dated as far back as Hirose's first snake robot from 1972 ([4]). More recent research robots include the JPL serpentine robot ([16]), the ACM-R5 ([21]) and the Perambulator-II ([22]), amongst others. Research has been aimed at mathematical formulation of snake motion, both kinematics ([14]) and dynamics ([12]), using robotic snakes with wheels ([21]), crawler

H. Bruyninckx et al. (Eds.): European Robotics Symposium 2008, STAR 44, pp. 13–22, 2008.

tracks ([9]) or only the snake body itself ([11]). Additionaly, simulation studies on adaptive snake robot locomotion have also been conducted using genetic programming techniques ([17]).

The presented sensor setup allows for detailed data collection from the environment, and the facility gives reproducible and reconfigurable experiments. The aim for these additions is to open a path leading to further insight into and more focused results on locomotion using obstacles as the main propulsive mechanism, possibly in combination with or as a modulation of already existing locomotion gaits. The locomotion approaches use only the snake body itself as the propulsive mechanism; no passive or active wheels or crawler tracks are attached, and friction against floor and walls is isotropic. Locomotion strategies are thus not restricted to specific conditions, but rather a general approach independent of underlying terrain.

Experiments outline the inherent possibilities in the benchmarking facility, showcasing a path planning and execution example, and readouts from force sensors from moving and connecting with obstacles.

The paper is organized as follows. Section 2 describes the robotic snake, Section 3 describes in more detail the sensor setup connected to the snake, and Section 4 the environment. In Section 5, results from experiments using the benchmarking facility are presented and discussed, and conclusions are given in Section 6.

2   The Snake Robot

The snake robot Aiko was originally created as a small-scale locomotion research model of another robotic snake developed at SINTEF; the water hydraulic fire-fighting snake robot Anna Konda ([11]). Earlier published work using Aiko for experimental verification includes 2D and 3D mathematical modeling of obstacle-aided locomotion by Transeth et al. ([19]).

2.1   Mechanical Design

Aiko consists of 10 identical segments, a special head segment, and 10 joints connecting the segments. The snake is shown in Fig. 1a. The joints act as cardan joints, using two 6 W DC-electromotors for the two degrees of freedom, as shown in Fig. 1b. Note that the motors are situated in the joints, and not in the segments. The length of each segment is $L_i = 0.122$ m, the radius $r_i = 0.115$ m, and the mass $m_i = 0.682$ kg, where $i \in [1, 10]$ enumerates the segments from head to tail, head segment not included. Each joint has a pitch and yaw angle limitation of $\pm 45^o$, and maximum joint moment is 2.5 Nm. The head segment is special in that it does not contain the same internals as the other segments, but rather makes room for a camera and three IR distance sensors. This gives the head segment a length of $L_h = 0.136$ m, and a radius of $r_h = 0.112$ m. The sensor setup is further described in Sect. 3.

2.2   Control System Design

This section describes hardware design, motion patterns and control concepts currently used in the robotic snake Aiko.

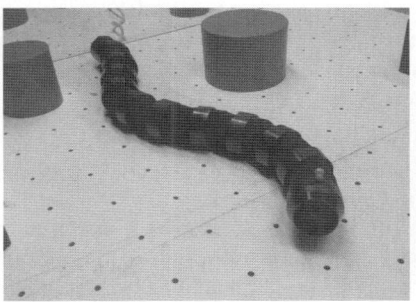
(a) The robotic snake Aiko

(b) Connected segments as cardan joints

Fig. 1. Physical apperarance of the robotic snake Aiko

Each segment is equipped with a microcontroller of type Atmel ATmega128, and necessary external components for data collection, communication and motor control. This makes each segment independently responsible for controlling its pitch and yaw in relation to the preceeding segment. The head segment does not contain any circuits for controlling its angles in relation to preceeding segments, as there are no segments preceeding it. Each segment communicates directly with an external cognition and pattern generation module, in the form of a 2.40 GHz, 1 GB RAM PC running Windows XP and Matlab/Java.

The cognition module handles higher level of reasoning, choice of action and autonomy, and HMI. Communication between the cognition module and the segments is handled by a type of field bus referred to as a CAN-bus, similar to the choice of bus for the ACM-R5 ([21]) and the GMD-Snake ([20]). The bus is a multidrop serial bus with automatic arbitration and broadcasting abilities. There is also a built-in bus-off mechanism for faulty communication and/or segment operation, making a robust system capable of single-segment shutdown. Fig. 2 shows how the total system is constructed conceptually.

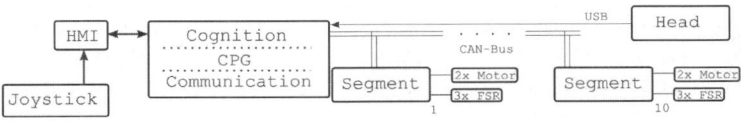

Fig. 2. Illustration of control system concept

Reference positions for the relative pitch- and yaw-angles are sent from the CPG (Central Pattern Generator) to each segment, which implements a local controller to ensure that the reference angle is reached. Currently implemented angle controllers include P- and PD-controllers. Current angles may also be returned from each segment on a request basis, which allows for an alternative centrally appointed controller.

The ATmega128 has a built-in 8 port 10 bit ADC which is used for data collection. All segments have three FSRs (Force Sensing Resistors) connected to ADC-ports, and segment 1 also has three IR distance sensing diodes of type Sharp 2D120X connected to ADC-ports, facing left, right and forward. Transmission of data samples are done in a request manner as for the angle measurements. A broadcasted request of samples will return the corresponding sample from all connected segments.

2.3   Motion Patterns

The CPG mentioned in Sect. 2.2, and shown in Fig. 2, constructs patterns corresponding to the wishes of the cognition module. The base pattern for which locomotion is achieved, is chosen in analogy to several earlier works, summarized by amongst others [5] and [12]; by partitioning the motion into a separate horizontal and vertical repeating pattern, more specifically a sine pattern. This is given in Eq. (1)

$$\begin{pmatrix} q_{i,\theta} \\ q_{i,\phi} \end{pmatrix} = \begin{pmatrix} A_\theta \sin(\omega_\theta t + (i-1)\delta_\theta + \delta_{0,\theta}) \\ A_\phi \sin(\omega_\phi t + (i-1)\delta_\phi + \delta_{0,\phi}) \end{pmatrix} + \begin{pmatrix} \psi_\theta \\ \psi_\phi \end{pmatrix} . \tag{1}$$

where $q_{i,\theta}$ is the yaw, or rotation about the horizontal axis for segment i, and $q_{i,\phi}$ is the pitch, or rotation about the vertical axis for segment i. A is the amplitude of the corresponding wave, ω the angular frequency, δ the phase offset, and ψ the angular offset. The difference between $\delta_{0,\theta}$ and $\delta_{0,\phi}$ gives the phase difference between the horizontal and vertical wave. Parameters δ_θ and δ_ϕ may also be seen as the sinusoidal frequency in segments for the two waves, as it can be given as

$$\delta_\theta, \delta_\phi = \frac{2\pi}{k} . \tag{2}$$

where k gives the wavelength in segments.

Lateral undulation

is a simple sine motion only in the horizontal plane, i.e. $A_\phi = 0$. This motion will not by itself produce any significant forward or sideways motion with isotropic friction.

Sidewinding

motion can be achieved by setting $\omega_\theta = \omega_\phi$, $\delta_\theta = \delta_\phi$, $\delta_{0,\theta} = 0$ and $\delta_{0,\phi} = \delta_{0,\theta} \pm \frac{\pi}{4}$. It was first described by [1], and makes the snake move laterally with respect to its initial straight heading by consecutive lifting and placing of contact points along the robotic snake. The parameter $\delta_{0,\phi}$ will direct the direction of lateral movement.

Sinus-lifting

motion is described in [15], and is achieved by setting $\omega_\phi = 2\omega_\theta$ and $\delta_\phi = 2\delta_\theta$. This gives a vertical wave with half the wavelength in segments travelling at double angular frequency. The parameter $\delta_{0,\theta} = 0$ and $\delta_{0,\phi} = \delta_{0,\theta} + \frac{\pi}{2}$ will produce ground contact at the snake's lateral center, whilst $\delta_{0,\phi} = \delta_{0,\theta} - \frac{\pi}{2}$ makes ground contact at the lateral extremities - in [15] this is referred to as a *3D pedal wave without stretching*. The resulting motion makes the robotic snake move forwards.

Rotation

is achieved by varying the sinus-lifting pattern, allowing for rotation around an arbitrary vertical rotation axis along the robotic snake corresponding to a segment.

3   Sensor Setup

The two main additions to the Aiko robotic snake, in relation to other robotic research snakes, are the force sensing snake skin and 3D sensing. Combining these measurements allows for a more detailed view of the snake's surroundings, which is advantageous in areas such as obstacle-aided locomotion and localization and mapping (e.g. for SLAM purposes).

3.1   Snake Skin

Earlier snake robot designs have had a tendency to focus more on articulation and actuation, and less on properties of the snake skin. There are several snake robots using wheels ([18],[22]), which achieves larger lateral than longitudinal friction. Other robotic snakes use only a rigid shell structure ([16]), possibly with contact switches alongside the segments ([20]).

As with Anna Konda ([11]), the snake robot Aiko uses FSRs for force measurements which alter their electrical resistance as a function of applied pressure. In the case of [11], a hard shell was rigidly secured outside the FSRs, which limited the FSR measurements to register contact forces. Differently from [11], the protective shell lining the FSRs is now floating, which means that any force acting on the shell laterally will displace it in relation to the underlying segment. Connecting the protective shell to the segment is a compliant material placed over the FSRs, and the protective shell is also pre-tensioned, keeping it in place and giving an initial bias to the FSRs.

Fig. 3 shows a concept illustration of the shell fastened outside the segment, and Fig. 1a shows the entire robotic snake equipped with these shells at the segments. The three locations of the FSRs were chosen to enable measurment of lateral forces, and simultaneously measure the force of pressure against the ground. Lateral forces are measured as a total difference of force increase on one side versus force decrease on the other, and vertical pressure as an increase on the bottom FSR.

3.2   3D Sensing

The head segment may include different sensor technologies to provide a test-bed for e.g. SLAM. The snake robot is currently equipped with a SwissRanger SR-3000 time-of-flight 3D camera. The camera provides 3D range measurements at each of its 176x144 pixels at a speed of up to 40 frames/second. Range images are measured by combining a 2D imaging sensor with a NIR light source modulated at 20 MHz. By sampling the modulated light at 4 temporally equidistant intervals, the range at each pixel location can be computed by analyzing the phase shift between the light emitted and the light returned to the pixel. The SR-3000 has already been used in several mobile robots ([2],[6]), in order to provide range data for 3D mapping. The attached IR sensors can also be used for redundancy and calibration for the 3D camera.

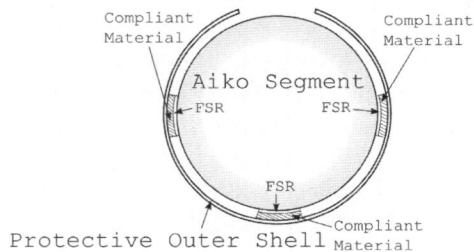

Fig. 3. Illustration of the protective outer shell connected via a compliant material to a segment and the FSRs

4   Research Environment

Little work has been done in the field of creating an artificial benchmarking environment in the case of snake robot locomotion. There have been some rudimentary ad hoc obstacle courses like in the instance of [3], as well as the simulation environment ([17]), but to the authors' knowledge no dedicated benchmarking facility exists.

Evaluation of motion patterns and path planning strategies requires a facility which should (1) give reproducible experiments, and (2) be reconfigurable. This paper presents a modular obstacle course and attached ceiling mounted camera system allowing an object's position to be determined at fixed or variable time intervals. The position fix is crucial to evaluate robot performance, as it gives the trajectory travelled, which in turn enables performance benchmarking parameters such as speed over ground (see e.g. [17]) and traversability.

4.1   Modular Obstacle Course

The obstacle course is shown in Fig. 4, and consists of a floor with holes spaced as equilateral triangles with side length 15 cm. This pattern is used both as placing points for obstacles, and position references for the camera system. Obstacles are structures

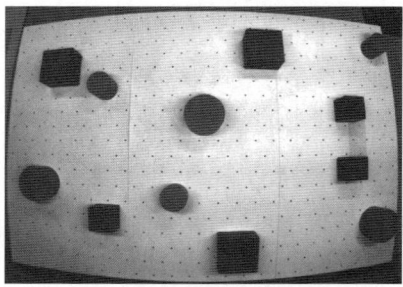

Fig. 4. Modular obstacle course as seen from ceiling mounted camera

of varying size and geometrical shape, with two rods underneath allowing for rigid placement on the obstacle course. The modular obstacle course is presently equipped with obstacles based on squares and circles, with side lengths 20 cm and 30 cm for the squares, and radius 10 cm and 15 cm for the circles. All obstacles presently have a height of 20 cm, and a third dimension may be added to the obstacle course by placing objects of height no greater than the climbing proficiency of a robotic snake allows. The obstacle shapes are not limited to the described shapes, but may be of arbitrary shape and material.

4.2   Ceiling Mounted Camera System

A 1.3 megapixel uEye camera, manufactured by IDS Imaging Development Systems, is mounted in the ceiling above the obstacle field. By detecting two reflective markers on the head segment of the snake robot, the camera tracks the position and heading 4 times each second. While tracking the head, image processing algorithms simultaneously detect the position of the snake robot along its entire length. These algorithms can assist in analysis of snake robot locomotion, and provide additional input to the cognitive module to determine segment angles and contact points of the snake robot.

5   Experimental Results

Two experiments were conducted to showcase the various aspects of the benchmarking facility, described as following.

5.1   Path Planning Using Forward Motion and Rotation

The snake robot was laid out according to Fig. 5, and a path planning algorithm based on Dijkstra's algorithm and a cost map encouraging straight movement and avoidance of obstacles was used to make the robot autonomously move to the desired location on the obstacle course. Position and orientation data from the ceiling mounted camera system gave feedback to the algorithm. Allowed motion primitives by the CPG were either direct forward motion in the form of sinus lifting, or a rotating motion around the fifth segment. Resulting path of the snake head and distance to target is shown in Fig. 5. This simple experiment shows the ease of which the benchmarking facility may be used for evaluating motion patterns and path planning strategies. In addition to the snake head, arbitrary points on the snake may be tracked in the same way, allowing for a detailed analysis of the snake's movements through the course.

   The aforementioned 3D camera has been tested in initial mapping experiments, and found to be insufficient to provide good 3D mapping in the research environment. An example of the accuracy problem occurs when bright objects near the camera distort the range of distant objects by reducing the range to these objects. Thus, a wall 1 meter distant from the camera might appear 70 cm distant if the camera is placed near or on a bright floor, which was the case in the environment. These problems with the camera accuracy have been described in [7] and [8], and ongoing work suggests that the accuracy of the 3D camera can be improved significantly ([10],[13]).

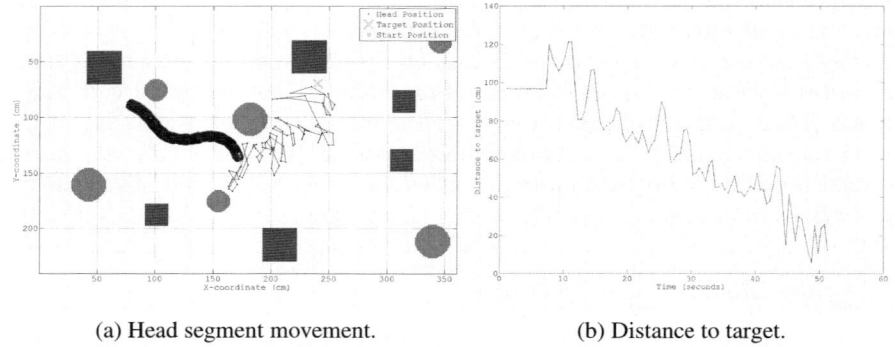

(a) Head segment movement.   (b) Distance to target.

Fig. 5. Path planning through obstacle course

5.2   Force Readings from Obstacle Interference

The snake robot was set up using lateral undulation motion with an amplitude of 0.61 radians. Two circular obstacles of radius 15 cm were introduced to the right and left of the snake at a distance of 10 cm at timestep 300, with centerline coinciding with the snake's lateral centerline at the third segment. Results from the FSRs of the third segment are shown in Fig. 6b, and corresponding relative angles of the third segment as opposed to the second are shown in Fig. 6a. Presented results from the FSRs have deducted the static bias, which is found by reading the corresponding FSRs at stand-still. The differential magnitude responses of the FSRs from friction and impulse forces can be seen in Fig. 6b. Each FSR sensor must be calibrated to remove the effect of pretensioning and nonlinear operation response. Calibrated sensor readouts may range from simple contact - no-contact responses to direct force measurements.

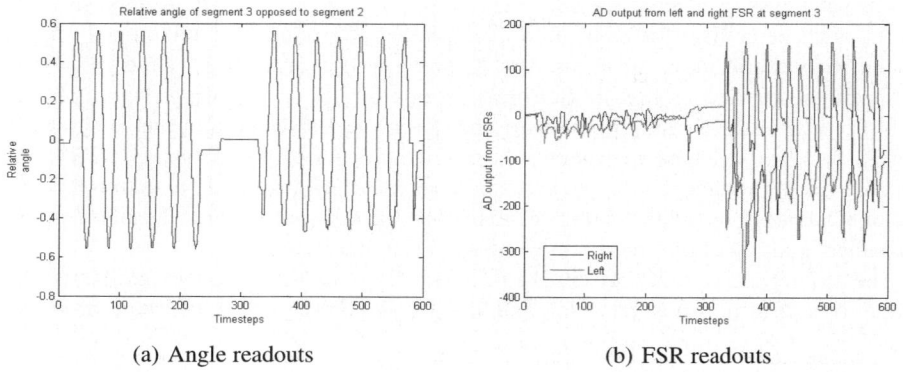

(a) Angle readouts   (b) FSR readouts

Fig. 6. Angle and FSR data from lateral undulation with and without obstacle interference, introducing the interference at timestep 300

6   Conclusions

This paper has presented a benchmarking facility allowing reproducible, reconfigurable experiments in a controlled environment with data collection possibilities both from the environment and the snake itself.

Future research on adaptive snake robot locomotion needs some platform to evaluate robustness and adaptibility of experimental designs, as well as detailed data from both snake and environment. This benchmarking facility is a step further in this direction, and the modularity of the setup enables easy interchanging of sensors and sensor setup to experiment on different configurations.

There is still much work to be done to discover some optimal sensory means in combination with a gait adapting to and using its close environment, and the presented facility enables a common reference on which to experiment and evaluate future ideas.

References

1. Burdick, J.W., Radford, J., Chirikjian, G.S.: A 'sidewinding' locomotion gait for hyper-redundant robots. In: Proc. IEEE Int. Conf. Robotics and Automation, May 1993, pp. 101–106 (1993)
2. Einramhof, P., Olufs, S., Vincze, M.: Experimental evaluation of state of the art 3d-sensors for mobile robot navigation. In: 31st AAPR/OAGM Workshop (2007)
3. Granosik, G., Borenstein, J.: Integrated joint actuator for serpentine robots. IEEE-ASME Trans. Mech. 10(5), 473–481 (2005)
4. Hirose, S.: Biologically Inspired Robots: Snake-Like Locomotors and Manipulators. Oxford University Press, Oxford (1993)
5. Hirose, S., Mori, M.: Biologically inspired snake-like robots. In: Proc. IEEE Int. Conf. Robotics and Biomimetics (2004)
6. Kadous, M., Sheh, R., Sammut, C.: Caster: a robot for urban search and rescue. In: Proceedings of the 2005 Australasian conference on robotics and automation, ARAA, pp. 1–10 (2005)
7. Kahlmann, T., Ingensand, H.: Calibration of the fast range imaging camera swissranger for use in the surveillance of the environment. In: Kamerman, G.W., Willetts, D.V., Steinvall, O.K. (eds.) Electro-Optical Remote Sensing II, SPIE, vol. 6396, p. 639605 (2006)
8. Kahlmann, T., Remondino, F., Ingensand, H.: Calibration for increased accuracy of the range imaging camera swissranger. In: Proceedings of the ISPRS Commision V Symposium Image Engineering and Vision Metrology, vol. XXXVI, pp. 136–141 (2006)
9. Kamegawa, T., Yarnasaki, T., Igarashi, H., Matsuno, F.: Development of the snake-like rescue robot 'Kohga'. In: Proc. IEEE Int. Conf. Robotics and Automation, vol. 5, pp. 5081–5086 (2004)
10. Kavli, T., Kirkhus, T., Thielemann, J.T., Jagielski, B.: Modelling and compensating measurement errors caused by scattering in time-of-flight cameras (to be published, 2008)
11. Liljebäck, P., Stavdahl, Ø., Beitnes, A.: SnakeFighter. development of a water hydraulic fire fighting snake robot. In: Proc. IEEE Int. Conf. Control, Automation, Robotics, and Vision (December 2006)
12. Liljebäck, P., Stavdahl, Ø., Pettersen, K.Y.: Modular pneumatic snake robot: 3D modelling, implementation and control. In: Proc. 16th IFAC World Congress (July 2005)
13. Mure-Dubois, J., Hügli, H.: Real-time scattering compensation for time-of-flight camera. In: Proc. Of the ICVS (2007)

14. Nilsson, M.: Serpentine locomotion on surfaces with uniform friction. In: Proc. IEEE/RSJ Int. Conf. Intelligent Robots and Systems, pp. 1751–1755 (2004)
15. Ohno, H., Hirose, S.: Design of slim slime robot and its gait of locomotion. In: Proc. IEEE/RSJ Int. Conf. Intelligent Robots and Systems, vol. 2, pp. 707–715 (2001)
16. Paljug, E., Ohm, T., Hayati, S.: The jpl serpentine robot: a 12-dof system for inspection. In: IEEE Int. Conf. on Robotics and Automation, vol. 3, pp. 3143–3148 (1995)
17. Tanev, I.: Genetic programming incorporating biased mutation for evolution and adaptation of snakebot. Genetic Programming and Evolvable Machines 8(1), 39–59 (2007)
18. Togawa, K., Mori, M., Hirose, S.: Study on three-dimensional active cord mechanism: Development of ACM-R2. In: Proc. IEEE/RSJ Int. Conf. Intelligent Robots and Systems, vol. 3, pp. 2242–2247 (2000)
19. Transeth, A.A., Liljebäck, P., Pettersen, K.Y.: Snake robot obstacle aided locomotion: An experimental validation of a non-smooth modeling approach. In: Proc. IEEE/RSJ Int. Conf. Intelligent Robots and Systems (to appear, 2007)
20. Worst, R., Linnemann, R.: Construction and operation of a snake-like robot. In: Proc. IEEE Int. Joint Symp. Intelligence and Systems, Rockville, MD, USA, November 1996, pp. 164–169 (1996)
21. Hiroya Yamada, S., Chigisaki, M., Mori, K., Takita, K., Ogami, K., Hirose, S.: Development of amphibious snake-like robot ACM-R5. In: Proc. 36th Int. Symp. Robotics (2005)
22. Ye, C., Ma, S., Li, B., Liu, H., Wang, H.: Development of a 3d snake-like robot: Perambulator-ii. In: Int. Conf. on Mechatronics and Automation, ICMA 2007, pp. 117–122 (2007)

Architecture for Neuronal Cell Control of a Mobile Robot

Dimitris Xydas[1], Daniel J. Norcott[1], Kevin Warwick[1], Benjamin J. Whalley[2],
Slawomir J. Nasuto[1], Victor M. Becerra[1], Mark W. Hammond[1,2],
Julia Downes[1], and Simon Marshall[2]

[1] School of Systems Engineering, University of Reading, UK
{D.Xydas,D.J.Norcott,K.Warwick,S.J.Nasuto,V.M.Becerra,
J.Downes}@reading.ac.uk
[2] School of Pharmacy, University of Reading, UK
{B.J.Whalley,M.W.Hammond,S.Marshall}@reading.ac.uk

Summary. It is usually expected that the intelligent controlling mechanism of a robot is a computer system. Research is however now ongoing in which biological neural networks are being cultured and trained to act as the brain of an interactive real world robot – thereby either completely replacing or operating in a cooperative fashion with a computer system. Studying such neural systems can give a distinct insight into biological neural structures and therefore such research has immediate medical implications. In particular, the use of rodent primary dissociated cultured neuronal networks for the control of mobile 'animats' (artificial animals, a contraction of animal and materials) is a novel approach to discovering the computational capabilities of networks of biological neurones. A dissociated culture of this nature requires appropriate embodiment in some form, to enable appropriate development in a controlled environment within which appropriate stimuli may be received via sensory data but ultimate influence over motor actions retained. The principal aims of the present research are to assess the computational and learning capacity of dissociated cultured neuronal networks with a view to advancing network level processing of artificial neural networks. This will be approached by the creation of an artificial hybrid system (animat) involving closed loop control of a mobile robot by a dissociated culture of rat neurons. This 'closed loop' interaction with the environment through both sensing and effecting will enable investigation of its learning capacity This paper details the components of the overall animat closed loop system and reports on the evaluation of the results from the experiments being carried out with regard to robot behaviour.

Keywords: Dissociated neurones, robotic animats, culture stimulation, neural plasticity.

1 Introduction

The cognitive and motor abilities of animals suggest that the nervous system is capable of solving sensory-motor tasks which pose considerable challenges for current computational approaches. Prior work has used simplified models of the nervous system in the form of artificial neural networks for tasks such as controlling a mobile robot. However, these models differ from their real world counterparts, not only in accuracy but also in flexibility and power.

For the purpose of the present research, it is necessary that the disembodied cell culture is provided with embodiment, since a dissociated cell culture growing in isolation

H. Bruyninckx et al. (Eds.): European Robotics Symposium 2008, STAR 44, pp. 23–31, 2008.
springerlink.com © Springer-Verlag Berlin Heidelberg 2008

and receiving no sensory input is unlikely to develop useful operation since sensory input significantly affects neuronal connectivity and is involved in development of meaningful relationships necessary for useful processing. As a result, the use of different animats seems the most logical solution to the re-embodiment problem.

Typically, *in vitro* neuronal cultures consist of thousands of neurones, hence signals generated by them are highly variable and multi-dimensional. In order to extract from such data components/features of interest which are representative of the network's overall state, appropriate pre-processing and dimensionality reduction techniques must be applied.

The electrically-evoked and spontaneous responses of the neuronal network are coupled to the robot architecture via a machine learning interface mapping the features of interest to specific actuator commands. Mapping sensory data from the robot to a set of suitable stimulation protocols delivered to the neuronal network closes the robot-culture loop. Thus, signal processing can be broken down into two discrete areas: 'culture to robot', an output machine learning procedure processing recorded neuronal activity and 'robot to culture', an input mapping process, from sensor to stimulus.

Several animats reported in the literature have been constructed in order to investigate the control capacity of hybrid systems. Notably, Shkolnik created a very interesting control scheme for a simulated robot [1]. Two channels of a Multi-Electrode Array (MEA) were selected for stimulation. A stimulating signal consisted of a +/-600mV, $400\mu s$ biphasic pulse delivered at varying intervals. The concept of information coding is formed by testing the effect of exciting with a given time delay called the Inter-Probe Interval (IPI) between two probes i.e. electrode pulses. This technique gives rise to a characteristic response curve which forms the basis for deciding the animat's direction of movement using basic commands (forward, backward, left and right).

Other groups have used a simulated rat [2] which moved inside a four-wall environment including barrier objects, or the physical robots such as 'Koala' and 'Khepera' robots [3]. The latter were used in an embodiment experiment wherein one of the robots ('hybrid living robot - hybrot') would attempt to maintain a constant distance from the other which moved under random control. The Koala robot managed to successfully approach the Khepera and maintain a fixed distance from it.

DeMarse investigated the computational capacity of cultured networks and introduced the idea of implementing the results in a "real-life" problem, such as that of controlling a simulated aircraft's flight path (e.g. altitude and roll adjustments) [4].

It is clear that even at such an early stage animat re-embodiments (real or virtual) have a prevailing role in the study of biological learning mechanisms. Our proposed physical and simulated animats provide the starting point for creating a proof-of-concept control loop around the neuronal culture and a basic platform for future more specific reinforcement learning experiments. As the fundamental problem is the coupling of the robot's goals to the culture's input output mapping the design of the animat's architecture discussed in this paper emphasises the need of flexibility and the use of machine learning techniques in search of such coupling.

The next section describes the main elements of the closed loop control of the animat and details the current system's architecture and section 3 describes the initial tests and our preliminary results. Finally, section 4 concludes with an overview of current progress and a discussion on planned future extensions.

2 Closing the Loop

Our animat system is constructed with a closed-loop, modular architecture in mind. Neuronal networks exhibit spatiotemporal patterns with millisecond precision [5], processing of which necessitates a very rapid response from neurophysiological recording and robot control systems. The software developed for this project runs on Linux-based workstations communicating over ethernet via fast server-client modules, thus providing the necessary speed and flexibility required when working with biological systems.

The study of cultured biological neurones has in recent years been greatly facilitated by the availability of commercially available systems such as the MEA's manufactured by MultiChannel Systems GmbH [7]. These consist of a glass culture petri-dish lined with an 8x8 array of electrodes as shown in Figure 1 next.

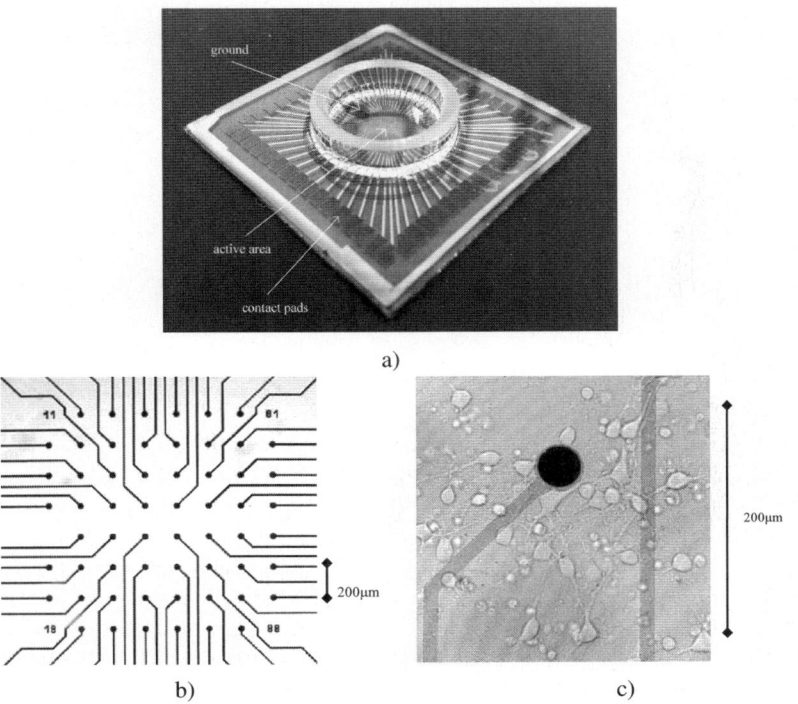

Fig. 1. a) Typical glass MEA, showing large contact pads which lead to the electrode column – row arrangement **b)** Electrode arrays in the centre of the MEA, as typically seen under an optical microscope, **c)** Single electrode close-up, showing a number of neuronal cells in close proximity along with a vast number of neural connections between them

A standard MEA measures 49 mm x 49 mm x 1 mm and its electrodes provide a bidirectional link between the culture and the rest of the system. The data acquisition hardware is provided by MCS and includes the head-stage (MEA connecting interface), amplifier, stimulus generator and PC data acquisition card. The remaining components of the system include the equipment manufactured by MCS such as the MEA petri-dishes which host the cultured networks and the recording hardware controlled by the MCCard (Peripheral Component Interconnect (PCI) – based); the Miabot Pro [7], a small mobile robot exhibiting very accurate motor precision (<0.5 mm) and speed (~3.5 m/s;) (Merlin Robotics [8]); a PCI-DAS1200 DAQ card (Measurement Computing Corp) [9]; and stimulating hardware. The primary software used for signal recording is the Linux-based MEABench toolkit [10]. The modular approach to the problem as discussed earlier, can be seen in more detail in Figure 2 below.

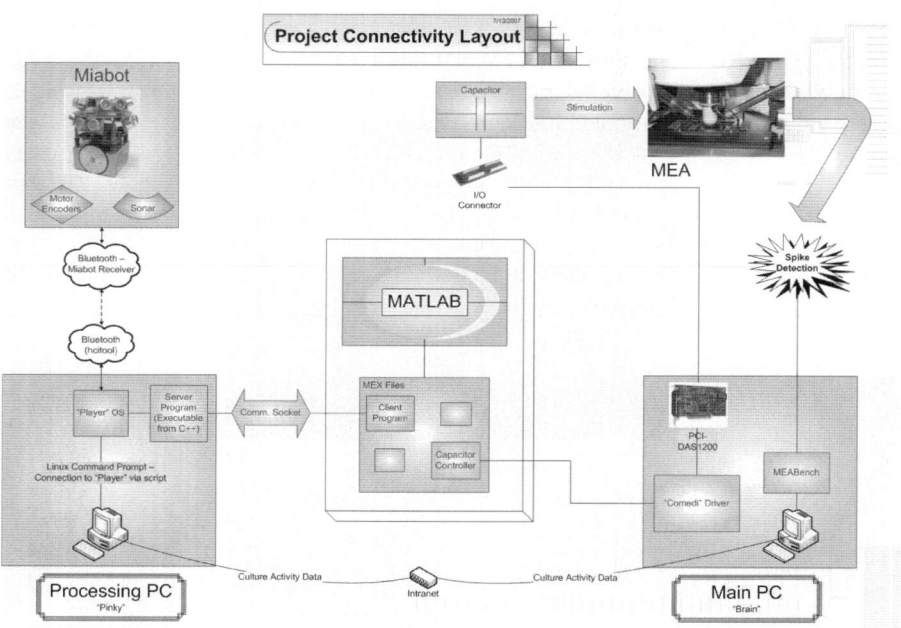

Fig. 2. Modular Layout of Animat/MEA System

The Miabot is wirelessly controlled via Bluetooth. Most Miabot development software is developed under Windows. However, we also utilised for the project the Linux-based 'Player' project [11], which provides an interface compatible with the Miabot. Player provides an appropriate set of interfaces such as position and sensor commands thus constituting a generic robot OS. To complete the robot control system, appropriate control drivers must be included. For this purpose we used an open source software, Miabot Plugin [12], [13], which interfaces with the Player project allowing control of most of the Miabot's functions. Player can connect to the Miabot wirelessly via Bluetooth, and client-side software written in C++ allows connecting to Player over a Transmission Control Protocol (TCP) socket in order to send motor commands and read sonar data. Subsequent control programs were written in C++ or Matlab "Mex" files (converted C++ code) which communicate over TCP ports.

In order to visualise the sonar data in a more effective way than the basic command prompt output and to provide a simulated environment for the experiments, the virtual reality Matlab toolbox [14] was employed to display a virtual Miabot, its sonars and the immediate environment around it, i.e. the wooden playpen. The 3D model, sensory representations and surrounding environment were designed under the 3DStudio Max modelling package (Autodesk Inc. [15]) and exported as a Virtual Reality Modelling Language (VRML) file. Also, the current simulation model can be used to generate course-plotting graphs of the robot's path inside its constrained space over any required period of time.

The main advantage of any simulated device over its real-life counterpart is of course the avoidance of all potential real-life noise sources and physical constraints imposed upon the experiment. The simulation enables rapid deployment of experiments without dealing with real-time problems, power consumptions etc. An added advantage is that noise parameters can be modelled to emulate real-life conditions if required by the experiment. However, a real-life robot is the ultimate aim of the project, since its principal objective is the evaluation of the potential of live cultures in real-life problem solving situations; with all the related noise and disturbances included.

Real-life and simulated representations of the animat embodiments can be seen in Figure 3 below. The Miabot may be extended with further multiple sensory inputs such as sonar arrays, camera modules etc.

 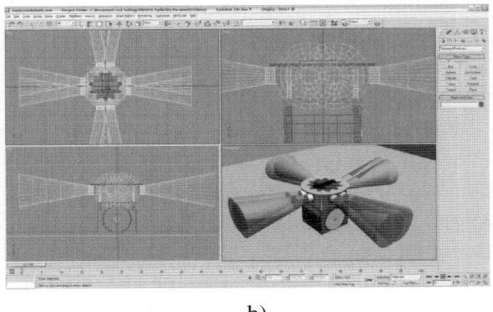

a)   b)

Fig. 3. a) Miabot Pro with 8-way sonar array pack, **b)** Virtual animat designed using mainly basic geometrical primitive shapes and exported as VRML file

In order to control the stimulus applied to the culture we have developed a custom set of libraries which interface with existing available low-level drivers, along with a flexible and easy-to-use GUI front-end.

The modular approach to the architecture has resulted in a system with easily reconfigurable components. The obtained closed-loop system can efficiently handle the information-rich data that is streamed via the recording software. A typical sampling frequency of 25 kHz of the culture activity recording leads to a very large required bandwidth and in consequence vast data storage space for raw data. For this reason, on-the-fly streaming of spike-detected data is the preferred method when investigating real-time closed-loop learning techniques.

A Linux-based system constitutes the core of our developed software incorporating freely available recording and data analysis toolkit, MEABench, a tried-and-tested software kit used by many leading research groups around the world (Georgia Tech [1], Caltech, [2] and the University of Florida [3]).

3   Preliminary Results

Initial proof-of-concept tests of the developed system were split in two parts. An initial interfacing test was set up in order to validate the animat's modular control architecture using pre-recorded neuronal data files. The activity of each electrode, in terms of its detected mean firing rate, is determined from the samples of the recorded set. The four most active electrodes are selected. The mean firing rate between the first two chosen electrodes is used to determine the Miabot's forward/backward direction of motion, with the analogous coupling of the remaining two electrodes activity for left/right direction.

A second set of tests used live streamed spike data as actuator commands. The robot's forward-backward and left-right movements were controlled by the activity levels of 4 of the most active MEA electrodes in terms of firing frequency. The electrode array was split into four quadrants mapped onto four directional robot commands. The later demo was used to test the control loop's stability and processing speed with respect to the recording software and possible lags due to data stream buffering.

Live data streaming and inclusion of the stimulation procedure all in an online manner have enabled us to consider how acquired data should be pre-processed and subsequently mapped into output commands.

Finding a mapping between the external goals set for the robot onto the mechanisms involved in information processing in the culture has been one of the most significant challenges facing animat development. The presently developed system permits analysis of culture activity patterns and its response to external stimuli which are instrumental in defining such mapping and identification of underlying plasticity mechanisms. Such investigations make possible the offline identification of electrodes used in conditioning experiments aimed at creating culture responses onto which the robot actions may be mapped.

The developed stimulus software allowed us to perform the preliminary experiments on Long Term Potentiation (LTP) and tetanisation protocols, with promising results.

4   Conclusions

4.1   Overview

The project, although still in early stages, has achieved a number of milestones and overcome a large array of technological challenges.

The culture preparation techniques are constantly being refined and have lead to successful and stable cultures that exhibit both spontaneous and induced spiking/bursting activity.

A stable robotic infrastructure has been set up, tested and is in place for future machine learning and culture behaviour experiments. The animat platform consists of both a hardware element and a software simulation element. The current rate of progress could be the inspiration for offspring projects investigating culture-mediated control of a wide array of additional robotic devices, such as robotic arms/grippers, mobile robot swarms and multi-legged walkers to name but a few.

4.2   Future Work

There are a number of ways in which the current system will be expanded in the future. The Miabot may be extended to include additional sensory devices such as extra sonar arrays, mobile cameras and other range-finding hardware. A considerable current limitation is the battery power supply of an otherwise autonomous robot. A main future consideration is the inclusion of a powered-floor, which would provide the robot with relative autonomy for a longer period of time while the suggested machine learning techniques are applied and the culture's behavioural responses are monitored. The future work will adapt a Miabot to operate on an in-house powered floor, so the animat can be provided with a constant power supply; this feature is necessary since machine learning and culture behaviour tests will be carried out for many minutes and even hours at a time. At present however the robotic simulation provides an alternative solution to continuous operation of the closed loop avoiding current hardware limitations.

A variety of custom stimulation techniques utilising the effect of weak electric fields on whole-network activity are also being constructed [19]. This will provide an interesting alternative stimulation method, particularly considering that neurons are much more sensitive to such fields than had been previously believed. A number of hardware approaches have been proposed, keeping in mind the hardware constraints of the MEA recording system. The stimulus generating electrodes are controlled by a PCI-DAS1200 Measurement Computing data acquisition card, which is in turn controlled by appropriate software through the Linux environment.

Using the full dimensionality of the culture-generated outputs will require the dimensionality reduction procedures extracting useful information for animat control. They will take inspiration from established data-processing algorithms such as data clustering techniques, Self Organising Maps (SOM's) [20], Generative Topographic Mapping [21] and Principal Component Analysis.

The current hardcoded mapping between the robot goals and the culture input output relationships will be extended to reinforcement learning techniques such as Q-Learning [22] and Temporal Difference Methods [23] which will reduce or even eliminate the need for a priori mapping choice.

Also, progression of the project will require benchmarking both the machine learning techniques and the results obtained by the culture. In order to achieve this, we will aim to develop a simulation of the neural network, based on the culture's observed connectivity density and activity. This behavioural evaluation model is likely to provide great insight into the workings of the neuronal network by comparing the model's performance versus the culture's performance as well as learning capabilities as expressed by changes in its neural plasticity.

Acknowledgements. This work is funded by the Engineering and Physical Sciences Research Council (EPSRC) under grant No. EP/D080134/1.

References

1. Shkolnik, A.C.: Neurally controlled simulated robot: applying cultured neurons to handle an approach / avoidance task in real time, and a framework for studying learning in vitro, Emory University. Mathematics and Computer Science (2003)
2. DeMarse, T.B., et al.: The Neurally Controlled Animat: Biological Brains Acting with Simulated Bodies. In: Autonomous Robots, vol. 11, pp. 305–310. Kluwer Academic Publishers, Dordrecht (2001)
3. Neurally-Controlled Animat. Potter Group. [Online] [Cited: 08 29, 07.], http://www.neuro.gatech.edu/groups/potter/animat.html
4. DeMarse, T.B., Dockendorf, K.P.: Adaptive flight control with living neuronal networks on microelectrode arrays. In: Proceedings of IEEE International Joint Conference on Neural Networks (IJCNN 2005), pp. 1548–1551 (2005)
5. Rolston, J.D., Wagenaar, D.A., Potter, S.M.: Precisely Timed Spatiotemporal Patterns of Neural Activity In Dissociated Cortical Cultures. Neuroscience 148, 294–303 (2007)
6. Multi Channel Systems Homepage. [Online] [Cited: 09 16, 07.], http://www.multichannelsystems.com
7. Miabot Pro Research Pack. Merlin Robotics - Creators of Robots for people. [Online] [Cited: 09 07, 07.] (1880), http://www.merlinrobotics.co.uk/merlinrobotics/miabot-research-pack-p-61.html?osCsid=2ba3af93a76f783161d2ef73b5018803
8. Merlin Robotics - Creators of Robots for people. Merlin Robotics. [Online] [Cited: 10 31, 07.], http://www.merlinrobotics.co.uk/merlinrobotics/
9. Measurement Computing Corp.: Product: 'PCI-DAS1200'. Data Acquisition from Measurement Computing. [Online] [Cited: 09 07, 07.], http://www.measurementcomputing.com/cbicatalog/cbiproduct_new.asp?dept_id=138&pf_id=659&mscssid=BTUKT461K5Q48KNQJL6JNK9T5TKXFN9O
10. Wagenaar, D.A.: Meabench. Wagenaar's Meabench Multi-electrode data acquisition and analysis. [Online] [Cited: 07 21, 07.], http://www.its.caltech.edu/~pinelab/wagenaar/meabench.html
11. The Player Project. [Online], http://playerstage.sourceforge.net/index.php?src=player
12. The Robot Soccer Project. In: Robot Lab - University of Nottingham. [Online] [Cited: 09 10, 07.], http://www.asap.cs.nott.ac.uk/řobots/wiki/index.php/The_Robot_Soccer_Project
13. Miabot Plugin Manual (Wiki). Robot Lab - University of Nottingham. [Online] [Cited: 11 04, 2007.], http://www.asap.cs.nott.ac.uk/řobots/wiki/index.php/Plugin_Manual_%28Wiki%29
14. Virtual Reality Toolbox. The MathWorks - MATLAB and Simulink for Technical Computing. [Online] [Cited: 11 05, 2007.], http://www.mathworks.com/access/helpdesk/help/toolbox/vr/

15. Autodesk 3ds Max. Autodesk. [Online] [Cited: 09 16, 07.],
    `http://usa.autodesk.com/adsk/servlet/`
    `index?id=5659302&siteID=123112`
16. Potter, S., et al.: NeuroLab. The Laboratory for Neuroengineering (NeuroLab). [Online]
    [Cited: 08 29, 07.], `http://neuro.gatech.edu/`
17. Pinelab at Caltech. The Pine Lab. [Online] [Cited: 09 17, 07.],
    `http://www.its.caltech.edu/p̃inelab/new_pinelab_page/`
    `pine_lab.htm`
18. Biomedical Engineering Homepage. Pruitt Family Department of Biomedical Engineering.
    [Online] [Cited: 09 17, 07.], `http://www.bme.ufl.edu/`
19. Francis, J.T., Gluckman, B.J., Schiff, S.J.: Sensitivity of Neurons to Weak Electric Fields.
    Part 19. Journal of Neuroscience 23, 7255–7280 (2003)
20. Vesanto, J., Alhoniemi, E.: Clustering of the self-organizing map. IEEE Computational In-
    telligence Society, IEEE Transactions on Neural Networks 11, 586–600 (2000)
21. Bishop, C.M., Svensén, M., Williams, C.K.I.: GTM: The Generative Topographic Mapping.
    Neural Computation 10, 215–234 (1998)
22. Watkins, C.J.C.H., Dayan, P.: Q-learning. In: Machine Learning, vol. 8, pp. 279–292.
    Springer, Netherlands (1992)
23. Tesauro, G.: Practical issues in temporal difference learning. In: Machine Learning, vol. 8,
    pp. 257–277. Springer, Netherlands (1992)

The Ares Robot:
Case Study of an Affordable Service Robot

Pedro Santana[1], Carlos Cândido[2], Paulo Santos[2], Luís Almeida[2],
Luís Correia[1], and José Barata[3]

[1] LabMAg, University of Lisbon, Portugal
   pfs@uninova.pt
[2] R&D Division, IntRoSys, S.A.
[3] UNINOVA, New University of Lisbon, Portugal

Summary. Robustness is pivot for robots operating in all-terrain environments. This demand comes mainly due to the highly heterogeneous and unstructured nature of the terrain. Two particular topics are sensitive to this problem: locomotion control and wheel odometry. A behaviour-based approach is proposed for the locomotion control system, whereas a heuristic exploiting the kinematic and dynamical constraints of the robot is used to enhance wheel odometry accuracy. Experimental results on the Ares robot, which is a $1.5\,m^2$ vehicle with four independently steered wheels, show the ability of the proposed methods to cope with all-terrain environments. In addition, the modules for localisation, mapping, and obstacle avoidance are also addressed in order to provide a global perspective over the Ares robot's control system.

1 Introduction

Robot mobility is a hot topic if robots are to be used in all-terrain. Wheeled, legged, tracked, and hybrid solutions have been considered, being the tracked solutions the most successful in domains like urban Search & Rescue and EOD (e.g. IRobot's[1] Packbot). Wheeled platforms have been the choice for interplanetary rovers (e.g. NASA MARS rovers[2]) though. In opposition to the success of wheeled and tracked solutions, the complexity and cost of legged solutions have been hampering their application in real-life applications. Addressing the topic of medium size affordable service robots capable of operating in all-terrain environments, the wheeled solution emerged as natural. Wheel-based robotic platforms require less maintenance, and are typically cheaper, more efficient, and lighter than their tracked counterparts.

The wheeled robot herein proposed, i.e. the Ares robot, has been developed to fill the gap between improved indoor robots for outdoor environments (e.g. MobileRobots Inc.[3] Pioneer 3-AT) and complex solutions as those considered in interplanetary missions. To our knowledge, no other robot is endowed with the mobility degree of the Ares robot at such a low cost, and with so little maintenance requirements. These features are essential to enable mass production robots for all-terrain environments.

Robustness is also key for robots operating in all-terrain environments. This demand arises mainly from the highly heterogeneous and unstructured nature of the terrains in

---

[1] http://www.irobot.com/
[2] http://marsrover.nasa.gov/
[3] http://www.activrobots.com/

H. Bruyninckx et al. (Eds.): European Robotics Symposium 2008, STAR 44, pp. 33–42, 2008.
springerlink.com   © Springer-Verlag Berlin Heidelberg 2008

question. Two particular topics are sensitive to this problem: (1) locomotion control and (2) wheel odometry. A behaviour-based approach is proposed for the locomotion control system, whereas an heuristic exploiting the kinematic and dynamical constraints of the robot is used to enhance wheel odometry accuracy. Experimental results on the Ares robot show the ability of the proposed methods to cope with all-terrain environments. In addition, the modules for localisation, mapping, and obstacle avoidance are also addressed in order to provide a global perspective over the Ares robot's control system.

This paper is organised as follows. The Ares robot platform is presented in Sec. 2, followed by the description of the major software components in Sec. 3. Afterwards, field trials are reported in Sec. 4. Finally, some conclusions and future work ideas are drawn in Sec. 5.

2   The Ares Robot

This section overviews the main characteristics of the physical robot considered in this paper. All-terrain robots must be able to adapt themselves to the unevenness of the terrains in which they will be deployed, and at the same time, be agile enough to dodge unexpected obstacles. To reduce mechanical stress and energy costs, the robot must be able to produce smooth trajectories avoiding slippage as much as possible. With its four independently steered wheels, the Ares robot (see Fig. 1)[8] has been developed to take all these issues into account.

(a)   (b)

Fig. 1. The Ares robot in a) and its four locomotion modes in b): turning-point (top-left), lateral (top-right), displacement (bottom-left), and double Ackerman (bottom-right)

The degrees of freedom Ares possesses allow it to displace in several modes. In particular, (1) in *double Ackerman mode* the robot is able to produce circular trajectories, (2) in *turning-point mode* it rotates around its own geometric centre, (3) in *displacement mode* it produces linear trajectories along a limited set of directions ($[-\pi/4, \pi/4]$ relative to robot's front), and (4) in *lateral displacement mode* it moves sideways. The robot is composed of two main blocks (with two wheels each), the front and the rear ones.

These blocks can freely and independently rotate around a longitudinal axis. By having this passive joint, the robot is capable of being compliant with respect to uneven terrain. The upper bounds of the volume occupied by the robot are 1.5 m x 1.4 m x 0.7 m. The actual volume varies according to the selected locomotion mode. By using bicycle wheels, the tires can be easily replaced in order to better comply with the needs of a given terrain. These two characteristics allied to the 40 cm height to the ground endow the Ares robot with great flexibility. Being able to surpass most natural obstacles, the perception requirements for safe navigation drop.

Due to its low-cost and versatility (e.g. useful for both localisation and environment perception), stereo vision has been selected as primary sensory modality. The key hardware components are a Videre Design STOC stereo head, a Honeywell HMR 3000 attitude sensor, a Novatel DGPS OEMV-1 system, four RoboteQ AX3500 boards for speed control of the eight Maxon 150 W motors, a Diamond Systems Hercules EBX PC-104 stack as on-board computer for direct robot control, and a Pentium M 2.0 GHz laptop running stereo vision, mapping, and navigation algorithms. Both computers run the operating system Linux.

3   Control Software

The architecture follows the behavioural paradigm [2], in which the robot's control is distributed over several perception-action loops (called behaviours). In addition to behaviours, the system also encompasses a set of functional modules. Briefly, localisation and attitude estimation (cf. Section 3.2) is the basis for the creation of the environment's map. Based on the obstacles present in the map and robot's estimated localisation/attitude, the obstacle avoidance behaviour directs the robot according to a desired direction and a desired speed (cf. Section 3.3). Two particular higher level behaviours sending requests to the obstacle avoidance behaviour are the recording and path following. At the bottom of the architecture lays the locomotion control system, behaviour-based as well, which controls the robot's actuators according to the output generated by the obstacle avoidance behaviour (cf. Sec. 3.1).

3.1   Locomotion Control

Classical control theory is usually concerned with optimal set-points following. When the set-points are highly dynamic (i.e. hardly reachable) and the robot's interactions with the environment highly non-linear and discontinuous, classical solutions collapse drastically. Bearing this in mind, in [9] a novel behaviour-based approach focused on robustness instead of optimality has been proposed for the locomotion control of the Ares robot. In such approach, each wheel has an independent controller composed of a set of behaviours (i.e. units linking perception to action) generating force vectors according to a given criterion. The resulting force is applied to the wheel's steering actuator. Besides the well known local minima problem, behaviour fusion alone has not enough expressiveness to handle exceptional situations, which typically have a discontinuous nature.

The following describes an enhanced version of the method, which in addition to behaviour fusion, also considers behaviour arbitration among other biologically inspired features. In the case of an arbitration node, the action generated by the behaviour with higher priority is passed on. In a fusion node a weighed average of each behaviour output is considered instead, being the weights related to each behaviour's priority. Therefore, arbitration nodes are more interesting to allow changing the global behaviour in a sudden and qualitative way. On the other hand, fusion nodes are better to enable cooperation among behaviours, and in particular, to have one behaviour modulating other behaviour's output.

Fig. 2 depicts the model of each wheel controller. Data flowing from perception to behaviours, from behaviours to coordination nodes, and from behaviours/coordination nodes to actuators are conveyed through information links. In order to allow hierarchical decomposition and run-time adaptation, behaviours can activate (excite) or deactivate (inhibit) other behaviours or information links through activation links.

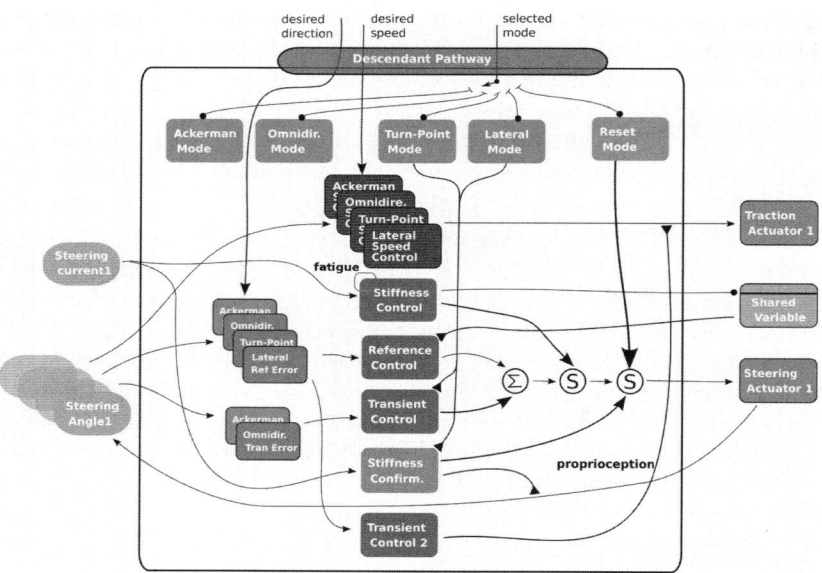

Fig. 2. The behaviour-based wheel controller. Arrows refer to information links. The thickest the arrow, the higher the priority of the behaviour. Lines with dotted and triangular end points refer to excitatory and inhibitory activation links, respectively. Arbitration and fusion nodes are denoted by circles with labels S and Σ, respectively.

Higher control layers can modulate the locomotion controllers for a given direction of motion, speed, and locomotion mode. Activating a locomotion mode reflects in the activation of the corresponding hierarchical higher behaviour ("Mode" labelled boxes). The desired speed feeds directly the behaviours responsible for controlling the robot's speed ("Speed" labelled boxes), whereas the desired direction feed into the perceptual entities ("Ref Error" labelled boxes) responsible for computing the wheel's steering error.

Due to space limitations, only the double Ackerman mode will be detailed. The `reference control` behaviour seeks to steer the wheel towards the angle that allows the robot to turn towards the desired motion direction. Hence, each wheel will seek to reach its target regardless of the existence of any other slower or faster wheel. The `transient control` behaviour adapts the output of the `reference control` behaviour so as to maintain the Ackerman geometry. If the controlled wheel is drifting from the Ackerman geometry (when compared to the other three wheels), it must be slowed down or speeded up, depending on the situation. As all wheels tend to do the same, eventually they all converge to the Ackerman geometry, provided that the system's parameters have been carefully tuned. Upon reaching an upper threshold of the current in the wheel steering actuator, the `stiffness control` behaviour suppresses the output resulting from the fusion of both `reference control` and `transient control`. Hence, if a too strong force is projected onto the wheel in such a way that contradicts the current steering speed, the wheel is asked to stop for a given period. Each time the current level reaches the aforementioned upper threshold before the stopping period expires, the period is incremented by a time constant and its counting restarted, implementing a kind of *fatigue* effect. Here fatigue results from an external behaviour, which progressively inhibits other behaviours' outputs for a longer period, whereas in [3] fatigue arises as an intrinsic property of each behaviour. When the period expires its default value is restored for the next time the behaviour gets activated. When this behaviour becomes active, it excites a `shared variable`, i.e. it sets a shared flag to one, which in turn will inhibit all wheels `reference control` behaviours. The outcome is that all other three wheels will no longer pursue the desired direction in order to focus all the effort in maintaining the Ackerman geometry.

As soon as the `stiffness control` behaviour becomes active, the wheel stops, and as a result the current applied to the steering motor drops. Therefore, the fatigue effect would never occur in this situation. In order to fix this, an *active perception* mechanism was developed and implemented in the `stiffness confirmation` behaviour. This behaviour is active a few seconds before the inactivation of the `stiffness control` behaviour. When active, it suppresses other behaviours output and asks the actuator to turn in the direction it was turning before the activation of the `stiffness control` behaviour for a while, and then to turn in the opposite direction for the same amount of time. If the obstacle is still present, the current will rise and the `stiffness control` behaviour will trigger the fatigue effect. Otherwise the wheel will turn in one direction, then in the opposite one, staying roughly in the same place before the `stiffness confirmation` behaviour was activated. Then, the `stiffness control` behaviour will timeout and the operations are resumed normally. This swing behaviour of the wheel induces other wheels to follow it in order to maintain the Ackerman geometry. To avoid this, the steering angle percept of the wheel in question is frozen (i.e. *gated*) during that period, meaning that the other wheels will not be sensitive to the swinging. The `transient control 2` behaviour compels the robot to stop when the steering actuators geometry error reaches a given threshold. This behaviour ensures that the mechanical structure does not collapse in extreme, unexpected, situations.

Active perception, perceptual gating, fatigue, and behavioural modulation are all ubiquitous mechanisms in nature, and herein exploited to build up a locomotion control system well adapted to handle exceptional situations.

3.2   Localisation

Despite being error prone, odometry is always a relevant aspect of a localisation system, and can in fact provide very useful results in extreme situations [7]. Thus, rather than using expensive inertial systems, odometry has been taken as main mechanism to cope with GPS dropouts. The state estimator first predicts the current state based either on visual odometry [1] or on wheel odometry in case the former fails to produce accurate results. Then, in the correction step the estimation approaches both GPS and attitude sensor outputs in order to guarantee global convergence. The convergence speed towards the global references is a function of the sensor's error estimates. The following proposes a wheel odometry method that exploits the kinematic and dynamical characteristics of the Ares robot.

Based on the kinematic model of the locomotion mode in question (e.g. the bicycle model [12] for Double Ackerman motion), the robot's displacement and heading variation, $(\Delta x_i, \Delta y_i, \Delta \psi_i)$, can be estimated by taking into account both steering angle and travelled linear distance of any wheel $i \in \{1, 2, 3, 4\}$. The redundancy introduced by the possibility of calculating wheel odometry based on a single wheel requires a decision on which wheel to consider. Observations on the dynamical behaviour of the robot suggest that all wheels affect each other strongly, meaning that wheel odometry computation benefits from considering all wheels rather than just a single one. Hence, all wheels estimates contribute to the global estimate, $(\Delta x, \Delta y, \Delta \psi) = (\Sigma_{i=1}^{4} w_i \cdot \Delta x_i, \Sigma_{i=1}^{4} w_i \cdot \Delta y_i, \Sigma_{i=1}^{4} w_i \cdot \Delta \psi_i)$, where the weight of each contribution, w_i, is,

$$w_i = \frac{1 - e_i / \Sigma_{j=1}^{4} e_j}{\Sigma_{k=1}^{4} (1 - e_k / \Sigma_{j=1}^{4} e_j)} \quad (1)$$

being $e_i = |\hat{\delta}_i - \delta_i|$ the difference between the expected angle for wheel i, $\hat{\delta}_i$, and its current steering angle, δ_i. $\hat{\delta}_i = \frac{1}{3} \cdot \Sigma_{j \neq i} \Phi(\delta_j)$, where $\Phi(\delta_j)$ returns the steering angle wheel i should have in order to be coherent (according to the locomotion mode in question) with the steering angle of wheel j.

Intuitively, the odometry computed based on each wheel is weighed for the global estimate according to a function of its estimated error, e_i, normalised with the other wheels estimated errors. Since it is impossible to determine which wheel is failing to meet the kinematic constraints of the locomotion mode, the error of each wheel is computed in relative terms by considering that all other wheels are in the correct position.

3.3   Mapping and Obstacle Avoidance

An obstacle detection algorithm for stereo vision based on [5] updates an Occupancy Grid (OG) [4] featuring an aging mechanism. A time-stamp is associated to each cell in the OG each time an obstacle is detected therein. In [11] the age of a cell is used to reduce the certainty of a cell containing an obstacle. Here a threshold on the age of a cell is used to avoid considering older obstacles in the obstacle avoidance algorithm, and not to remove them from the map. By considering only the newest obstacles, the problems associated to erroneous localisation in the map are diminished.

A set of arcs of trajectory are considered when selecting the next action to perform. The utility of each arc is computed by weighing the following criteria: (1) the distance it is possible to travel along the arc before hitting an obstacle; (2) the predicted heading error (relative to the desired one) the robot will have after travelling for a given small period of time along the arc; (3) the arc's index distance to the previously selected; and (4) the distance of the closest obstacle in a given arc's neighbourhood (i.e. neighbour arcs). The first component promotes arcs with farther obstacles, the second one directs the robot towards a given desired direction, the third one avoids intermittent switching between similarly good arcs, whereas the fourth one pushes the robot away from lateral obstacles.

The arc of trajectory, a_c^o, that would lead a point of the robot's contour, $c = (x_c, y_c)$, to collide against a point obstacle $o = (x_o, y_o)$, is computed according to a closed-form solution [6]. The distance, d_c^o, the robot can travel before hitting the obstacle is also computed with a closed-form solution. Being the contour of the robot composed of a set of contour points $c \in C$, computing the effect obstacles have in all arc trajectories requires performing the previous computations $c \cdot o$ times, with $o \in O$, being O the set of obstacles considered for obstacle avoidance. To avoid this computational load in runtime, a set of lookup tables are computed at the mission's onset, similar to the approach taken in [10].

In runtime each time an obstacle is detected, the possible distance to traverse in each arc of trajectory without colliding against the obstacle is obtained with a direct lookup in the table. Obstacles are obtained directly from the map. As mentioned, only the obstacles below a given age (e.g. 30 sec.), are considered. The speed is then selected according to the terrain's roughness, closest obstacle to the robot, desired speed, and dynamical constraints of the robot (i.e. taking into account that the robot should be able to stop before hitting the closest obstacle in the selected arc).

4   Field Trials

In a first experiment (see Fig. 3) the robot was manually driven in a highly slopped and rough terrain. The goal was to test the locomotion control under demanding conditions. The amount of times the `stiffness control behaviour` triggered (see figure bottom-left) reflects the complexity of the terrain as well as the ability of the method to cope with it. The forces projected onto the robot's wheels were so high that the steering motors could not maintain the pure Ackerman geometry most of the time. See for instance (figure bottom-right) the largest error the front-left wheel has relative to the other three wheels at each instant, while trying to maintain the Ackerman geometry.

Fig. 4 illustrates a second experiment in which the *obstacle avoidance behaviour* is modulated by a *path following behaviour*. The goal of the latter is to maintain the robot over a given path at an approximate desired speed of $0.5\,\mathrm{ms}^{-1}$. This path is previously recorded as a set of 1 m spaced waypoints by manually driving the robot. Only

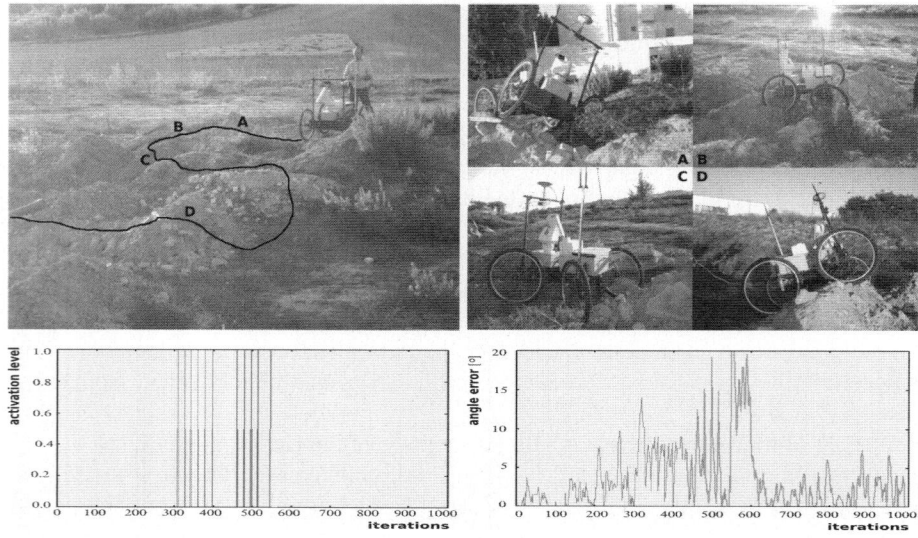

Fig. 3. Experimental setup to test the locomotion control system in which the robot was manually driven along a given path (represented as a black line in top-left). Snapshots of situations A, B, C, and D (top-right) illustrate the complexity of the considered terrain, which resulted in several activations of the `stiffness control behaviour` (bottom-left), and considerable error on front-left wheel steering angle (bottom-right).

(a)   (b)

Fig. 4. Resulting 33 m x 27 m map produced by the robot while autonomously following a given path (see text) in a). Only wheel odometry and compass information were considered. Both starting and ending points of the path are in location S. The travel direction is represented by the arrow. Black, white, and dark gray pixels represent obstacles, non-obstacle regions, and unmapped regions, respectively. The closed-loop line represents the estimated location of the robot during the run. The discontinuous path segment (top-left) represents a portion of the recorded path. The thicker segment of the followed path illustrates the effect of an external magnetic field on the compass. Snap-shots of situations A, B, C, and D in b).

wheel odometry and compass information is considered in order to test the ability of the wheel odometry method to handle a terrain where wheel friction is high, and consequently the wheels geometry is harder to maintain. In order to check the ability of the robot to avoid unexpected obstacles, the waypoints composing the recorded path were translated by 3 m in the direction of the arrow. This way the path to be followed is no longer free of obstacles. A sketch of a portion of the recorded path before the translation is presented to highlight the differences to the followed path. In *A* the robot faced a too close obstacle and consequently had to move backwards before aligning between that obstacle and another one. In *B*, *C*, and *D* the robot also had to dodge unexpected obstacles. After arriving its final destination, the robot's offset relative to the position where it has departed was measured to be roughly 2 m.

5   Conclusions and Future Work

Major software components of a four independently steered wheels robot for off-road environments have been presented. The most significant contributions of this paper are three folded. First, a behaviour-based solution for the locomotion problem has been used, allowing the control system designer to focus on robustness on an experimental basis. The heterogeneous and unstructured nature of all-terrain environments makes this approach more suited than one mostly concerned with optimal behaviour under a constrained set of operating ranges, as pursued in classical control theory. The second contribution is the heuristic employed to reduce wheel odometry errors, which exploits in a simple way the kinematic and dynamical constraints of the robot. Finally, the experimental results show Ares robot to be a successful case-study of a middle size affordable all-terrain robot. Calibration and adaptation of both locomotion control and wheel odometry parameters will be the main focus of future work. In addition, other heuristics for improvement of wheel odometry will be pursued.

Acknowledgements

The authors wish to acknowledge Vasco Santos and Mário Salgueiro for the support on preparing the experimental setup, João Lisboa, Hildebrando Cruz, Rui Maltez, and Hélder Monteiro for the development of the robot's mechanical platform, and Carlos Grilo and António Manso for their fruitful comments. This work was partially supported by FCT/MCTES grant No. SFRH/BD/27305/2006.

References

1. Agrawal, M., Konolige, K.: Real-time Localization in Outdoor Environments using Stereo Vision and Inexpensive GPS. In: Proc. of the 18th Int. Conf. on Pattern Recognition (ICPR 2006), Washington, DC, USA, vol. 3, pp. 1063–1068. IEEE Computer Society, Los Alamitos (2006)
2. Arkin, R.C.: Behavior-Based Robotics, May 1998. MIT Press, Cambridge (1998)
3. Correia, L., Abreu, A.: Forgetting and fatigue in mobile robot navigation. In: Bazzan, A.L.C., Labidi, S. (eds.) SBIA 2004. LNCS (LNAI), vol. 3171, pp. 434–443. Springer, Heidelberg (2004)

4. Elfes, A.: Sonar-based real-world mapping and navigation. Int. J. Rob. Autom. 3(3), 249–265 (1987)
5. Manduchi, R., Castano, A., Talukder, A., Matthies, L.: Obstacle detection and terrain classification for autonomous off-road navigation. Autonomous Robot 18, 81–102 (2003)
6. Minguez, J., Montano, L., Santos-Victor, J.: Abstracting Vehicle Shape and Kinematic Constraints from Obstacle Avoidance Methods. Autonomous Robots 20(1), 43–59 (2006)
7. Ojeda, L., Reina, G., Cruz, D., Borenstein, J.: The FLEXnav precision dead-reckoning system. Int. Journal of Vehicle Autonomous Systems 4(2–4), 173–195 (2006)
8. Santana, P.F., Barata, J., Correia, L.: Sustainable robots for humanitarian demining. Int. Journal of Advanced Robotics Systems (special issue on Robotics and Sensors for Humanitarian Demining) 4(2), 207–218 (2007)
9. Santana, P.F., Cândido, C., Santos, V., Barata, J.: A motion controller for compliant four-wheel-steering robots. In: Proc. of the IEEE Int. Conf. on Robotics and Biomimetics (ROBIO 2006), Kunming, China (December 2006)
10. Schlegel, C.: Fast local obstacle avoidance under kinematic and dynamic constraints for a mobile robot. In: Proc. of the IEEE/RSJ Int. Conf. on Intelligent Robots and Systems (IROS 1998), Victoria, B.C. Canada, pp. 594–599 (1998)
11. Singh, S., Simmons, R., Smith, T., Stentz, A., Verma, V., Yahja, A., Schwehr, K.: Recent progress in local and global traversability for planetaryrovers. In: Proc. of the IEEE Int. Conf. on Robotics and Automation (ICRA 2000), San Francisco, April 2000, vol. 2 (2000)
12. Wang, D., Qi, F.: Trajectory planning for a four-wheel-steering vehicle. In: Proc. of the IEEE Int. Conf. on Robotics and Automation (ICRA 2001), vol. 4 (2001)

Balancing the Information Gain Against the Movement Cost for Multi-robot Frontier Exploration

Arnoud Visser and Bayu A. Slamet

Intelligent Systems Laboratorium Amsterdam, Universiteit van Amsterdam (UvA) Kruislaan
403, 1098 SJ, Amsterdam, The Netherlands
arnoud@science.uva.nl

Summary. This article investigates the scenario where a small team of robots needs to explore a hypothetical disaster site. The challenge faced by the robot-team is to coordinate their actions such that they efficiently explore the environment in their search for victims.

A popular paradigm for the exploration problem is based on the notion of frontiers: the boundaries of the current map from where robots can enter yet unexplored area. Coordinating multiple robots is then about intelligently assigning frontiers to robots. Typically, the assignment of a particular frontier to a particular robot is governed by a cost measure, e.g. the movement costs for the robot to reach the frontier. In more recent approaches these costs are traded off with the potential gain in information if the frontier would be explored by the robot.

In this paper we will further investigate the effect of balancing movement costs with information gains while assigning frontiers to robots. In our experiments we will illustrate how various choices for this balance can have a significant impact on the exploratory behavior exposed by the robot team.

1 Introduction

This paper will investigate a multi-robot exploration approach that was designed with the disaster sites of the RoboCup Rescue Competitions in mind. These scenarios can be simulated in the real world [7] or virtually within the USARSim simulator [1]. In either case, the team of robots is challenged to explore the site and locate victims in a constrained amount of time. Afterwards, the efforts of the robot team are evaluated on the amount of covered area, the quality of the produced map and most importantly the number of located victims. See [2] for a more detailed discussion on this scoring process.

The exploratory efforts exposed by our robots have so far been governed by strictly reactive behavior (2006) and tele-operation (2007). Although the autonomous behavior has demonstrated good robustness and obstacle avoidance, any seemingly 'intelligent' autonomous exploration effort was due to randomizations that were inherent to the behavior control design [11].

A well-known paradigm to address the multi-robot exploration challenge in a more intelligent fashion is *frontier*-based exploration. The frontiers are typically defined as the boundaries of the currently mapped free area where the robot can enter yet unexplored area [14]. Collaborating robots can use these frontiers to coordinate their actions

H. Bruyninckx et al. (Eds.): European Robotics Symposium 2008, STAR 44, pp. 43–52, 2008.

[3, 14, 15], i.e.: assign robots to frontiers such that the robots simultaneously explore multiple yet unexplored parts of the environment. This shifts the exploration problem to a frontier assignment problem.

Most approaches use a *cost* measure to evaluate the utility of a frontier. The anticipated traveling distance to reach a frontier or the associated motion costs are examples of such cost measures. In [6] and [13] however, frontier evaluation approaches were presented that focus on the opposite measure: the *information gain* that can be expected if the frontier would be explored. This gain is expressed as an estimate for the amount of area that lies beyond the frontier. While [6] uses a sampling method that extrapolates the current map to estimate the information gain, the approach of [13] directly measures the expected gain from the *current* map.

This paper will look more carefully at the balance between the cost and the gain. In the following Section we will present our multi-robot exploration strategy. Subsequently, the balance between information gain and movement costs is varied in a number of the experiments in Section 3. This will lead to the conclusions in Section 4.

2 Multi-robot Rescue Site Exploration

Exploration addresses the challenge of directing a robot through an environment such that its knowledge about the environment is maximized [12]. A mobile robot typically maintains its knowledge about the external world in a map m. Increasing the knowledge represented by m is achieved by either reducing the uncertainty about current information, or by inserting new information. The latter occurs when the map coverage is extended as the robot visits areas in the external world not yet covered by m before.

The approach in [11] was to passively acquire the information to store in the map, i.e. while the robot was wandering around pursuing other objectives like finding victims. In this work however, the focus is on *active* exploration: to explicitly plan the next exploration action a which will increase the knowledge about the world the most. In this paradigm victim finding becomes the side-effect of efficient exploration.

Occupancy grids [8] are a convenient representation for m in order to address the exploration challenge as they lend themselves excellently for storing probabilistic information about past observations. Each cell x corresponds to a region in the external world and holds the value $p(x)$ that denotes the aggregated probability that this region in the real world is 'occupied', i.e. is (part of) an obstacle. The objective of active exploration can then be seen as to minimize the *information entropy* $H(m)$ [5] of the probability distribution defined over all $x \in m$:

$$H(m) = -\sum_{x \in m} p(x) log(p(x)) \tag{1}$$

Initially each grid cell has unknown occupancy, so $p(x) = 0.5$ for all $x \in m$ and the entropy of the map $H(m)$ is maximum. For exploration purposes the absolute value of $H(m)$ is not of interest, what is relevant is the difference in entropy before $H(m)$ and after $H(m|a)$ a particular exploration action a: the *information gain* $\Delta I(a)$ [9, 10].

$$\Delta I(a) = H(m|a) - H(m) \tag{2}$$

Note that the exploration action a could be a complex maneuver, consisting of a number of controls u_i and observations z_i that spans multiple time steps i. Hence, for predictions about $\Delta I(a)$ that lie multiple timestamps in the future, the set of possible exploration actions can grow rather fast. In many current exploration strategies this issue is addressed by evaluating only a limited set of future states. These approaches consider only the situations where a robot actually enters yet unexplored area, which are by definition the locations where open area borders on unknown area: the *frontiers* [14]. The borders can be easily derived from the occupancy grid map m; the unknown area involves all the cells x for which the occupancy $p(x)$ is still at its initial value $p(x) = 0.5$ and the open area involves all the cells for which $p(x)$ is sufficiently close to zero.

2.1 Estimating Beyond Frontiers on Occupancy Grids

A good autonomous exploration algorithm should navigate the robot to optimal target observation points. The approach presented in [13] enables a robot to distinguish these locations using a method that is based on 'safe regions'. The idea is that the robot simultaneously maintains two occupancy grids: one based on the maximum sensing range r_{max} of the range sensing device and another one based on a more conservative safety distance r_{safe}. Typical values for r_{max} and r_{safe} are 20 meters and 3 meters respectively. The result is that the safe region is a subset of the open area. Frontiers can then be extracted on the boundaries of the safe region where the robot can enter the free space. Subsequently, the area beyond the frontier can be estimated directly from the current map by measuring the amount of free space beyond the safe region.

Greedy exploration could continuously focus the robot to the frontier f with largest area $A(f)$ and which will ultimately lead to a complete coverage of the environment. More efficient exploration can be expected when also the distance $d(f)$ is considered in a utility function $U(f)$ that trades off the costs of moving to the frontier with the expected information gain. In our experiments we used the equation:

$$U(f) = A(f)/d^n(f) \tag{3}$$

The parameter n can be tuned to balance the costs against the gains, simular to the constant λ used in the utility function of [6]:

$$U(f) = A(f)e^{-\lambda d(f)} \tag{4}$$

2.2 Multi-robot Coordination

After the frontier extraction method illustrated in Section 2.1 and the utility function from Equation (3) we are left with the challenge to intelligently assign frontiers to the members of a multi-robot team.

Given the set of frontiers $F = \{f\}$ and team of robots $R = \{r\}$ the full utility matrix $U = [u_{ij}]$ can be computed that stores the utility u_{ij} for all possible assignment of robots $r_i \in R$ to frontiers $f_j \in F$. This matrix U is calculated with an Euclidian distance measure $d_{eu}(f)$. The Euclidean distance d_{eu} gives a lower bound of the actual distance to be

---

Algorithm 1. The algorithm for the assignment of frontiers to robots

---

Data: the identity of the current robot $r_c \in R$ and the map m as known by r_c.

Data: the set of robots r_i in R. Each r_i consist of the tuple $(x_{r_i}, y_{r_i}, \theta_{r_i})$.

Data: the set of frontiers f_j in F. Each f_j consist of the tuple $(x_{f_j}, y_{f_j}, A_{f_j})$.

Result: the pair r_c, f_c and the path p_c to the location (x_{f_c}, y_{f_c})

for *each robot r_i in R* **do**

    for *each frontier f_j in F* **do**

        $d_{eu} = \sqrt{(x_{f_j} - x_{r_i})^2 + (y_{f_j} - y_{r_i})^2}$;

        $u_{ij} = A(f_j)/d_{eu}^n$;

    end

end

$u_{max} = \max u_{ij}$;

repeat

    for *robot r_i and frontier f_j of u_{max}* **do**

        p=PathPlanning from (x_{r_i}, y_{r_i}) to (x_{f_j}, y_{f_j}) on map m;

        d_{pp}=length of path p;

        $u_{ij} = A(f_j)/d_{pp}^n$;

    end

    if $\max u_{ij} = u_{max}$ **then**

        Assign f_j to r_i;

        Prune U from i and j;

    end

    $u_{max} = \max u_{ij}$;

until *robot r_c is Assigned*;

p_c=last path p

---

traveled. This means that the utility values $[u_{ij}]$ are optimistic when $n > 1$. The actual distance to be traveled can be calculated by performing a path-planning operation, but this is typical a computation intensive operation. The efficiency of our algorithm is optimized by recalculating only a few elements of the matrix; only the highest utility $u_{ij} \in U$ is recalculated. Thereafter it is checked if this value is still the maximum value. Otherwise the new maximum value is recalculated with a distance measure based on path-planning. When the path-planned value is the maximum, the frontier f_j is assigned to robot r_i and the rows and columns of the utility matrix U are pruned. This process continues until a frontier is assigned to the current robot r_c. The pseudo code of this algorithm is given in Algorithm 1.

Note that the algorithm makes no assumption about the numbers of frontiers and robots. When this algorithm is applied in real time the computational consequences of a large number of frontiers and robots have to be studied, but typically the two step approach described above limits the path-planning operation to one or two frontiers per robot. So, this algorithm scales nearly linearly with the number of robots. Also, when there are less frontiers than robots some robots will not be assigned a frontier. However, in our experience these occasions are rather rare as robots usually find more frontiers than they can close.

2.3 Planning Safe Paths

In the second part of the algorithm a check is made if an obstacle free path exists to a frontier. The same occupancy grid that was used to extract the frontiers can also be used to plan safe paths that avoid obstacles. If paths would planned on the free-space robots may be guided to locations that are dangerously close to obstacles. This is a well-studied problem in robotics [4] and several solutions exist.

Because path-planning has to be performed for several robot-frontier combinations, a simple method has been applied that gives fast reasonable path. The occupancy map is converted into a safety map by convoluting the obstacles with the shape of the robot. When the shape of the robot is non-holonomic, the convolution can be performed by taking the Minkowski sum. For holonomic robots, as used in this study, the convolution can be approximated by employing a Gaussian convolution kernel. On this approximated safety map path-planning is performed with a breath-first algorithm.

This completes the set of tools necessary to enable coordinated frontier-based exploration by a team of robots. In the subsequent section the method will be applied to guide robots through to a virtual disaster site from two start positions.

3 Experiments and Results

Our experiments will be based on the 'Hotel Arena' that was used extensively during the RoboCup Rescue competitions of 2006[1]. Figure 1 shows a blue-print of this office-like environment. The wide vertical corridor in the center connects the lobby in the south with several horizontal corridors that go east and west. Numerous rooms border on all

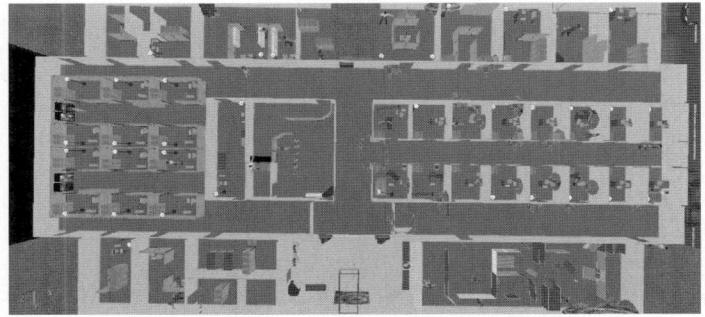

Fig. 1. A blue-print of the Hotel Arena, the virtual environment used for our experiments

---

[1] USARSim simulator, the simulated worlds used in these experiments and documentation are available on http://www.sourceforge.net/projects/usarsim The maps used in 2007 are unfortunately not publicly available yet.

the corridors. For the competition runs, robots would typically be spawned in the lobby or at the far ends of corridors, e.g. in the north-east, north-west or south-east corners.

Following the same setup as in the RoboCup competition each experiment will involve a run of 20 minutes. So the comparisons will focus on the amount of area that the robots were able to explore in this fixed time-window.

3.1   Results Based on Coordinated Exploration

These experiments involve multi-robot exploration using the presented approach. We used two spawn positions which were frequently used in the competitions of 2006. For each set of spawn positions a number of runs with a team of two collaborating robots were performed, each time with another formula for the utility function $U(f) = A(f)/d^n(f)$. In Figure 2 and 3 the resulting maps are given, for an inverse linear ($n = 1$), quadratic ($n = 2$) and cubic ($n = 3$) dependence on the distance $d(f)$.

On the maps the following color-coding is used:

- blue (dark grey) indicates unknown terrain,
- shades between light-blue (light grey) and white indicate the probability that the area is free from obstacles
- black indicates obstacles
- solid grey indicates 'safe region', as introduced in Section 2.1
- red indicates a victim
- light-green (light grey) line inside safe region indicates the path of the robots

In Figure 2 the robots started in the north-east corner. In Figure 2(a) the first robot (called Hercules) starts exploring the north-east corridor, while the second robot (called Achilles) explores the first two rooms. Hercules arrives at the T-junction, and decides to go south in the direction of the lobby. In the mean time Achilles is ready with the second room, and favors the unexplored corridors above exploring the nearby third room. The east corridor is chosen, which is explored till the end. Hercules explores firstly the west side and secondly the east side of the lobby. When the lobby is well covered, Hercules continues the exploration in the south-east corridor. This corridor is explored till the end, and the robot has time enough to enter a last chamber (a maze called the Yellow arena). This was a very successful run. More area can only be covered when Achilles would not have explored the first two rooms, but would instead explore the north-west corridor. Such a decision should be based on heuristics, because the existence of this corridor can only be known when a robot arrives at the T-junction. Until that time, the exploration of the nearby rooms is a good choice.

With an inverse quadratic dependence, the robot behavior is more tuned to nearby frontiers. As can be seen in Figure 2(b), with $n = 2$ the first three rooms are explored (one by Hercules, two by Achilles). Hercules goes again towards the lobby, while Achilles explores the north-west corridor. Hercules enters first the Yellow arena, before he continues with the south-east corridor. Notice in the north-west corridor that the localization is a few degrees off, because the robot Achilles had a lot of trouble to navigate in a confined space near the victim in the third room.

With an inverse cubic dependence, even the fifth room along the north-east corridor is entered (the entrance to the fourth room is blocked by a victim). Unfortunately, Achilles got stuck at that location, and the exploration has to be continued by the other robot. Hercules enters the large room west of the north-south corridor, and finds two victims. Afterwards, the north-south corridor and even the lobby is explored.

It should be clear from these examples that for a smaller n the robots have a tendency to stay in the corridors. This is a good strategy to cover a large area. On the other hand, a larger n can be used to direct the robots into the rooms. In those rooms victims can be found, but the confined space is more difficult to navigate, and the chance exists that the robot will not be able to leave the room anymore.

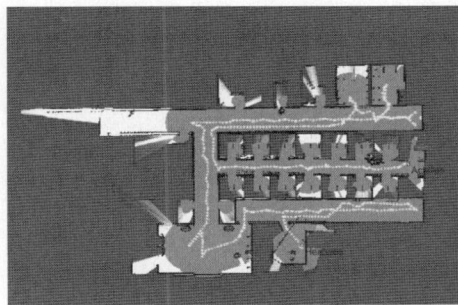
(a) $A(f)/d(f)$ (629 m^2, 6 victims)

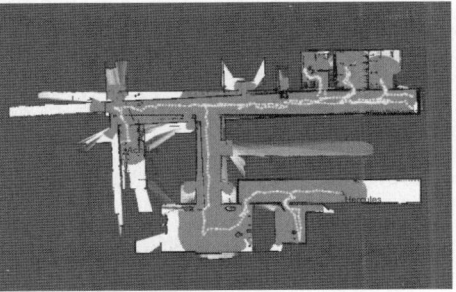
(b) $A(f)/d^2(f)$ (619 m^2, 8 victims)

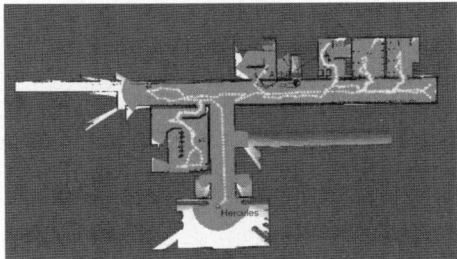
(c) $A(f)/d^3(f)$ (486 m^2, 8 victims)

Fig. 2. Exploration from the north-east corner with two robots

The experiments from the other start location show that tuning the parameter n not always leads to different behavior. For instance, in Figure 3(a-c) it can be seen that Hercules always explores the lobby, enters the south-east corridor and part of the Yellow arena. The south-east corridor is a dead-end, with enough free space to explore. Against the time that one is finished with this corridor, no time is left for any other choices.

The difference between the maps is mainly due to Achilles. In Figure 3(a) Achilles enters the north-south corridor and explores the east corridor. At the end of the corridor (again an dead-end), Achilles turns and goes back to the north-south corridor. In Figure 3(b) Achilles explores the north-east corridor and three rooms at the last corridor. In Figure 3(c) Achilles first checks the south-west corridor, before it heads towards the north-east corridor. Due to the detour there is only time left to explore a single room at the north-east corridor, but the area seen in the south-west corridor compensates more than enough. During its exploration, Achilles mainly encounters large corridors. Increasing the parameter n resulted in different behavior, because different corridors are chosen. This illustrates that difference in behavior doesn't always have the same effect. In Figure 3 an increase of n has as result more area and fewer victims, but this was mainly due to the particular layout of the corridors.

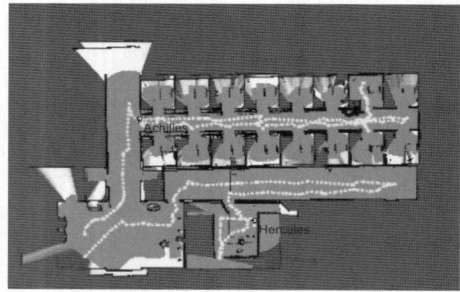
(a) $A(f)/d(f)$ (454 m^2, 6 victims)

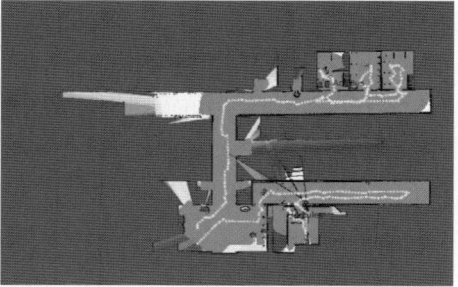
(b) $A(f)/d^2(f)$ (512 m^2, 6 victims)

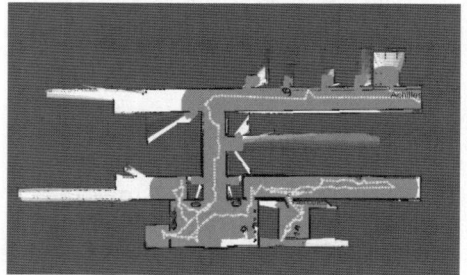
(c) $A(f)/d^3(f)$ (570 m^2, 4 victims)

Fig. 3. Exploration from the lobby with two robots

Notice that the number of victims found by the two robots in the previous experiments is comparable to the number of victims found in the semi-final by teams with four to eight robots [2]. Experiments with only two robots are not enough to make strong claims about the scalability of the algorithm to larger teams, but detailed analysis of the experiments indicated that in our current implementation the performance bottleneck is in merging the observations of multiple robots into a shared map. Planning the exploration on this shared map goes rather efficiently and is more a function of the length of the resulting path (e.g. when backtracking from a long dead-end corridor) than the number of robots or frontiers. Future experiments have back up this claim.

4   Conclusions

This paper investigated a frontier-based exploration approach that can be used to coordinate a team of robots. The approach assigns utilities to frontiers using a measure of the information gain that can be estimated directly from the *current* map. This information gain is balanced by the movement costs. In a first approximation these movement costs are estimated by the Euclidian distance. The actual movement costs are checked by performing path-planning on the map. Subsequently, a frontier with the highest utility is assigned to the members of a robot team.

The information gain and the movement costs can be balanced by a parameter. In the presented experiments is shown that tuning this parameter can change the overall behavior from exploring mainly corridors towards exploring nearby rooms. This parameter could be further tuned towards the scoring-function as applied in the RoboCup [2], but before this tuning is applied, the underlying navigation should be optimized. Currently, corridors and rooms are explored with the same care and speed. Inside the corridors the speed can be increased, to allow fast coverage of large areas. Inside the rooms the care could be increased, to guarantee that a room once entered could also be leaved.

In our experiments we have shown that this approach leads to efficient rescue site coverage. In future work we would like to investigate the possibilities for multi-robot coordinated exploration with more than two robots, study the influence of a-priori data, the effect of distributed decision making and the conditions where only limited communication is possible.

Acknowledgments

A part of the research reported here is performed in the context of the Interactive Collaborative Information Systems (ICIS) project, supported by the Dutch Ministry of Economic Affairs, grant nr: BSIK03024.

References

1. Balakirsky, S., Scrapper, C., Carpin, S., Lewis, M.: Usarsim: providing a framework for multi-robot performance evaluation. In: Proceedings of PerMIS 2006, August 2006. pp. 98–103 (2006)
2. Balakirsky, S., Carpin, S., Kleiner, A., Lewis, M., Visser, A., Wang, J., Ziparo, V.A.: Towards heterogeneous robot teams for disaster mitigation: Results and performance metrics from robocup rescue. Journal of Field Robotics 24(11-12), 943–967 (2007)
3. Burgard, W., Moors, M., Stachniss, C., Schneider, F.: Coordinated multi-robot exploration. IEEE Transactions on Robotics 21(3), 376–378 (2005)
4. Choset, H., Lynch, K.M., Hutchinson, S., Kantor, G.A., Burgard, W., Kavraki, L.E., Thrun, S.: Principles of Robot Motion: Theory, Algorithms, and Implementations, June 2005. MIT Press, Cambridge (2005)
5. Fox, D., Burgard, W., Thrun, S.: Active markov localization for mobile robots. Robotics and Autonomous Systems 25, 195–207 (1998)
6. González-Baños, H.H., Latombe, J.-C.: Navigation Strategies for Exploring Indoor Environments. The International Journal of Robotics Research 21(10-11), 829–848 (2002)
7. Jacoff, A., Messina, E., Weiss, B.A., Tadokoro, S., Nakagawa, Y.: Test arenas and performance metrics for urban search and rescue robots. In: Proceedings of the 2003 IEEE/RSJ International Conference on Intelligent Robots and Systems (October 2003)

8. Moravec, H.: Sensor fusion in certainty grids for mobile robots. AI Magazine 9, 61–74 (1988)
9. Sim, R., Roy, N.: Global a-optimal robot exploration in slam. In: Proceedings of the IEEE International Conference on Robotics and Automation (ICRA), Barcelona, Spain (2005)
10. Simmons, R.G., Apfelbaum, D., Burgard, W., Fox, D., Moors, M., Thrun, S., Younes, H.: Coordination for multi-robot exploration and mapping. AAAI/IAAI, 852–858 (2000)
11. Slamet, B., Pfingsthorn, M.: ManifoldSLAM: a Multi-Agent Simultaneous Localization and Mapping System for the RoboCup Rescue Virtual Robots Competition. Master's thesis, Universiteit van Amsterdam (December 2006)
12. Thrun, S., Burgard, W., Fox, D.: Probabilistic Robotics (Intelligent Robotics and Autonomous Agents), September 2005. MIT Press, Cambridge (2005)
13. Visser, A., Xingrui-Ji, v.I.M., Jaime, L.A.G.: Beyond frontier exploration. In: Proceedings of the 11th RoboCup International Symposium (July 2007)
14. Yamauchi, B.: Frontier-based exploration using multiple robots. In: AGENTS 1998: Proceedings of the second international conference on Autonomous agents, pp. 47–53. ACM Press, New York (1998)
15. Zlot, R., Stentz, A., Dias, M., Thayer, S.: Multi-robot exploration controlled by a market economy. In: Proceedings of the IEEE International Conference on Robotics and Automation (2002)

Compiling POMDP Models for a Multimodal Service Robot from Background Knowledge

Sven R. Schmidt-Rohr, Rainer Jäkel, Martin Lösch, and Rüdiger Dillmann

University of Karlsruhe, Karlsruhe, Germany
{srsr,jaekel,loesch,dillmann}@ira.uka.de

Summary. This paper presents an approach to create POMDP models, used for decision making by an autonomous service robot, from background knowledge. This allows the power of POMDP decision making to be applied on multimodal service robots with quite distinct stochastic dynamics in different modalities and a generally high model complexity. The two tiered approach presented allows both fine grained model adaptation as well as semantically more transparent knowledge representation and easy composition of new scenarios. The application of the process on a realistic mission scenario performed by a physical service robot is presented as an example.

1 Introduction

When building control systems which operate service robots, autonomy and robustness are primary goals. However, the complexity of the real world, especially limitations of perceptive methods and the dynamic, not completely deterministic course of events, make it a tough challenge. Thus, real world environments in which service robots operate can be described formally as partially observable, stochastic, dynamic and sequential. Methods and systems which employ autonomous control and decision making on service robots therefore need to take into account these characteristics explicitly. Autonomous behavior of a service robot can be accomplished by utilizing the measurements of multiple sensors of the robot and combining them with existing knowledge about properties of the world to derive actuator commands which are expected to lead to mission success.

This paper centers around this most abstract level of control, where decisions are made on a scenario level, including perceptions of all available sensor complexes and utilizing all available actuator complexes. On this level, abstract and symbolic *partially observable Markov decision processes* (POMDPs) are a framework for decision making for which algorithms with the ability to calculate approximately optimal decisions exist. However, to calculate decision policies, models are needed which describe the stochastic nature of the environment, the uncertainty of measurements and the mission objectives.

The use of POMDPs for decision making has shown promising results in simplified mono-modal domains like autonomous vehicles, dialog systems and primitive grasping where simple models can be used. Yet, in the complex mission environment of multimodal service robots the question of how to acquire models which describe the scenario settings is mostly unsolved.

H. Bruyninckx et al. (Eds.): European Robotics Symposium 2008, STAR 44, pp. 53–62, 2008.
springerlink.com   © Springer-Verlag Berlin Heidelberg 2008

In this paper, a system utilizing POMDP decision making for a real, highly multi-modal service robot is outlined while methods are presented to compile POMDP models from more abstract and reusable background knowledge.

2   State of the Art

The control systems of multi-modal autonomous robots are usually organized in layouts of some kind of architecture to manage the complexity of many different components and the amount of data. Three layer architectures have proven to be an efficient design [4]. The first layer processes the observations of individual sensor measurements into more general perceptions and controls actuators closely. The second layer sequences and supervises individual tasks of a mission while the third layer deliberatively selects those symbolic, abstract tasks.

In the scope of partially observable, stochastic and dynamic environments as encountered by physical service robots, methods like classical planning are unfeasible as those characteristics are not regarded. However, probabilistic decision theory models decision making of rational agents when facing uncertainty. An already usable framework within general probabilistic decision theory are partially observable Markov decision processes (POMDPs) [1], especially the class of discrete, model based POMDPs.

A POMDP is an abstract environment model for decision making under uncertainty [12], [2]. A POMDP models a flow of events in abstract states and discrete time. A specific POMDP model is represented by the 8-tupel $(S, A, M, T, R, O, \gamma, b_0)$. S is a finite set of symbolic states, A is a discrete set of actions and M is a discrete set of measurements. The transition model $T(s', a, s)$ describes the probability of a transition from state s to s' when the agent has performed action a. The observation model $O(m, s)$ describes the probability of a measurement m when the intrinsic state is s. The reward model $R(s, a)$ defines the numeric reward given to the agent when being in state s and executing action a. The parameter γ controls the time discount factor for possible future events. The initial belief state is marked by b_0. As POMDPs handle partially observable environments, there exists only an indirect representation of the intrinsic state of the world. In POMDPs, the belief state, a discrete probability distribution over all states in a scenario model, forms this representation. At each time step, the belief state is updated by Bayesian forward-filtering.

A decision about which action is most favorable for the agent when executed next, can be retrieved from a policy which contains information about the most favorable action for any possible belief distribution. The policy incorporates balancing the probabilities of the course of events into the future with the accumulated reward which has to be maximized.

While there exist different classes of approaches to compute a policy, those which utilize an explicit model have shown superior performance so far. Whereas computing exact, optimal policies from explicit models by value iteration is computationally intractable [6], approximate solutions as PBVI [10], discrete PERSEUS [13] or HSVI2 [11] deliver satisfying results while taking a reasonable amount of computation time for mid-sized scenarios.

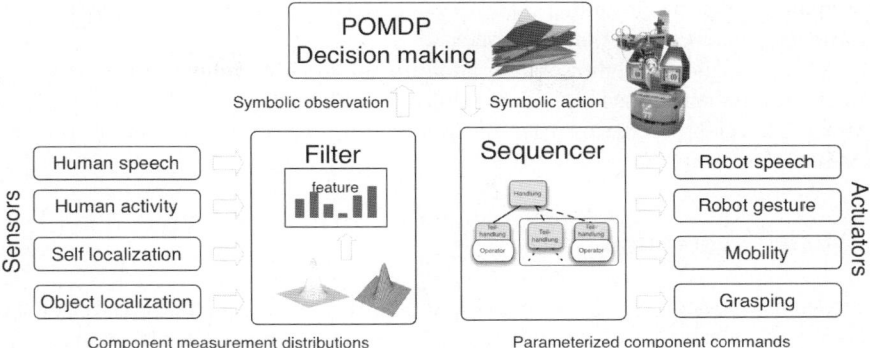

Fig. 1. The diagram shows the three layer architecture. Low level components processing perception are on the left, controlling actuators on the right. The midlevel perception filtering and fusion as well as the sequencer are at the center. The deliberative layer is at the top.

Explicit POMDP models used by those algorithms have been quite simple, even when having a large number of states and limited to mono-modal domains so far. Most research has dealt with navigation of autonomous vehicles [3] but there has also been work with dialog systems [14] and recently grasping by robot manipulators [5]. However, models for more abstract decision making of multi-modal service robots are less homogenous, making the acquirement of those models more difficult.

3   Application Domain

The methods for compiling models from abstract background knowledge as described in section 4 are based on the requirements arising from the capabilities and layout of the control system of a multi-modal service robot. Thus, our service robot with its decision and control system will be described shortly, first. The control architecture, shown in fig. 1, is designed along the classical structure of a three-layer architecture, with low-level control at the lowest, measurement filtering and execution supervision at midlevel and decision making at the top level. Three different capability domains exist at the control level: mobility, human-robot interaction and manipulation. Low level modules control the corresponding hardware directly, closely managing low level commands while processing sensor readings and delivering measurements, including uncertainty, to the layer above. The available capabilities in the mobility domain are driving and self-localization. Human-robot interaction is possible by speech output, speech recognition, robot arm/hand gestures and human activity recognition by a state-of-the-art procedure [8, 9]. Manipulation capabilities include arm movement, hand grasping as well as force-torque measurements.

The midlevel perceptive component filters and fuses perception into symbolic, abstract observations, usable by POMDP decision making. At the top level, the decision process chooses a symbolic, abstract action, based on the abstract observation and a POMDP policy, computed for the scenario. The abstract, symbolic action is interpreted as a sequential, flexible program [7] and is processed by the sequencer at midlevel which coordinates low level actuation components.

In typical assignments, the service robot utilizes all different capability domains to interact with humans, grasp and release objects and to move around between different locations. Especially scenarios where the robot performs waiter duties employ all three domains intensively.

4   Model Generation

The mission scenarios have to be modeled as POMDPs which includes observation uncertainties, stochastic world dynamics and mission objectives. In flat representations for which the mentioned, fastest policy computation methods exist, the size of the models grows quickly with the number of states and actions. For grid based navigation, the most popular POMDP application, transitions and observations are limited to local regions, leading to uniform and sparsely populated models which are easy to create even for large numbers of states.

When using POMDP reasoning in a multi-modal context on an abstract level, however, the models are far more complex, even with small state spaces. Different kinds of stochastic world dynamics like human behavior, navigation glitches or slippery objects overlap or are linked in the transition and observation models while there may be many mission objectives influencing the reward model.

Because superior, straight forward policy calculation methods exist for non-factored POMDPs, the presented approach maps several subspaces, representing different modalities into a single, unified state space:

$$s = (s_0, s_1, ..., s_n) \in S = S_1 \times S_2 \times ... \times S_n$$

A state s is hereby defined by a mapping from a unique set of sub-states in the respective S_i single-modality state-spaces. Apart from being able to utilize the mentioned algorithms, this approach maintains flexibility in modeling transition cross-dependencies between different modalities as well as combining unused and therefore redundant sub-state combinations.

In the following sections, a two tiered approach for compiling the observation, transition and reward models from simple, abstract and reusable descriptions of the scenario and environment properties is presented. This is crucial as the size of the transition model, a third grade tensor, grows cubically with state and action space. Additionally, by following this approach, the knowledge of the robot is organized in more meaningful semantics, covering different modalities which helps comprehensibility.

4.1   Direct Model Programming

The first layer, which can also be used standalone, is formed by a direct rule based programming system, which compiles the model matrices from compact, parameterized functions. A central element of the process are tables which keep temporary results of calculations based on those functions. All tables, whether designated for observation model, reward model or individual actions within a transition model, have a unified layout:

$$T : \{1,...,k\} \times \{1,...,m\} \times \{1,...,n\} \mapsto \mathbb{R} \tag{1}$$

Where k and m are row and column size of the corresponding model matrix and n is the number of sub-spaces which correspond to explicit modalities. By using a unified layout, calculation rules can be applied to all models equally well. After final results for all tables have been calculated, the model can be calculated by forming the product of all sub-states and then scaling for row or column sum 1.0 where necessary:

$$M : \{1,...,k\} \times \{1,...,m\} \mapsto \mathbb{R}, M(i,j) = \prod_{l=1}^{n} T(i,j,l) \tag{2}$$

The tables themselves are built and calculated step by step. First, the single-modality state-spaces have to be defined. This is either done by declaring states directly *state(modality,name)* or by loading an externally defined graph. This graph e.g. represents a driving graph of the mobile platform for navigation, including nodes and distances or e.g. a dynamic Bayes network modeling a dialog. In this case, states are automatically generated from the nodes, while the edges are kept in a data structure for use by calculation rules later on. Actions are declared explicitly on this level: *action(name,tags)*. After the tables have been created, their values are then computed by the application of rules. Rules can alter individual entries, rows, columns or whole tables while performing different kinds of arithmetic operations on them. A generic rule is defined as:

rulename(tableid,k,m,n,n/action,op,parameters). By using wildcards for k, m or *tableid*, rows, columns and whole tables can be modified. In the transition model this makes it possible to apply rules to all transitions from one state, all transitions to one state, all transitions of an action or even the whole model. As the tables cover the whole (product) state-space, transitions, observations or rewards depending on several modalities can be calculated. Yet, by calculating the rules in a modality-specific entry before calculating the final values, rule application is kept in semantically coherent groups.

Simple rules
directly set values or perform arithmetic operations on the addressed target set. In the following only the rule names and specific parameters are explained, although rules also contain the address parameters explained previously. Some selected simple arithmetic rules, where v denotes a value of an entry in the target set, are:

- *absolute(x)* : sets v to x.
- *equals(x)*: sets v to x on the main diagonal of a symmetric table.
- *relative(x, op)* : performs arithmetic operation $v = v\{+|-|*|/\}x$.
- *param(s, op, i_1, ..., i_n)*: where s addresses row or column and i_j are indices of it with op is $+$, $-$, $*$ or $/$, sets $v = s_{i_1}, \forall k \geq 2 : v = op(v, s_{i_k})$
- *offsetscale(i, a, b, c, d)*: with row r, col l, sets $v = a + b * r_i + c * l_i + d * (l_i - r_i)$.
- *linear_v(a, b, s, e)*: with i the row-index of the current v, calculates weighted interpolation $a + (b-a) * (i-s)/(e-a)$
- *linear_h(a, b, s, e)*: as previous, but on columns.

Simple rules can be used to set the values of whole rows, diagonals or columns efficiently, to model simple arithmetic relations or to fine-tune a model. Simple rules are generated by more abstract second tier knowledge (see sec. 4.2).

Special rules
perform calculation tailored to more specific needs. Some selected special rules are:

- *loadedges(g)*: directly sets individually addressed transitions to values from an externally defined graph.
- *gaussrelative(l, s, o, σ_1, ... , σ_n)*: calculates gaussian distribution on a graph in its n-dim space defined by indices l and s for node o with additional scaling $\phi()$: $v =$
  $$\prod_{i=1}^{l} \frac{1}{\sqrt{2\pi\phi(\sigma_i)^2}} \exp\left(-\frac{1}{2}\left(\frac{m_{s+i}-o_{s+i}}{\phi(\sigma_i)}\right)^2\right)$$
- *gaussfixed(l, s, o, σ_1, ... , σ_n)*: as previous, but with $\phi = id$.
- *gauss(l, s, σ_1, ... , σ_n)*: as previous, but with $o = k$.
- *linearrelative(o, p, d)*: calculates the transition probabilities on a graph using geometric distance instead of nodes to the goal with scaling $\phi()$:
  $$v = \begin{cases} \max\left(0, p+i\phi(d)\right), & \text{if m = ith node on the shortest path from o to k} \\ 0, & \text{otherwise} \end{cases}$$
- *linearfixed(o, p, d)*: as previous, but based on the number of nodes passed on the way.

Linearfixed/relative are an example for the usage of modality specific calculation. As the navigation of the service robot is realized by a graph based system in the low level component, navigational glitches and failures relate closely to that structure. Specific behavior of that system is modeled by the linear fixed rule: the more nodes a driving action has to pass towards the target state, the less likely will it arrive there. When not arriving, it will most likely end up in the node directly before the target node. less likely one before and so on. The rule linearrelative can include the length of connecting edges into the consideration. Together the rules model the real stochastic behavior of the navigation of the real robot quite closely for arbitrary input graphs, which exist for navigation.

4.2   Generation from Abstract Knowledge

While only direct model programming rules and functions make the creation of models with more than a few states feasible, their manual use is still intricate. New rules have to be created for any new scenario and they do hardly contain semantically sound and reusable knowledge about the world. Therefore a second, more abstract tier of rules and world modeling is necessary. Instead of arithmetic rules, knowledge domains are the main conceptual structure at this second tier. Knowledge from this conceptual structure is used for both transition and reward models.

The observation model on the other hand is derived from more specific analysis of the measurement uncertainties of individual modalities. For self-localization the uncertainty is calculated from distances, both topological on the graph and geometric distance, between states. For speech recognition it is calculated from utterance similarities based on the speech recognizer methodology. Considering human activity recognition, the uncertainty is calculated from an activity similarity metric which takes into account angles and velocities of body parts. By these means, a realistic observation model can be calculated easily for each modality set used in a specific scenario without any necessary manual modeling and no further knowledge based approach is necessary:

$$O(i, j) = \prod_{l=1}^{n} (1 - \langle i, j \rangle_l) \tag{3}$$

With single-modality l distance metric $\langle ., . \rangle_l$.

For transition and reward model building different knowledge domains can be utilized. For practical application, direct model programming rules will be generated from the abstract knowledge. Important domains are:

1. General descriptive attributes of robot, humans and environment.
2. Modality specific state knowledge.
3. Action type specific knowledge.
4. Robot independent environment dynamics.
5. General interaction knowledge.

Examples for domain 1 are *robot::explorative* or *robot::efficient* which will influence the state specific reward model rules. The attribute *human::friendly* on the other hand can serve as a global parameter to general interaction knowledge (domain 5) in the transition model. An example for domain 2 is assigning location states a *danger* and a *benefit* value which are then scaled by global attributes for the reward model. In domain 3 cost metrics for an action can be scaled, e.g. the penalty per execution time where the action penalty is calculated from the complexity of the sequential program for grasping actions or the navigation graph for driving actions. Domain 4 contains dynamic environment behavior independent of the actions of the robot, e.g. how often a new person enters the scene and thus contains reusable transition knowledge for the *idle/NOP* action. Domain 5 organizes specific interaction elements into abstract classes, e.g. *informal protocol* like greetings, acknowledgments, facings, *commands* or *descriptive attributes*. Generally, some knowledge expressions do not only modify transition and reward model, but can also lead to the inclusion of additional actions, e.g. dialog stimulating or specific information gain actions.

Second tier knowledge items have a rule description containing the first tier rules which are created when utilizing the knowledge and the parametrization taking place. When applied in practice, the knowledge is addressed in rules looking quite similar to first tier rules, but without arithmetic parameters.

5   Experiments and Results

While the approach is quite general, the main goal is practicability, as creating a POMDP model for a mission scenario of a real, multi-modal service robot without any compilation mechanism is totally infeasible. Therefore, the system has been used for creating a model of a realistic mission scenario, which was then utilized by the decision and control system to manage the behavior of the physical service robot in the modeled scenario.

As basic setting, a simple waiter mission was chosen for evaluation with common aspects of domestic service robot scenarios: verbal and non-verbal human-robot interaction, navigation and object delivery. Although being still somewhat limited in size,

the scenario was designed as being mostly natural: the robot waits for potentially interested persons, engages in interaction while keeping awareness about body postures and
offers its services to persons who are considered interested. When requested, it fetches
a cup from a fixed position, then addresses the human again to request to which of two
destinations it shall it shall be delivered, expecting an instruction with both gesture and
speech. If the robot is unsure about the perception, it may request again, if it is sure
enough, it delivers the cup and returns to a position, waiting for new "clients". Only
onboard sensors are used, speech recognition is using an onboard microphone, activity
recognition the onboard time-of-flight camera and navigation the onboard laser scanner.
Thus, measurement uncertainties are high and POMDP reasoning makes perfect sense,
modeling human behavior and navigation glitches as stochastic.

Grasping was handled in the sequential programs in this case, thus the modality
domains utilized were navigation, dialog and human body activity. For navigation, 8
nodes were chosen as relevant states from the navigation graph, for dialog 5 states
were used, as were for body activity. As in most nodes, not all interaction stages are
performed, there are many redundant product states, which were combined to speed up
policy computation. In the end, there were 28 unique POMDP states.

The following shows the second tier rules for the model compilation process ("Labor*"
are names of navigation nodes/states):

```
Idle("Environment::standard", "Human::friendly");
Explore("Robot::cautious");
Interact("LaborOstVoodoo", "Robot::patient", "Human::unfriendly",
    "Environment::UserUnknown") -> Target("Bring cup/FaceRobot");
Interact("LaborPlanck", "Robot::patient", "Human::unfriendly",
    "Environment::UserUnknown") -> Target("To Fermi/PointFront"),
    Target("To West/PointBack");
On("LaborOstVoodoo/Bring cup/FaceRobot", "Important") ->
    PickUp("LaborPlanck", "Cup");
On("LaborPlanck/To Fermi/PointFront", "LessImportant") ->
    PutDown("LaborFermi", "Cup");
On("LaborPlanck/To West/PointBack", "LessImportant") ->
    PutDown("LaborWestTisch", "Cup");
```

It should be noted, that much information is present in the navigation graph, which
however already exists for the navigation component. Additional background information is present with dialog transition probabilities for different kinds of people
(friendly/unfriendly, familiar/strangers etc.) for simple dialog elements like greetings
and commands. Reassurance and navigation actions are added automatically by second tier rule processing, leading to 11 distinct POMDP actions, which are mapped to
compound sequential programs. This leads to a reward model with 308 entries, an observation model with 784 probabilities and a transition model with 8624 probabilities.
Superimposing effects in reward, observation and transitions are managed by the process. Creating such a model manually would be infeasible.

The policy was calculated from the model by the PBVI algorithm and then used
on the physical robot by the system presented in sec. 3. A controlled experiment was
made, where a human supervisor, distinct from the interacting person in the experiment,
recorded true states of the world, true requests made and the behavior of the robot. The
robot performed the waiter duties for scheduled 30 minutes while interacting persons
were to behave on average according to the modeled stochastic behavior template. Although the measurements were quite noisy as indicated by automated recordings, the

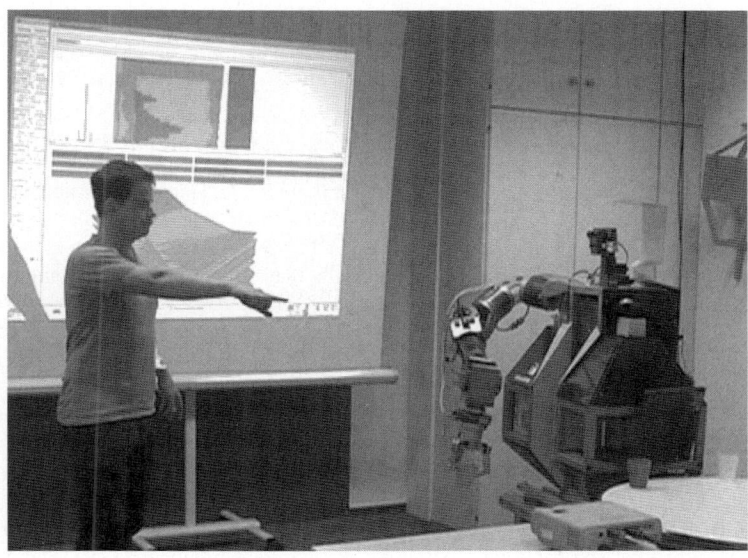

Fig. 2. A picture from an experiment run. The robot *Albert* fetched the cup and requested the destination. At the background a live projection of the POMDP decision making process is visible as transmitted via wireless network from the robot. The belief state can be seen plotted on top, while a 3D cut of the 28D policy is visible below.

robot was able to balance annoying reassurance questions and executing a bring/fetch action quite optimally. All fetch operations were performed when requested while 7 out of 9 times, the cup was brought to the correct location. Figure 2 shows the robot interacting with a human while performing the presented mission scenario.

6 Conclusion and Outlook

The presented approach makes creating POMDP models for service robot mission scenarios feasible. A scenario model can be composed out of semantically meaningful building blocks and at the same time put emphasis on the important stochastic properties in each modality. Next, methods to organize and utilize much larger sets of knowledge to build very different kinds of scenarios will be investigated as well as using a knowledge based approach for incremental, online model learning.

References

1. Aström, K.J.: Optimal control of markov decision processes with incomplete state estimation. Journal of Mathematical Analysis and Applications 10 (1965)
2. Cassandra, A.R., Kaelbling, L.P., Littman, M.L.: Acting optimally in partially observable stochastic domains. In: Proceedings of the Twelfth National Conference on Artificial Intelligence (1994)

3. Foka, A., Trahanias, P.: Real-time hierarchical pomdps for autonomous robot navigation. Robot. Auton. Syst. 55(7), 561–571 (2007)
4. Gat, E.: On three-layer architectures. In: Artificial Intelligence and Mobile Robots, MIT/AAAI Press, Menlo Park (1997)
5. Hsiao, K., Kaelbling, L.P., Lozano-Pérez, T.: Grasping pomdps. In: ICRA, pp. 4685–4692 (2007)
6. Kaelbling, L.P., Littman, M.L., Cassandra, A.R.: Planning and acting in partially observable stochastic domains. Artif. Intell. 101(1-2), 99–134 (1998)
7. Knoop, S., Schmidt-Rohr, S.R., Dillmann, R.: A Flexible Task Knowledge Representation for Service Robots. In: The 9th International Conference on Intelligent Autonomous Systems (IAS-9), Kashiwa New Campus, The University of Tokyo, Tokyo, Japan, März (2006)
8. Knoop, S., Vacek, S., Dillmann, R.: Sensor fusion for 3d human body tracking with an articulated 3d body model. In: Proceedings of the 2006 IEEE International Conference on Robotics and Automation (ICRA), Orlando, Florida (2006)
9. Lösch, M., Schmidt-Rohr, S., Knoop, S., Vacek, S., Dillmann, R.: Feature set selection and optimal classifier for human activity recognition. In: Robot and Human Interactive Communication 2007 (ROMAN 2007), 16th IEEE International Symposium on, Jeju Island, Korea, August 26-29, 2007 (2007)
10. Pineau, J., Gordon, G., Thrun, S.: Point-based value iteration: An anytime algorithm for POMDPs. In: International Joint Conference on Artificial Intelligence (IJCAI), pp. 1025–1032 (August 2003)
11. Smith, T., Simmons, R.: Focused real-time dynamic programming for mdps: Squeezing more out of a heuristic. In: Nat. Conf. on Artificial Intelligence (AAAI) (2006)
12. Sondik, E.J.: The optimal control of partially observable Markov Decision Processes. PhD thesis, Stanford university (1971)
13. Spaan, M., Vlassis, N.: Perseus: Randomized point-based value iteration for pomdps. Journal of Artificial Intelligence Research 24, 195–220 (2005)
14. Williams, J.D., Poupart, P., Young, S.: Using factored partially observable markov decision processes with continuous observations for dialogue management. Technical report, Cambridge University Engineering Department Technical Report: CUED/F-INFENG/TR.520 (March 2005)

Constraint Based Object State Modeling

Daniel Göhring, Heinrich Mellmann, and Hans-Dieter Burkhard

Institut für Informatik, LFG Künstliche Intelligenz Humboldt-Universität zu Berlin,
Unter den Linden 6, 10099 Berlin, Germany
goehring@informatik.hu-berlin.de
http://www.aiboteamhumboldt.com

Summary. Modeling the environment is crucial for a mobile robot. Common approaches use Bayesian filters like particle filters, Kalman filters and their extended forms. We present an alternative and supplementing approach using constraint techniques based on spatial constraints between object positions. This yields several advantages: a) the agent can choose from a variety of belief functions, b) the computational complexity is decreased by efficient algorithms. The focus of the paper are constraint propagation techniques under the special requirements of navigation tasks.

1 Introduction

Modeling the world state is important for many robot tasks. But usually robots have a limited field of view, which makes it hard to acquire the whole surrounding from one image. Bayesian filters [3] have been very successful in solving this problem by incorporating sensor data over time. A very famous member of the Bayesian filter family is the Kalman filter [6] using Gaussian distribution functions. But many distributions can neither be processed by a Kalman filter nor by one of its extensions. For non gaussian distributions particle filters have become very popular. But the calculation of the sample set can become very costly, making it inappropriate for real time applications.

Given an image of a scene, we have constraints between the objects in the image and the objects in the scene. Object parameters, image parameters and camera parameters are dependent by related constraints. Given odometry (or control) data, subsequent positions are constrained by measured speed and direction of movements. They can be combined with sensor measurements [5].

We have to deal with incomplete or with noisy measurements. With incomplete measurements, the result of constraint propagation will be ambiguous, while noisy measurements may lead to inconsistent constraints. Related quality measures have been discussed in our paper [4]. In this paper we discuss constraint propagation methods for solving navigation problems.

The main difference to classical propagation is due to the fact that navigation tasks do always have a solution in reality. For that, inconsistencies have to be resolved e.g. by relaxing constraints. Moreover, navigation tasks are not looking for a single solution of the constraint problem. Instead, all possible solutions are interesting in order to know about the ambiguity of the solution (which is only incompletely addressed by

H. Bruyninckx et al. (Eds.): European Robotics Symposium 2008, STAR 44, pp. 63–72, 2008.

particle filters). For that, the notion of conservative propagation functions is introduced. It can be shown that this notion coincides to some extend to the classical notion of local consistency (for maximal locally consistent intervals).

The paper is structured as follows: Section 2 gives an introduction and an example for description and usage of constraints generated from sensor data. In Section 3 we present the formal definitions and the backgrounds for usage of constraints. Basics of constraint propagation in the context of navigation tasks are discussed in Section 4, and an efficient algorithm is presented.

2   Perceptual Constraints

A constraint C is defined over a set of variables $v(1), v(2), ..., v(k)$. It defines the values those variables can take:

$$C \subseteq Dom(v(1)) \times ... \times Dom(v(k))$$

We start with an example from RoboCup where the camera image of a robot shows a goal in front and the ball before the white line of the penalty area (Figure 1). It is not too difficult for a human interpreter to give an estimate for the position (x_B, y_B) of the ball and the position (x_R, y_R) of the observing robot. Humans can do that, regarding relations between objects, like the estimated distance d_{BR} between the robot and the ball, and by their knowledge about the world, like the positions of the goalposts and of the penalty line.

The program of the robot can use the related features using image processing. The distance d_{BR} can be calculated from the size of the ball in the image, or from the angle of the camera. The distance d_{BL} between the ball and the penalty line can be calculated, too. Other values are known parameters of the environment: (x_{Gl}, y_{Gl}), (x_{Gr}, y_{Gr}) are the coordinates of the goalposts, and the penalty line is given as the set of points $\{(x, b_{PL})| - a_{PL} \leq x \leq a_{PL}\}$. The coordinate system has its origins at the center point, the y-axis points to the observed goal.

The relations between the objects can be described by constraints. The following four constraints are obvious by looking to the image, and they can be determined by the program of the observing robot:

C_1:   The view angle γ between the goalposts (the distance between them in the image) defines a circle (periphery circle).

C_2:   The ball lies in the distance d_{BL} before the penalty line.

C_3:   The distance d_{BR} between the robot and the ball defines a circle such that the robot is on that circle around the ball.

C_4:   The observer, the ball and the left goal post are on a line.

The points satisfying the constraints by C_1 (for the robot) and by C_2 (for the ball) can be visualized immediately on the playground as in Figure 1.

The constraint by C_3 does not give any restriction to the position of the ball. The ball may be at any position on the playground, and then the robot has a position somewhere

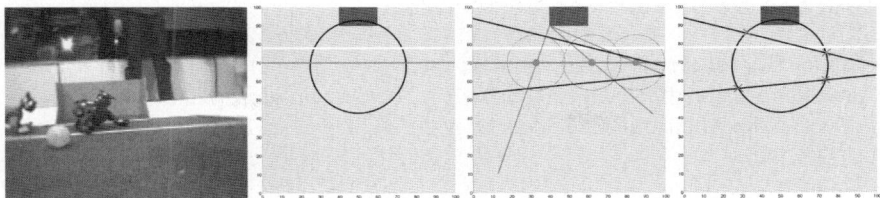

Fig. 1. Example from RoboCup (Four legged league): A robot is seeing a goal and the ball before the penalty line. The first Figure on left side illustrates the scene from the view of the robot. The next three figures show the related constraints that can be used for localization. Left: The picture shows a part of the field with the goal and the white penalty line, the periphery circle according to C_1, and the line of the Ball-Line-Constraint C_2. Middle: The picture shows the Constraint C_2 for the ball, some of the circles according to constraint C_5, some of the lines according to C_4, and the resulting two lines for C_6. Right: Constraints according to C_7: The position of the robot is one of the four intersection points between the periphery circle (C_1) and the lines according to C_6.

on the circle around the ball. Or vice versa for reasons of symmetry: The robot is on any position of the playground, and the ball around him on a circle. In fact, we have four variables which are restricted by C_3 to a subset of a four dimensional space. The same applies to constraint C_4.

The solution (i.e. the positions) must satisfy all four constraints. We can consider all constraints in the four dimensional space of the variables (x_B, y_B, x_R, y_R) such that each constraint defines a subset of this space. Then we get the following constraints:

$$C_1 = \{(x_B, y_B, x_R, y_R) \mid \arctan \frac{y_{Gl} - y_R}{x_{Gl} - x_R} - \arctan \frac{y_{Gr} - y_R}{x_{Gr} - x_R} = \gamma\} \qquad (1)$$

$$C_2 = \{(x_B, y_B, x_R, y_R) \mid (x_B \in [-a_{PL}, a_{PL}], y_B = b_{PL} - d_{BL}\} \qquad (2)$$

$$C_3 = \{(x_B, y_B, x_R, y_R) \mid (x_B - x_R)^2 + (y_B - y_R)^2 = d_{BR}^2\} \qquad (3)$$

$$C_4 = \{(x_B, y_B, x_R, y_R) \mid \frac{x_R - x_B}{y_R - y_B} = \frac{x_B - x_{Gl}}{y_B - y_{Gl}}\} \qquad (4)$$

Then the possible solutions (as far as determined by C_1 to C_4) are given by the intersection $\bigcap_{1,...,4} C_i$. According to this fact, we can consider more constraints $C_5, ..., C_n$ as far as they do not change this intersection, i.e. as far as $\bigcap_{1,...,n} C_i = \bigcap_{1,...,4} C_i$. Especially, we can combine some of the given constraints.

By combining C_2 and C_3 we get the constraint $C_5 = C_2 \cap C_3$ where the ball position is restricted to any position on the penalty line, and the player is located on a circle around the ball. Then, by combining C_4 and C_5 we get the constraint $C_6 = C_4 \cap C_5$ which restricts the positions of the robot to the two lines shown in Figure 1 (middle).

Now intersecting C_1 and C_6 we get the constraint C_7 with four intersection points as shown in Figure 1 (right). According to the original constraints C_1 to C_4, these four points are determined as possible positions of the robot. The corresponding ball positions are then given by C_2 and C_4.

3   Formal Definitions of Constraints

We define all constraints over the set of all variables $v(1), v(2), ..., v(k)$ (even if some of the variables are not affected by a constraint). The domain of a variable v is denoted by $Dom(v)$, and the whole universe under consideration is given by

$$U = Dom(v(1)) \times \cdots \times Dom(v(k))$$

For this paper, we will consider all domains $Dom(v)$ as (may be infinite) intervals of real numbers, i.e. $U \subseteq \mathbb{R}^k$.

Definition 3.1. *(Constraints)*

   1. A **constraint** *C over $v(1), ..., v(k)$ is a subset $C \subseteq U$.*
   2. An assignment β of values to the variables $v(1), ..., v(k)$, i.e. $\beta \in U$, is a **solution** *of C iff $\beta \in C$.*

Definition 3.2. *(Constraint Sets)*

   1. A **constraint set** *$\mathscr{C}$ over $v(1), ..., v(k)$ is a finite set of constraints over those variables: $\mathscr{C} = \{C_1, ..., C_n\}$.*
   2. An assignment $\beta \in U$ is a **solution** *of $\mathscr{C}$ if β is a solution of all $C \in \mathscr{C}$, i.e. if $\beta \in \bigcap \mathscr{C}$*
   3. A constraint set $\mathscr{C}$ is **inconsistent** *if there is no solution, i.e. if $\bigcap \mathscr{C} = \emptyset$*

The problem of finding solutions is usually denoted as solving a constraint satisfaction problem (CSP) which is given by a constraint set $\mathscr{C}$. By our definition, a solution is a point of the universe U, i.e. an assignment of values to all variables. For navigation problems it might be possible that only some variables are of interest. This would be the case if we are interested only in the position of the robot in our example above. Nevertheless we had to solve the whole problem to find a solution.

In case of robot navigation, there is always a unique solution of the problem in reality (the positions in the real scene). This has an impact on the interpretation of solutions and inconsistencies of the constraint system (cf. Section 4).

The constraints are models of relations (restrictions) between objects in the scene. The information is derived from sensory data, from communication with other robots, or from knowledge about the world – as in the example from above. Since information may be noisy, the constraints may not be as strict as in the introductory example from Section 2. Instead of a circle we get an annulus for the positions of the robot around the ball according to C_3 in the example. In general, a constraint may concern a subspace of any dimension (e.g. the whole penalty area, the possible positions of an occluded object, etc.). Moreover, constraints need not to be connected: If there are indistinguishable landmarks, then the distance to such landmarks defines a constraint consisting of several circles.

Other constraints are given by velocities: Changes of locations are restricted by the direction and speed of objects. This means that a position cannot change too much within a short time.

There are many redundancies which are due to all available constraints. Visual information in images usually contain lots of useful information: Size and appearance of observed objects, bearing angles, distances and other relations between observed objects, etc. Only a very small part of this information is usually used in classical localization algorithms. This might have originated in the fact, that these algorithms have been developed for range measurements. Another problem is the large amount of necessary calculation for Bayesian methods (grids, particles). Kalman filters can process such large amounts, but they rely on additional presumptions according to the underlying statistics.

Like Kalman filters, the constraint approach has the advantage, that it can simultaneously compute positions of different objects and the relations between them. Particle filters can deal only with small dimensions of search spaces.

For constraint methods, we have the problem of inconsistencies. According to the noise of measurements, it may be impossible to find a position which is consistent with all constraints. In our formalism the intersection of all constraints will be empty in such a case. Inconsistency in constraint satisfaction problems means usually that there does not exist a solution in reality. But in our situation, the robot (and the other objects) do have their coordinates, only the sensor noise corrupted the data. Related quality measures for constraint sets have been investigated in [4].

4   Constraint Propagation

Known techniques (cf. e.g. [1] [2]) for constraint problems produce successively reduced sets leading to a sequence of decreasing restrictions

$$U = D_0 \supseteq D_1 \supseteq D_2, \supseteq \cdots$$

Restrictions for numerical constraints are often considered in the form of k-dimensional intervals $I = [\mathbf{a}, \mathbf{b}] := \{\mathbf{x} | \mathbf{a} \leq \mathbf{x} \leq \mathbf{b}\}$ where $\mathbf{a}, \mathbf{b} \in U$ and the $\leq$-relation is defined componentwise. The set of all intervals in U is denoted by $\mathscr{I}$. A basic scheme for constraint propagation with

- A constraint set $\mathscr{C} = \{C_1, ..., C_n\}$ over variables $v(1), ..., v(k)$ with domain $U = Dom(v(1)) \times ... \times Dom(v(k))$.
- A selection function $c : \mathbb{N} \to \mathscr{C}$ which selects a constraint C for processing in each step i.
- A propagation function $d : 2^U \times \mathscr{C} \to 2^U$ for constraint propagation which is monotonously decreasing in the first argument: $d(D, C) \subseteq D$.
- A stop function $t : \mathbb{N} \to \{true, false\}$.

works as follows:

Definition 4.1. *(Basic Scheme for Constraint Propagation, BSCP)*

Step(0) Initialization: $D_0 := U$, $i := 1$
Step(i) Propagation: $D_i := d(D_{i-1}, c(i))$.
       If $t(i) = true$: Stop.
       Otherwise $i := i + 1$, continue with Step(i).

We call any algorithm which is defined accordingly to this scheme a BSCP-algorithm.

The restrictions are used to shrink the search space for possible solutions. If the shrinkage is too strong, possible solutions may be lost. For that, backtracking is allowed in related algorithms.

To keep the scheme simple, the functions c and t depend only on the time step. A basic strategy for c is a round robin over all constraints from $\mathscr{C}$, while more elaborate algorithms use some heuristics. A more sophisticated stop criterion t considers the changes in the sets D_i. Note that the sequence needs not to become stationary if only $D_i = D_{i-1}$. Actually, the sequence $D_0, D_1, D_2, \ldots$ needs not to become stationary at all.

For localization problems with simple constraints it is possible to compute the solution directly:

Corollary 4.1. *If the propagation function d is defined by $d(D,C) := D \cap C$ for all $D \subseteq U$ and all $C \in \mathscr{C}$, then the sequence becomes stationary after $n = card(\mathscr{C})$ steps with the correct result $D_n = \bigcap \mathscr{C}$.*

For simpler calculations, the restrictions D_i are often taken in simpler forms (e.g. as intervals) and the restriction function d is defined accordingly.

Usually constraint satisfaction problems need only some but not necessarily all solutions. For that, the restriction function d does not need to regard all possible solutions (i.e. it need not be conservative according to definition 4.3 below). A commonly used condition is local consistency:

Definition 4.2. *(Locally consistent propagation function)*

1. A restriction D is called **locally consistent w.r.t. a constraint** *C if*

$$\forall d = [d_1, ..., d_k] \in D \quad \forall i = 1, ..., k \quad \exists d' = [d'_1, ..., d'_k] \in D \cap C : d_i = d'_i$$

i.e. if each value of a variable of an assignment from D can be completed to an assignment in D which satisfies C.
2. A propagation function $d : 2^U \times \mathscr{C} \to 2^U$ is **locally consistent** *if it holds for all D, C: $d(D,C)$ is locally consistent for C.*
3. The **maximal locally consistent** *propagation function $d_{maxlc} : 2^U \times \mathscr{C} \to 2^U$ is defined by $d_{maxlc}(D,C) := Max\{d(D,C) | d$ is locally consistent$\}$.*

Since the search for solutions is easier in a more restricted the search space (as provided by smaller restrictions D_i), constraint propagation is often performed not with d_{maxlc}, but with more restrictive ones. Backtracking to other restrictions is used if no solution is found.

For localization tasks, the situations is different: We want to have an overview about all possible poses. Furthermore, if a classical constraint problem is inconsistent, then the problem has no solution. In localization problem, there does exist a solution in reality (the real poses of the objects under consideration). The inconsistency is caused e.g. by noisy sensory data. For that, some constraints must be relaxed or enlarged in the case of inconsistencies. This can be done during the propagation process by the choice of even a larger restrictions than given by the maximal locally consistent restriction function.

Definition 4.3. *(Conservative propagation function)*
A propagation function $d : 2^U \times \mathscr{C} \to 2^U$ is called **conservative** *if $D \cap C \subseteq d(D,C)$ for all D and C.*

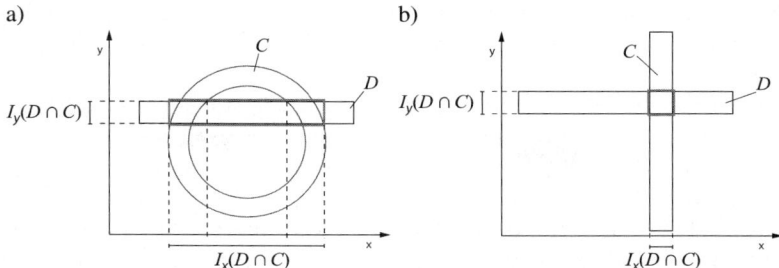

Fig. 2. Constraint propagation with intervals D for a) a circular constraint C b) a rectangular constraint C. *Intervals of Projection* w.r.t. $C \cap D$ are illustrated.

Note that the maximal locally consistent restriction function d_{maxlc} is conservative. We have:

Proposition 4.1. *Let the propagation function d be conservative.*

1. *Then it holds for all restrictions D_i : $\bigcap \mathscr{C} \subseteq D_i$.*
2. *If any restriction D_i is empty, then there exists no solution, i.e. $\bigcap \mathscr{C} = \emptyset$.*

If no solution can be found, then the constraint set is inconsistent. There exist different strategies to deal with that:

- enlargement of some constraints from $\mathscr{C}$,
- usage of only some constraints from $\mathscr{C}$,
- computation of the best fitting hypothesis according to $\mathscr{C}$.

We have discussed such possibilities in the paper [4].

As already mentioned above, intervals are often used for the restrictions D, since the computations are much easier. Constraints are intersected with intervals, and the smallest bounding interval can be used as a conservative result. Examples are given in Fig. 2.

Definition 4.4. *(Interval Propagation)*

1. *A propagation function d is called an **interval propagation function** if the values of d are always intervals.*
2. *The **minimal conservative** interval propagation function $d_{minc} : 2^U \times \mathscr{C} \to \mathscr{I}$ is defined by $d_{minc}(D,C) := Min\{I | I \in \mathscr{I} \wedge D \cap C \subseteq I\}$ for all D and C.*

The results by minimal conservative interval propagation functions can be computed using projections.

Definition 4.5. *(Interval of projection)*

*The (one-dimensional) **Interval of projection** w.r.t. to a set $M \subseteq U$ for a variable v is defined as the smallest interval containing the projection $\Pi_v(M)$ of M to the variable v: $I_v(M) = Min\{I | I \subseteq \mathbb{R} \wedge \Pi_v(M) \subseteq I\}$. It can be computed as $I = [a,b]$ with $a := Min(\Pi_v(M))$ and $b := Max(\Pi_v(M))$.*

Both, maximal local consistency and minimal conservatism leads to the same results, and both can be computed using the projections (Figure 2):

Proposition 4.2

1. $d_{maxlc}(D,C) = d_{minc}(D,C)$
2. $d_{minc}(D,C) = I_{v(1)}(D \cap C) \times .... \times I_{v(k)}(D \cap C).$

While local consistency is the traditional approach (to find only some solutions), the approach with conservative intervals is more suited for localization tasks because it can be modified w.r.t. to enlarging constraints during propagation for preventing from inconsistency. In case of inconsistencies, the algorithm below is modified accordingly in step 6. The related work is still under investigation.

The following simple and practicable algorithm is used for propagation. The stop condition compares the progress after processing each constraint once. Since stabilization needs not to occur, we provide an additional time limit. Note that the step counting s is not identical to the steps i in the basic scheme BSCP (but could be arranged accordingly).

Algorithm 1. Constraint Propagation with Minimal Conservative Intervals, MCI-algorithm

---

> **Input**: constraint set $\mathscr{C} = \{C_1, ..., C_n\}$ with variables $\mathscr{V} = \{v_1, ..., v_k\}$ over domain U
>   and a time bound T
> **Data**: $D \leftarrow U,\, s \leftarrow 1,\, D_{old} \leftarrow \emptyset$
> **Result**: minimal conservative k-dimensional interval D

1   **while** $s < T$ & $D \neq D_{old}$ **do**
2 | $D_{old} \leftarrow D;$
3 | **foreach** $C \in \mathscr{C}$ **do**
4 | | **foreach** $v \in \mathscr{V}$ **do**
5 | | | $D(v) \leftarrow I_v(D \cap C);$
            | **end**
6 | | $D \leftarrow D(v_1) \times \cdots \times D(v_n);$
            | **end**
7 | $s \leftarrow s + 1;$
    end

4.1 Experimental Results

In our experiments within the RoboCup soccer domain (see section 2), we compared a standard implementation of a Monte-Carlo particle filter with the algorithm described above.

We used constraints given by fixed objects like goalposts, flags and field lines identified in the images by the camera of the robot. It was easy to derive the related constraints: distances to landmarks are defined by circular rings in a generic form, where only the distances derived from the vision system of the robot have to be injected.

Constraints given by observed field lines are defined by rectangles and angles, the distances and the horizontal bearings are sufficient to define these constraints. All this can be done automatically.

While the particle filter used data from odometry, the constraint approach was tested with only the actual vision data. Since we were able to exploit various redundancies for the MCI, the accuracy of the results were comparable.

Our experiments showed that the MCI algorithm works several times faster than a related particle filter. We performed experiments with a different number of particles. Even with very small sample sets (about 50 particles) the computational costs for the MCPF were several times higher than for MCI. A disadvantage of particle based approaches is that many particles are necessary to approximate the belief which comes at high computational costs.

In further experiments we investigated more ambiguous data (i.e. when only few constraints are available). In this case, the MCI provided a good estimation of all possible positions (all those positions which are consistent with the vision data). The handling of such cases is difficult for MCPF because many particles would be necessary. Related situations may appear for sparse sensor data and for the kidnapped robot problem. Odometry can improve the results in case of sparse data (for MCPF as well as with additional constraints in MCI). But we would argue that the treatment of true ambiguity by MCI is better for the kidnapped robot problem.

5   Conclusion

Constraint propagation techniques are an interesting alternative to probabilistic approaches. From a theoretical point of view, they could help for better understanding of navigation tasks at all. For practical applications they permit the investigation of larger search spaces employing the constraints between various data. Therewith, the many redundancies in images can be better used. This paper has shown how sensor data can be transformed into constraints. We presented an algorithm for constraint propagation and discussed some differences to classical constraint solving techniques. In our experiments, the algorithm outperformed classical approaches like particle filters.

The different strategies for dealing with inconsistencies have to be investigated in more detail. This will be done by connecting the results from this paper with our results from [4]. In further work we will analyze constraint based approaches for cooperative object modeling tasks as well as very dynamic situations with quickly changing object states.

References

1. Davis, E.: Constraint propagation with interval labels. Artificial Intelligence 32 (1987)
2. Goualard, F., Granvilliers, L.: Controlled propagation in continuous numerical constraint networks. ACM Symposium on Applied Computing (2005)
3. Gutmann, J.-S., Burgard, W., Fox, D., Konolige, K.: An experimental comparison of localization methods. In: Proceedings of the 1998 IEEE/RSJ International Conference on Intelligent Robots and Systems (IROS), IEEE, Los Alamitos (1998)

4. Göhring, D., Gerasymova, K., Burkhard, H.-D.: Constraint based world modeling for autonomous robots. In: Proceedings of the CS&P (2007)
5. Jüngel, M.: Memory-based localization. In: Proceedings of the CS&P (2007)
6. Kalman, R.E.: A new approach to linear filtering and prediction problems. Transactions of the ASME - Journal of Basic Engineering 82, 35–45 (1960)

A COTS-Based Mini Unmanned Aerial Vehicle (SR-H3) for Security, Environmental Monitoring and Surveillance Operations: Design and Test

G. Belloni[2,3], M. Feroli[3], A. Ficola[1], S. Pagnottelli[1,3], and P. Valigi[2]

[1] Department of Electronic and Information Engineering, University of Perugia, Italy
{ficola,pagnottelli}@diei.unipg.it
[2] Polo Scientifico Didattico di Terni, University of Perugia, Italy
{belloni,valigi}@diei.unipg.it
[3] Siralab Robotics srl, Terni, Italy
michele.feroli@siralab.com

Summary. This paper presents the development of a mini unmanned aerial vehicle (UAV) with wingspan of 2.5 m. A flight control system is constructed using small and light components. The logical interconnection and schematic layout of the Automatic Flight Control are presented. The UAV has been successfully tested carrying a high resolution camera, and was able to acquire the images and the video of the fly zone and transmit them back to the ground station.

1 Introduction

This work is an in-depth examination of others [11, 12, 13] and pertains to a mini UAV design.

The availability of unmanned autonomous systems is a key issue in rescue and security operations and, among the class of unmanned vehicles, UAV's are of special interest in view of their flexibility and efficiency. Literature on UAV's is quite large, with contributions ranging from complete control of a single UAV to team coordination.

Accurate description of the control scheme for a single UAV is given, among many others, in [1, 2, 3] and the references therein, while team coordination is addressed, among many others, in [4, 5] and the references therein. The case of rescue applications has been addressed, among others, in [6].

In this paper, attention is focused on the description of a single UAV, comprising the three main management layers, namely Mission Planning, Guidance and Control. Also, the system has a bidirectional telemetry communication system, allowing continuous monitoring of on-board signals and real-time interaction with the vehicle. Such a feature is of paramount importance in rescue and security, where missions adjustment and re-planning is the normal way of operation.

Section 2 describes system requirements and design, Section 3 describes system realization, while some experimental results are reported in Section 4. Finally, Section 5 draws some conclusions and outlines future developments.

H. Bruyninckx et al. (Eds.): European Robotics Symposium 2008, STAR 44, pp. 73–82, 2008.
springerlink.com

2   System Design

The purpose of the system described in this paper is to pursue autonomous aerial monitoring and surveillance of a given region. System requirements comprise: about one hour flight endurance and 10 km data link range, a cargo bay suitable to lodge a video cam and a high resolution camera weighting 0.5 kg, electric propulsion. Automatic landing is carried out with a low speed spiral descent.

To achieve these requirements, an integrated design approach has been used, whereby the mechanical and aerodynamic design have been carried out together with the design of the electronics and of the automatic control system. The total weight resulted less than 4 kg (payload 0.5 kg, battery 1.2 kg, avionics 0.5 kg, structure, motor and propeller 1.8 kg) with 2.5 m wingspan, allowing either catapult or hand launching. The aircraft is shown in Fig.1.

(a) SH-H3 vehicle   (b) Launcher

Fig. 1. The aircraft

The realized complete aerial unmanned system is composed of a fixed-wing aircraft, an autopilot, a launcher (Fig. 1) and a Ground Control Station (GCS). The GCS is implemented on a PC through which it is possible to monitor the state of the aircraft flight and send mission data to the autopilot. The mission data consist of a list of waypoints (WPs) that the aircraft must follow. Every WP is described by GPS position, altitude and approach speed. Moreover, through the GCS it is possible to control the payload (a digital video-camera or a thermocamera) and to program an automatic photo relief session at the engaging of a desired WP.

The whole control system of any unmanned aerial vehicle comprises three main layers, namely Mission Planning (MP), Flight Guidance (FG), and Automatic Flight Control (AFC).

The Mission Planning level is mostly based on the use of suitable Man Machine Interfaces (see Fig. 3), and therefore requires some form of human intervention. Mission Planning is usually carried out in advance, off-line. Depending on the mission objectives, continuous re-planning may be required on-line, such as in the case of rescue and security applications. In this case, the MP layer could benefit from additional utilities for decision support. The MP consists in defining the list of the coordinates of the points

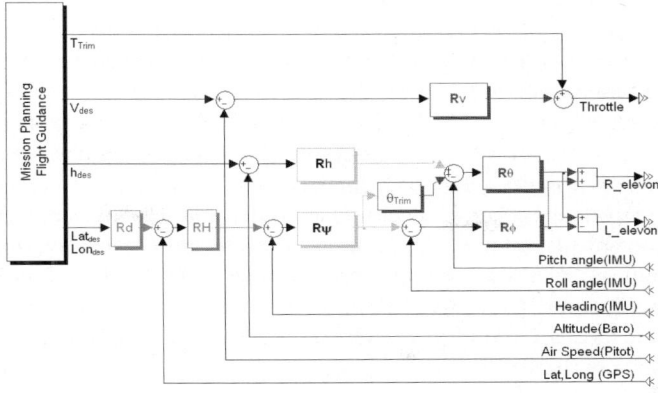

Fig. 2. The control system

that must be reached by the UAV. We distinguish among four kinds of points: take off, landing, way-points and orbits. The take-off point is automatically set by the GPS measurement; the landing point is the second point that must be defined. The way-points are the positions that must be reached with an assigned error, while the orbit points specify the positions that must be maintained for a certain time, with the UAV orbiting around. The path can be modified in real time: in particular it is possible to insert new points, modify them, specify if the orbit must continue or stop. As for the speed, it is defined in correspondence of each point for the subsequent route.

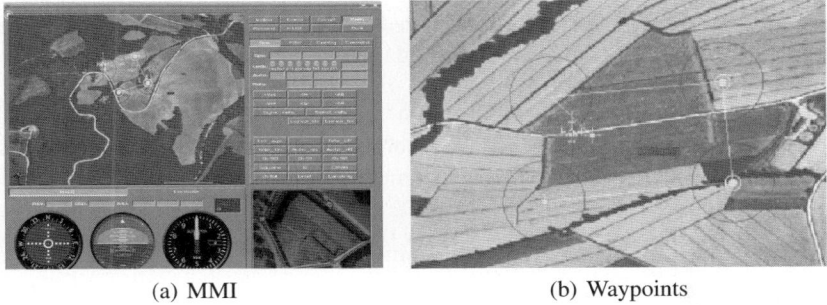

(a) MMI   (b) Waypoints

Fig. 3. Man Machine Interface

The Flight Guidance layer is responsible for converting the planned mission into a suitable trajectory that can be actually followed by the vehicle. The route, defined during the mission planning, is a sequence of segments characterized by the angle they form with the geographic north, and the length. The transition from a segment to the subsequent one is determined by the following rules: i) way-point: the distance is smaller than a prescribed value; ii) orbit-point: a certain number of orbits was completed or by a command from the ground station. The desired heading angle is discontinuous at the transition instants; to avoid possible actuator saturation, the heading angle function is smoothed by a first order low pass filter. Both mission planning and flight guidance layers run on a base station PC.

The Automatic Flight Control layer is responsible for ensuring system stability and performance, i.e., the tracking of the assigned trajectories. The vehicle described in this paper is a fixed wing tail-less aircraft. This architecture has been chosen, because it offers some advantages with respect to a more traditional one (wing, fuselage and tail). In fact, the tail surfaces and the fuselage generate additional drag; the fuselage is a weak point in case of crash-landing; finally, removing tail and fuselage reduces the weight. On the other hand, the tail absence gives rise to a certain lack of maneuvrability; for instance, the "de-crabbing" during the landing or the sideslip are not possible, because they require the vertical rudder. These restrictions are not critical for the application, because the required maneuvers are only level flight, steady climb/descent and level/climb/descent bank-to-turn. The main task of the tail surfaces, namely the longitudinal and lateral stabilization of the aircraft, can be ensured by a proper wing design and suitable control algorithms, using only the two wing elevons. The tail absence requires that the spiral and dutch-roll modes must be asymptotically stable, because they can be controlled mainly by the rudder and the vertical tail fin. To this purpose, the tip winglet design plays an important role. The other aircraft modes (phugoid, short-term and roll modes) can be controlled by the elevons, if stability augmentation is required.

The AFC layer relies on a number of sensors on board: an Inertial Measurement Unit (IMU) including a magnetic compass for attitude measurement, a Pitot probe for the indicated air speed, a baroaltimeter and a GPS receiver. The AFC is organized into three partly independent regulators, responsible for speed, altitude and direction control (Fig. 2).

Speed control. The speed control subsystem consists of a single linear regulator R_v and employs the indicated air speed measured by the Pitot probe. The measurements are filtered by a 4th order FIR filter. The regulator is

$$R_v(z) = c_{0v}(1 - c_{1V})/(1 - c_{1V}z^{-1}) \tag{1}$$

The low pass term is introduced to prevent current spikes that are non effective for the speed control and increase the power consumption. A trim command is introduced to improve the climbing maneuver:

$$T_{trim} = mg \sin(V_c/V) \tag{2}$$

where V_c and V are respectively the climbing and aircraft speeds, and mg is the aircraft weight.

Altitude control. The altitude control is performed by the elevons, commanded symmetrically. The scheme consists of an inner loop for the stability augmentation around the pitch axis and an external loop for the altitude regulation. The inner loop employs the pitch angle provided by the IMU; the pitch angle reference is computed by the regulator R_h in function of the desired altitude and the measured one, provided by the baroaltimeter. The pitch controller R_θ is a lead-lag network.

The altitude control is performed by a PI regulator. The altitude measurement is filtered by a 4th order FIR filter. A trim command is introduced in order to compensate the possible altitude decrease during the bank-to-turn maneuvers. In fact, during a banked turn, the lift force must increase to balance the weight; this is obtained by increasing the pitching angle with a term θ_{trim}, computed in function of the desired roll angle ϕ_{des}

$$\theta_{trim} = k(1 - cos(\phi_{des}))$$ (3)

Direction control. The direction control is also performed by the elevons, commanded antisymmetrically. The scheme consists of an inner loop for the stability augmentation around the roll axis and an external loop for the heading regulation. The inner loop employs the roll angle provided by the IMU; the roll angle reference is computed by the regulator $R_d(z)$ in function of the desired heading and the measured one, provided by the IMU. In particular, the desired heading is computed in function of the position of future target and the present one, expressed with latitude and longitude. If the distance between the waypoints is sufficiently small, the positions can be specified on a tangent plane to the earth at a certain latitude: therefore, assuming a local earth frame with $(x, y, z) = (North, East, Down)$, the position errors are

$$\Delta x = R(Lat_{des} - Lat_{meas})$$
$$\Delta y = R(Lon_{des} - Lon_{meas}) \cos(Lat_{meas})$$ (4)

where R is the local earth radius; the desired heading is $H = arctan(\Delta y / \Delta x)$. The heading regulator R_ψ is a simple gain. The roll controller is a lead-lag network; the roll angle is measured by the IMU. The desired roll angle is also employed to compute the trim command for the pitch control system.

3   System Realization

The architecture of the AFC layer is described in Fig.4. It is based on a modular organization, centered around the data and power bus, which allows easy communications among the several modules. The CPU module uses the bus to exchange data with the measurement and sensor modules, i.e., AFC layer signals, while connections with the aileron servomotors and with the propulsion brushless motor occur by means of dedicated connections; this is required also to increase electromagnetic disturbance rejection. The data bus is also used to exchange additional information between the UAV and the ground station, employed by the guidance and mission planning layers. Video stream is transmitted to the ground station by a 2.4 GHz video transmitter, still to increase disturbance rejection. Photo snaps can be taken on request from the ground station or automatically in function of the aircraft coordinates; the images are recorded on a memory stick on-board.

   To increase flexibility, the run-time selectable parameters of all the modules and subsystems are under control of the on-board CPU and therefore, through the radio link, they can be modified and adapted during mission execution under supervisor command.

   As mentioned in [7], the use of modular COTS-based (Commercial Off-The-Shelf) systems enabled major breakthroughs in Unmanned Aerial Vehicles performance. The innovative COTS systems are very useful to UAV developers through their ability to combine superior performance components within smaller, light-weight chassis. In this

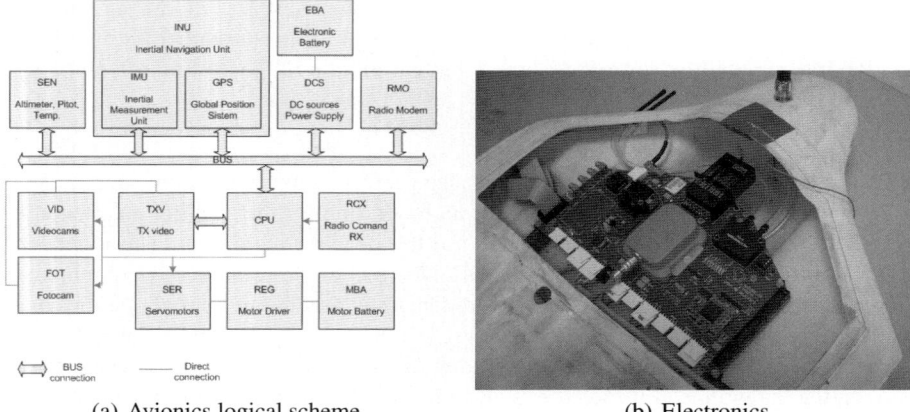

(a) Avionics logical scheme   (b) Electronics

Fig. 4. Autopilot

work this approach was followed mainly to have the possibility to test different modules and to substitute them quickly, to match and improve performance. To this aim the electronic project was organized in modular way, in which a central section, shared from all the modules, supplies signals and powers by a 34 data lines BUS. The CPU module is able, through the bus, to communicate with all the on board sensors and to calculate the proper commands for the actuators. The CPU is a module as well and therefore it can be replaced easily. In fact, to experience the unlike potentialities, we carried out tests with different CPUs processor and operating systems. The realized and replaceable electronic sections are five: BUS, CPU, Radio modem (RMO), GPS receiver and Sensors (SEN). In the BUS board we implemented also DCS (DC sources), for several power supplies, and the IMU (Inertial Measurement Unit). A logical outline, that shows how the avionics is connected to the aircraft, is reported in Fig. 4(a). This architecture allows either automatic on board control (autopilot) or remote control; a commercial radiocommand can control directly the elevons by means of a radio receiver (RCX). This characteristic is very important during the system tests and allows to accelerate experiments, carrying the aircraft in the wished conditions by the radiocommand, which can command the insertion of the autopilot in the desired moment.

Nevertheless this architectural approach has the disadvantage of larger size and weight; therefore it was utilized only to carry out tests, which last one year. The experiments were oriented to select appropriate sensors and validate the modules performances. Finally the various modules were integrated in a single mainboard that is visible in Fig. 4(b). It was obtained a significant weight and size reduction. The prototype takes up a surface of 221 cm^2 and has a mass of 370 g.

In order to process the control algorithms shown in Fig. 2, the main low level loop represented in Fig. 5(a) was implemented by the CPU. It takes about 2.61 ms for a complete cycle execution (data acquisition and filtering and computing the actuator's commands). The loop frequency execution is 50 Hz and is temporized by an external signal that produce an ISR (interrupt service routine) call.

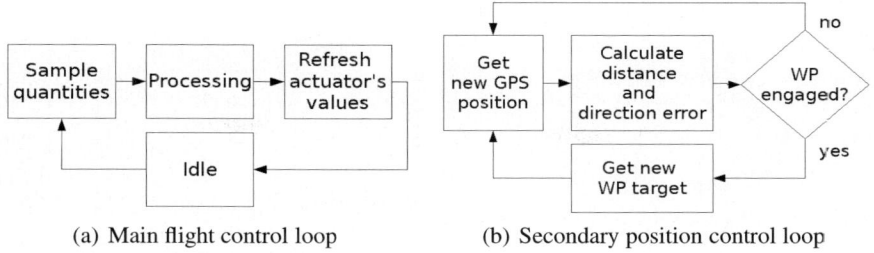

(a) Main flight control loop   (b) Secondary position control loop

Fig. 5. Control processing loops

In the idle state the primary loop releases time-machine resources to perform the secondary loop process (Fig. 5(b)) in which the CPU gets a new position from the GPS chip, calculates target's distance and direction error, and checks if the actual waypoint is engaged. In this case the new GPS target position is fetched from the memory. The secondary loop has a frequency of 1 Hz (the same of the GPS module). This is possible as the program runs on a scheduled preemptive multitasking operative system.

4   Experiments

To verify and evaluate the IMU's performance in like-real conditions, we built a simple centrifugal machine that was utilized for carried out some simulation in aircraft bank. This was possible by changing the roll angle of the avionics and measuring radius and angular velocity.

The proposed UAV has been extensively tested, to exploit its capabilities, and to reveal problems and weak points. Experimental tests have been performed to study the accuracy of the localization system, based on the proper joint use of data from heterogeneous sensors. The integration approach used so far is based on heuristic rules, while the use of data fusion techniques, based on suitable extension of classical approaches (see [8, 9, 10]) is subject of ongoing activities. An example of test mission is described. A four WPs path was defined and sent to the autopilot (Fig. 3(b)) and is composed of GPS positions: 1) 12.317187°E, 43.004414°N, 2) 12.320230°E, 43.004662°N, 3) 12.320409°E, 43.002823°N, 4) 12.317057°E, 43.002502°N. The altitude was programmed at 130 m.

The UAV was programmed for execute the path for an endless-loop. Fig. 6 shows a mission carried out for 32 minutes. The measures in meters are obtained by a cylindrical projection of GPS's measure differences around the Ground Control Station position (12.318906°E 43.003048°N). There are visible some trajectories farther than others; this is due to the wind influence. A tridimensional view of the followed path shows how the aircraft maintains the altitude at 130 m over the ground.

In some cases the aircraft did not reach the desired waypoint at the first attempt. This happens when the wind influence perturbs significantly the normal air route.

To test the vehicle ability to perform critical rescue tasks, an experimental champaign with a thermocamera by FLIR has been also conducted.

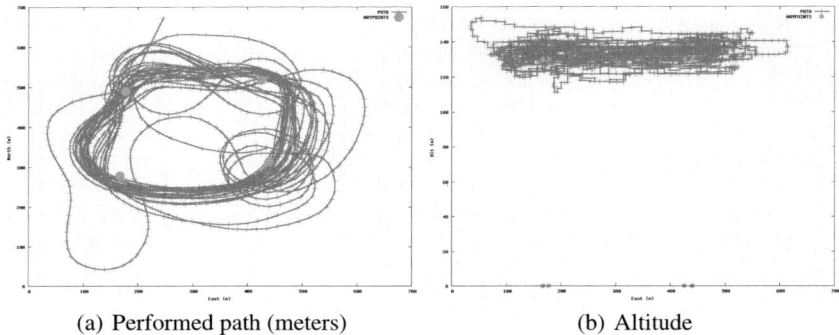

(a) Performed path (meters)   (b) Altitude

Fig. 6. Performed paths

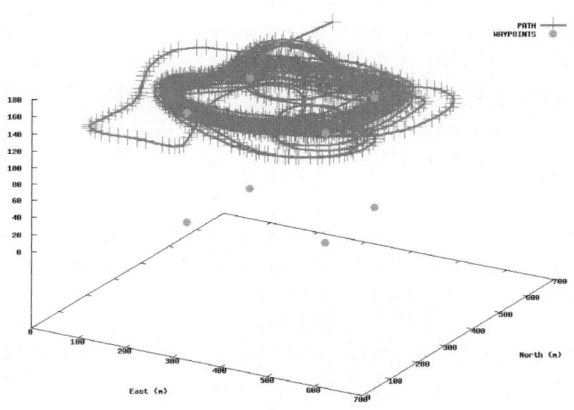

Fig. 7. 3D performed paths

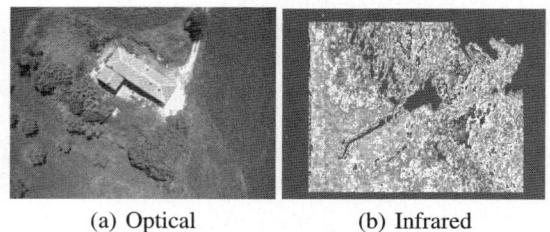

(a) Optical   (b) Infrared

Fig. 8. Images

An example picture taken from the video stream is reported in Fig. 8(b). Additional tests have been carried out with a 5 MPixels digital camera. An example picture is reported in Fig. 8(a). One of the advantages of the UAV presented in this paper is that highly detailed images can be taken, because the aircraft can fly also at low altitudes (20-50 m); in this case resolutions of about 3 cm can be achieved. Finally, the low noise electrical propulsion makes the UAV suitable for surveillance and security applications.

5   Conclusions

The development of an electrically powered UAV has been described, and experimental results have been presented. The flight control system is based on small and light-weight components. The vehicle can carry vision systems, with a real-time data link with the ground station. The UAV prototype has been successfully tested carrying a high resolution camera, and was able to acquire the images and the video of the fly zone and transmit them back to the ground station. Future activities will cover more advanced navigation schemes, adaptive controllers, and team cooperation.

Acknowledgments

The work has been supported by funds from Polo Scientifico Didattico di Terni, under grant Progetto di Sviluppo 2006 and from Siralab Robotics Srl. The authors wants to thank FLIR Systems Italia, which kindly made available a thermocamera for the experimental tests.

References

1. Johnson, E.N., Proctor, A.A., Ha, J., Tannenbaum, A.R.: Visual Search Automation for Unmanned Aerial Vehicles. IEEE Trans. Aerospace and Electronic Systems 41(1), 219–232 (2005)
2. Campbell, M., D'Andrea, R., Lee, J.W., Scholte, E.: Experimental Demonstrations of Semi-Autonomous Control. In: Proc. of the 2004 American Control Conference, Boston, Massachusetts, June 30-July 2, 2004, pp. 5338–5343 (2004)
3. Quigley, M., Goodrich, M.A., Griffiths, S., Eldredge, A., Beard, R.W.: Target Acquisition, Localization, and Surveillance Using a Fixed-Wing Mini-UAV and Gimbaled Camera. In: Proc. of the 2005 IEEE International Conference on Robotics and Automation, Barcelona, Spain, pp. 2600–2605 (2005)
4. Beard, R., McLain, T., Nelson, D.B., Kingston, D., Johanson, D.: Decentralized Cooperative Aerial Surveillance Using Fixed-Wing Miniature UAVs. Proc. of the IEEE 94(7), 1306–1324 (2006)
5. Giulietti, F., Pollini, L., Innocenti, M.: Autonomous Formation Fligth. IEEE Control System Magazine, 34–44 (2000)
6. Ryan, A., Hedrick, J.K.: A mode-switching path planner for UAV-assisted search and rescue. In: Proc. of the 44th IEEE Conference on Decision and Control and the European Control Conference 2005, Seville, Spain, pp. 1471–1476 (2005)
7. Eder, J.: COTS-based systems enable breakthroughs in unmanned aerial vehicles. In: Boards and Solutions (June 2006)
8. Martinelli, A., Martinelli, F., Nicosia, S., Valigi, P.: Sensor Fusion for Robot Localization. In: Siciliano, B., Bicchi, A., Nicosia, S., Valigi, P. (eds.) Articulated and Mobile Robotics for Services and Technologies (RAMSETE). Lecture Notes in Control and Information Sciences, Springer, London (2001)

9. Pagnottelli, S., Taraglio, S., Valigi, P., Zanela, A.: Visual and Laser Sensory Data Fusion for Outdoor Robot Localisation and Navigation. In: Seattle, W.U. (ed.) Proc. of the 12th Int. Conf. on Advanced Robotics (ICAR 2005), Seattle, Washington, USA, July 18-20 (2005)
10. Pagnottelli, S., Valigi, P.: SARA: a Flexible Framework for Rapid Prototyping of Mobile Robotics Applications. In: Proc. of the 14th Mediterranean Conf. on Control and Automation, Ancona, Italy (July 2006)
11. Belloni, G., Feroli., F., Ficola, A., Pagnottelli, S., Valigi, P.: An autonomous aerial vehicle for unmanned security and surveillance operations: design and test. In: Proc. of IEEE International Workshop on Safety, Security, and Rescue Robotics (SSRR 2007), Rome, Italy (September 2007)
12. Belloni, G., Feroli., F., Ficola, A., Pagnottelli, S., Valigi, P.: A small electric Unmanned Aerial Vehicle (SR-H3 UAV) for application in environmental coastal monitoring. In: IFAC Conference on Control Applications in Marine Systems, Bol, Croatia (September 2007)
13. Venanzoni, R., Gigante, D., Ferranti, F., Reale, L., Belloni, G., Feroli, M., Pagnottelli, S.: Il supporto delle foto aree da piccolo velivolo (SR-H3) nella rappresentazione cartografica di Habitat dell'Allegato I alla Dir 92/43/EEC. Congresso Nazionale di Fitosociologia, Ancona, Italy (July 2007)

Eyes-Neck Coordination Using Chaos

Boris Durán[1], Yasuo Kuniyoshi[2], and Giulio Sandini[1]

[1] Italian Institute of Technology - University of Genova, Via Morego 30, 16145 Genova, Italy
boris@unige.it, giulio.sandini@iit.it
[2] University of Tokyo, Laboratory for Intelligent Systems and Informatics,
Eng. Bldg. 2. 7-3-1 Hongo, Bunkyo-ku, Tokyo, Japan
kuniyosh@isi.imi.i.u-tokyo.ac.jp

Summary. The increasing complexity of humanoid robots and their expected performance in real dynamic environments demand an equally complex, autonomous and dynamic solution. Our approach for the creation of real autonomy in artificial systems is based on the use of nonlinear dynamical systems. The purpose of this research is to demonstrate the feasibility of using coupled chaotic systems within the area of cognitive developmental robotics.

Using a robotic head, we demonstrate that the visual input coming into the head's eyes is enough for the self-organization of the axes controlling the motion of eyes and neck. No specific coding of the task is needed, which results in a very fast adaptation and robustness to perturbations. Another equally important goal of this research is the possibility of having new insights about how the coordination of multiple degrees of freedom emerges in human infants. We show that the interaction between body and environment modifies the inner connections of the controlling network resulting in the emergence of a tracking behavior.

1 Introduction

Most of today's humanoid platforms follow an almost 50-year-old tradition of control theory that started with the industrial automation at the beginning of the 1960s. The methodology followed by this approach is based on modeling as precise as possible both the plant and the controller; and filtering or processing as noise the different unexpected circumstances that could occur during the operation of the system. This approach has worked pretty well when the system is in a fixed framework and the environmental conditions are known and controlled; however, this will not be the case for humanoid robots of the future. It is absolutely necessary to start working on a different approach if we want to design and build systems that move and act in the same kind of dynamic environments where humans move and act. A more adaptive and flexible theory is needed in order to 'control' a device that is supposed to move within an ever-changing environment. These are our first steps towards the design and implementation of a real autonomous cognitive architecture based on nonlinear dynamical systems.

Although the study of nonlinear dynamical systems and chaos has also a long history, real applications that make direct use of chaos theory have not been fully developed. The purpose of this research is to demonstrate the feasibility of using coupled chaotic systems [1] within the area of cognitive developmental robotics. Based on the model of behavior emergence introduced by Kuniyoshi et al. [2], we study the coordination of multiple degrees of freedom in humanoid robots.

H. Bruyninckx et al. (Eds.): European Robotics Symposium 2008, STAR 44, pp. 83–92, 2008.
springerlink.com © Springer-Verlag Berlin Heidelberg 2008

The task of tracking an object has been fully studied and many solutions presented before. Based either in position errors or velocity mismatches, some approaches try to control the activation of motors by means of robust PID controllers [3, 4, 5], while others base their controllers in fuzzy logic [6] or neural networks [7]. In any case, the common methodology in these approaches is to compute expensive Jacobian and kinematic expressions thinking in all the possible circumstances the system could encounter.

All these works comprehend the state of the art in motor control for tracking systems; therefore it would not be necessary to develop new solutions. However, the tracking problem represented the simplest test bed for the study of coupled chaotic systems, both in a simulated environment and for its implementation in a real platform. Our approach differs from previous work mainly in two aspects: first, our system does not need to deal with complex equations of kinematics and dynamics; second, the main goal behind our research is not to improve the performance of existing algorithms but, through our experiments, start building the basis of a dynamic model for motion emergence that embrace as a single entity body and environment. Following Esther Thelen and Linda Smith's suggestion that "action and cognition are also emergent and not designed" [8], another equally important goal of this research is the possibility of having new insights about how the coordination of multiple degrees of freedom emerges in human infants.

The following section contains a short introduction on chaos and coupled chaotic systems; as well as a description of the model of behavior emergence proposed in [2]. Section III describes the experimental setup and the results of our experiments from the implementation of our model when working with constant parameters. In Section IV it is presented the results of a developmental process in a five degree of freedom implementation of our approach. Finally, conclusions and guidelines for future work are summarized in section V.

2   Coupled Chaotic Systems

2.1   A Short Introduction to Chaos

The word 'chaos' has been used to represent a part of nonlinear dynamical systems theory that deals with the unpredictable behavior of a system governed by deterministic rules, [9]. One of the most common, and probably the simplest, deterministic rule that generates chaos is the logistic map (1). This second-order difference equation was studied by the biologist Robert May as a model of population growth [10]. In this equation, the parameter α controls the nonlinearity of the system. In order to keep the system bounded between -1 and 1, α takes values between 0 and 2, Fig. 1.

$$f(x_n) = 1 - \alpha x_{n-1}^2 \tag{1}$$

A stand-alone logistic map (internal feedback whitout external influences) stabilizes in an specific behavior depending on its initial condition and the value of α. This very simple rule can generate fixed points, Fig. 1a; periodic oscillations of period two, Fig. 1b; period four, Fig. 1c; and following the period doubling path until reaching a choatic behavior, Fig. 1d.

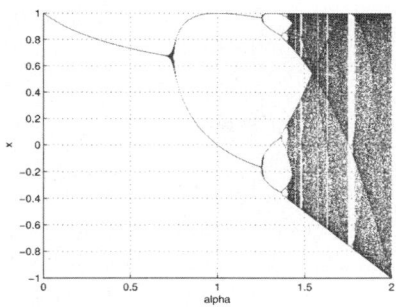
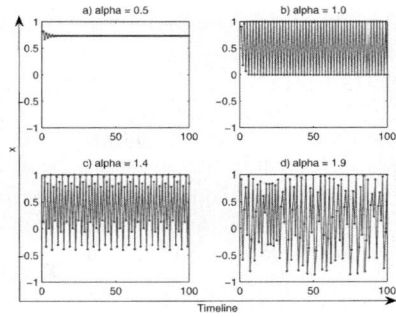

Fig. 1. Left, bifurcation plot for logistic map. Right, different outputs for Logistic Map depending on α.

2.2 Coupled Maps with Adaptive Connections

Coupled Map Lattices (CML) and Globally Coupled Maps (GCM), were introduced by Kunihiko Kaneko in the middle of the 1980's as an alternative for the study of spatiotemporal chaos [1]. In short, this kind of dynamical systems use discrete partial difference equations to study the evolution of a process described by discrete steps in space and time but with continuous states. Two parameters control the dynamics of these maps: a chaoticity factor and the strenght of connections among their elements.

Due to the chaotic nature of the system, it is possible to see one of the main properties of chaotic systems: two slightly different initial conditions amplify their difference through time. On the other hand, the system tries to synchronize the activations of all its chaotic elements by coupling them. In between these two states of complete chaos and complete synchronization, interesting states emerge like the formation of clusters oscillating in different phases and amplitudes.

The study of dynamically varying the connections among the elements in a GCM was done by Ito and Kaneko [11, 12]. The model is described by the set of equations in (2). The first equation correspond to a GCM, where f represents a chaotic map; (2b) updates each unit's connections coming from other units in the network; and (2c) specifies the hebbian rule governing the relationship between all units.

$$x_n^i = f\left((1-\varepsilon)x_{n-1}^i + \varepsilon \sum_{j=1}^{N} w_n^{ij} x_{n-1}^j \right), \tag{2a}$$

$$w_{n+1}^{ij} = \frac{\left[1 + \delta g\left(x_n^i, x_n^j\right)\right] w_n^{ij}}{\sum_{j=1}^{N} \left[1 + \delta g\left(x_n^i, x_n^j\right)\right] w_n^{ij}}, \tag{2b}$$

$$g(x,y) = 1 - 2|x-y| \tag{2c}$$

In (2b), δ represents the degree of plasticity of the connections and ranges from 0 to 1. The weights w^{ij} in (2b) refer to the influence from unit j going into unit i. All self-connections were set to 0; and the initial condition for all remaining connections are equal to $1/(N-1)$, N being the number of chaotic units.

2.3   A Model for Behavior Emergence

The states of each of the elements in a GCM, or a CML, depend only on the internal dynamics of these systems; they are not influenced in any moment by an external force. When taking these concepts to robotic applications it is necessary to think in a way of including the environment within the dynamics of the system.

The model used in this project is based on the approach followed by Kuniyoshi and Suzuki [2]. Their model uses both, the local interaction (CML) and the global interaction (GCM) but with the environment as the external force influencing the internal dynamics of the network. In our case, only GCM was used since no extra benefit was seen when including local connections; nevertheless the overall approach is the same, Fig. 2.

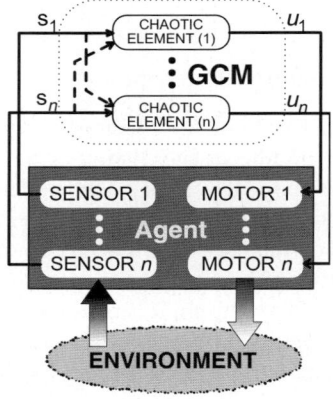

Fig. 2. Body-environment interaction through coupled chaotic fields

3   Implementation

A copy of the iCub's head, the humanoid platform of the Robotcub's project [13], was used in the present work. The head's hardware and software components will be described in the following subsections together with the implementation of the algorithms used to create a dynamic smooth pursuit.

3.1   Hardware and Software

The head has six degrees of freedom: yaw, pitch and roll for the neck, a single pitch motion for both eyes and independent yaw motors for each eye. DC-micromotors are used for moving the different joints; each motor contains an incremental encoder that provides the position of the joint at any time. All motors and sensors are controlled by a suite of DSP chips which channel data over a CAN bus to a computer in charge of iCub's high-level behavioral control [14].

Due to the large amount of sensori-motor information generated within the platform the iCub's software was configured to run in parallel on a distributed system of computers. An open-source framework for robotics named YARP (Yet Another Robot Platform) [15] was used for the implementation of the algorithms. It is important to mention that the focus of this project is not the extraction of saliencies from moving images, which is in itself a hard problem in computer vision. A tracking algorithm available in the YARP repository was used as the visual component in charge of providing us with the horizontal and vertical coordinates of a moving object. With this information we focus our efforts on the motor control problem.

3.2   Methodology

Each camera provides two quantities: the position of the target in vertical and horizontal directions. These values modify the position of each motor; thus generating a coupled chaotic system with 6 logistic maps, Fig. 3. The algorithm governing the dynamics of the system is governed by (3).

$$u_n^i = f\left((1 - \varepsilon)s_{n-1}^i + \varepsilon \sum_{j=1}^{N} w_n^{ij} s_{n-1}^j\right) \tag{3a}$$

$$\begin{aligned} m_n^i &= G_m(u_n^i + O_m) \\ s_n^i &= G_s(r_n^i + O_s) \end{aligned} \tag{3b}$$

Where m is the output applied to each motor as speed values, s and u are inputs and outputs respectively of the chaotic field block, and r is the raw value coming from the sensors. Finally, G_m, G_s, O_m, and O_s are gains and offsets of the sensors and motors respectively; these values are applied in the same magnitude to all elements in the system.

The methodology for tuning offsets was done by approximating the average of the raw output from the logistic map towards a zero average of the motor activation values. In other words, offsets should be chosen in such a way that the activations from the logistic map oscillate around zero. Gains G_m were chosen depending on the speed

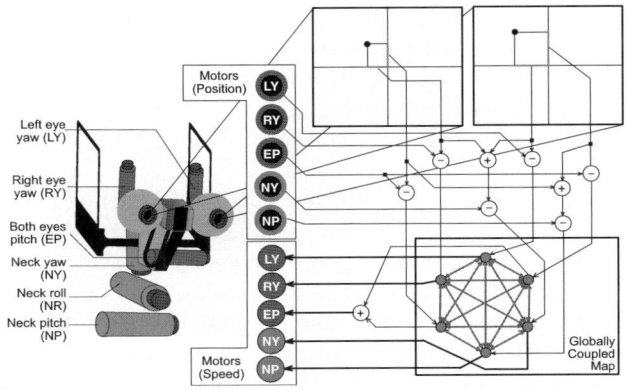

Fig. 3. iCub's sensorimotor diagram, 5dof actuation

limits of the motors. The following parameters were fixed during all experiments: G_s=1.0, O_s=-0.8, $G_{LY} = G_{RY} = G_{EP}$=25.0, G_{NY}=70, G_{NP}=35, and O_m= 0.0; α = 1.9, and $\varepsilon = 0.1$.

3.3   Results

The motion of both eyes and the motion of the head is shown in Fig. 4. This plot shows the motion of the eyes relative to the head and the motion of the head relative to its fixed position. In this plot is possible to see the coordination between eyes and neck. The target was moved in random directions and at different speeds. Since the joints of the neck give approximately an extra 60 degrees on each side and on each direction, an object can be tracked in a wider space. It was also observed an increase of the tracking speed when compared to the 3dof case (2-eye tracking). The motors in the neck help the motors in the eyes to follow the object in a faster way, especially in the yaw direction.

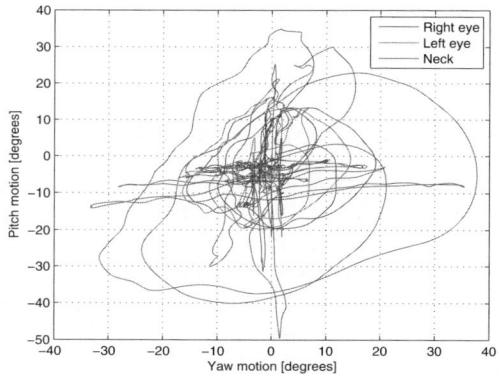

Fig. 4. Motion of both eyes and neck

The coordination between both eyes and between eyes and neck in each direction can be more easily appreciated in Fig. 5. Since the tracking algorithm works on independent threads in each camera, different points in space are delivered to the GCM. This 'computer vision' problem creates the errors observed during some points during the experiments.

The activations of all units grouped in yaw, Fig. 6, and pitch Fig. 6 directions show the dynamics of the system. Here is also possible to see the coordination of chaotic units since all activations are gathered along the diagonal of each plot. The nonlinearity of the chaotic units give them enough freedom to use the rest of the space when needed but always staying and returning back to this diagonal.

The development of weak and strong connections among the chaotic units depend on the level of interaction they have through time, Fig. 7. Even though all connections

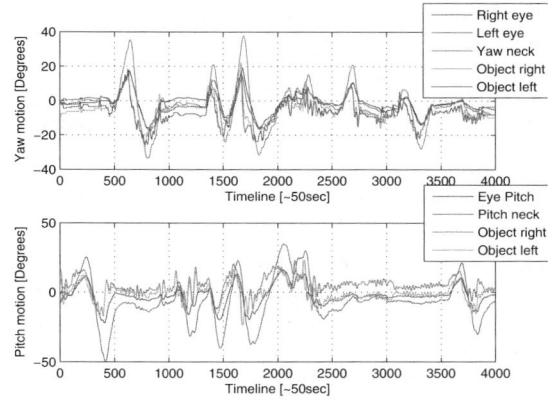

Fig. 5. Position of target w.r.t. center of eye: yaw motion, top; and pitch motion, bottom

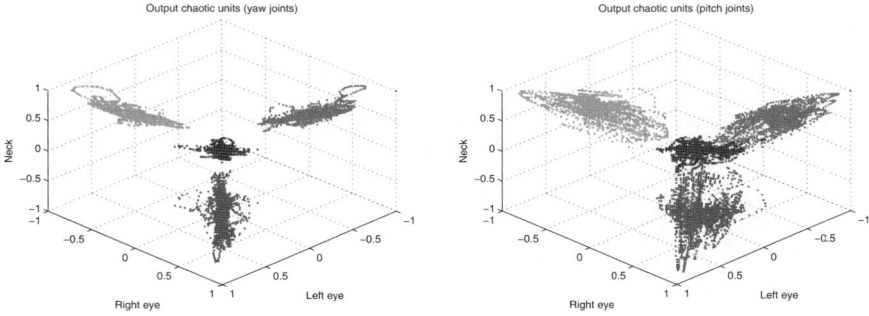

Fig. 6. Left, phase space (yaw). Right, phase space (pitch).

start with the same value, the system takes only a few time steps to separate in groups of strong and weak connections. A very interesting observation from this plot is that after approximately 500 steps, the connections arriving to any unit oscillate around the middle of the permitted strength. Extreme cases are with pitch units in each eye LP and RP which develop a very strong influence from the pitch motion of the neck NP but a zero influence from one to another. Yaw units develop a more balanced influence in their network, oscillating always around 0.5.

At time step 3500 the system has entered in an almost fully developed state where its internal connections vary very little. In the end, each unit is influenced by no more than two other units within the whole network, Fig. 8. As expected, two independent sub networks emerge after approximately 20 seconds. In one side all chaotic units fed by yaw motions strengthen their connections while weakening those towards and from 'pitch' units; and the same happens with those units fed by pitch motions when compared to 'yaw' units.

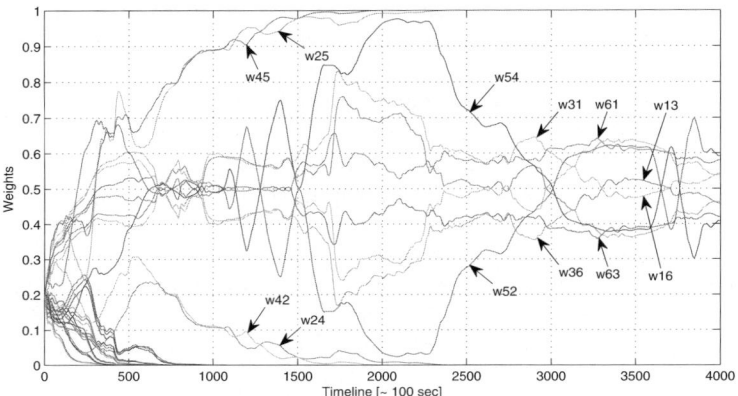

Fig. 7. Development of connections in time

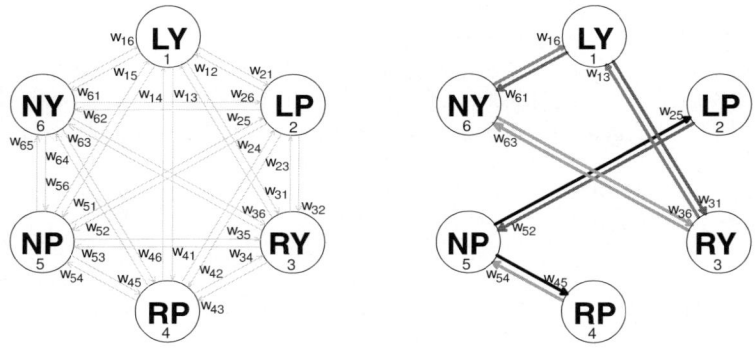

Fig. 8. Initial (left), and final (right) configurations of the GCM

4   Conclusions and Future Work

Conclusions

A very simple experiment for demonstrating the feasibility of applying coupled chaotic systems in the area of cognitive developmental robotics has been shown in this project. Tracking an object moving in front of a camera has been solved in several ways previously, from using very simple trigonometric solutions to advanced control algorithms. However, this task represented the simplest test bed for the study of emergence of a reactive behavior in a real platform.

A copy of the iCub's head [13], a 6 DOF robotic platform, was used to replicate the sensori-motor configuration of a real head. The tracking algorithm used in all experiments was taken from the YARP repository [15]. The experience obtained in previous experiments with the simulation and implementation of a single eye tracking [16] gave us enough confidence to increase the complexity of our model. The present work contains the results on the development of connections in the eyes-neck coordination problem (5 DOF).

We have demonstrated that a visual input is enough for the self-organization of a globally coupled map whose outputs are used as speed values activating each of the joints of our device. No specific coding of the task is needed, which results in a very fast reactive behavior. A very simple Hebbian rule was used to study the development of connections within the core of the system, a globally coupled map. From normalized initial connections we saw them changing through time, restructuring the 'brain' according to the experiences with the environment. In the final stage, two independent sub networks were formed, one containing yaw-related chaotic units only and the other pitch-related chaotic units only. The smooth pursuit behavior emerged during this process.

Future Work

The iCub's head includes also an inertial sensor which will be used in the future as another element influencing the chaotic field. Several questions should be addressed regarding the correspondences between this research and the biological counterpart; for example, if a smooth pursuit behavior emerged from the interaction of chaotic units, could it be possible to obtain other visual behaviors like vestibulo-ocular reflex (VOR), vergence or saccades in the same way?

The tracking algorithm used in all experiments does not focus on the same point in both cameras; consequently a displacement is observed when comparing the centers of both images. Therefore, this algorithm will be modified in order to visually track the same point in space.

Acknowledgments

The work presented in this paper has been in part supported by the ROBOTCUB project (IST-2004-004370), funded by the European Commissionthrough the Unit E5 - Cognitive Systems.

References

1. Kaneko, K., Tsuda, I.: Complex Systems: Chaos and Beyond. Springer, Heidelberg (2001)
2. Kuniyoshi, Y., Suzuki, S.: Dynamic emergence and adaptation of behavior through embodiment as a coupled chaotic field. In: Proceedings of 2004 IEEE/RSJ International Conference on Intelligent Robots and Systems, pp. 2042–2049 (2004)
3. Metta, G., Gasteratos, A., Sandini, G.: Learning to track colored objects with log-polar vision. Mechatronics 14(9), 989–1006 (2004)
4. Bernardino, A., Santos-Victor, J.: Binocular visual tracking: Integration of perception and control. IEEE Transactions on Robotics and Automation 15(6), 1080–1094 (1999)
5. Coombs, D., Brown, C.: Real-time binocular smooth pursuit. International Journal of Computer Vision 11(2), 147–164 (1993)
6. Aja Fernandez, S., Alberola Lopez, C., Ruiz Alzola, J.: A fuzzy-controlled kalman filter applied to stereo-visual tracking schemes. Signal Processing 83(1), 101–120 (2003)

7. Kumarawadu, S., Watanabe, K., Kiguchi, K., Izumi, K.: Self-adaptive output tracking with applications to active binocular tracking. Journal of Intelligent and Robotics Systems 36(2), 129–147 (2003)
8. Thelen, E., Smith, L.: A dynamic systems approach to the development of cognition and action. The MIT Press Inc, Cambridge (1994)
9. Strogatz, S.: Nonlinear dynamics and chaos. Addison Wesley, New York (1994)
10. May, R.: Simple mathematical models with complicated dynamics. Nature 261, 459–467 (1976)
11. Kaneko, K.: Relevance of dynamic clustering to biological networks. Physica D: Nonlinear Phenomena 75(1), 55–73 (1994)
12. Ito, J., Kaneko, K.: Spontaneous structure formation in a network of dynamic elements. Phys. Rev. E. 67(4), 46–226 (2003)
13. Robotcub project, http://www.robotcub.org
14. Sandini, G., Metta, G., Vernon, D.: The icub cognitive humanoid robot: An open-system research platform for enactive cognition. Journal of Bionics Engineering 1(3), 191–198 (2004)
15. Metta, G., Fitzpatrick, P., Natale, L.: Yarp: Yet another robot platform. International Journal of Advanced Robotic Systems 3(1), 43–48 (2006)
16. Duran, B., Sandini, G.: Towards the implementation of a "chaotic" smooth pursuit. In: IEEE/RAS International Conference on Humanoid Robots (accepted, 2007)

Formation Graphs and Decentralized Formation Control of Multi Vehicles with Kinematics Constraints

Ufuk Y. Sisli and Hakan Temeltas

Istanbul Technical University, Department of Electrical Engineering, Robotics Lab., Istanbul, 34409, Turkey
sisli@itu.edu.tr, temeltas@elk.itu.edu.tr

Summary. In this work, decentralized formation control of a multi vehicle system is investigated. Each vehicle model considers kinematic constraints of differential drives which is a principal approach for various application for mobile robotics in 2D space. A virtual leader is assigned to navigate the whole cluster in a certain formation via predefined paths. Each vehicle produces its own control signal via communication with other vehicles and interactions with virtual leader. These interactions and communications are modeled with formation graphs. Formation graphs are widely used in multi vehicle decentralized formation control area. They provide a robust and scalable control approach and tools for designing stable systems.

1 Introduction

Multi-vehicle system control is currently an attractive research area due to the increasing usage of unmanned autonomus vehicles in a wide variety of applications. However, these systems introduce new problems that do not exist in single-vehicle systems, like communication, coordinated path planning, sensor fusion and formation control. Formation control is one of the most researched of these problems, since almost all multi vehicle missions require vehicles to shape a desired formation.

Early works on the subject are mostly focued on centralized control methods where vehicles in the system are directed by a central controller and are not required to communicate with each other. Generally speaking, these systems collect the state information from all units, calculate globally optimum control signals and transmit back to the corresponding vehicles. Examples of centralized systems are [1] using various optimization techniques and [2, 3] using mixed-integer programming.

On the other hand, it is possible to extend the vehicles with communication capabilities. In this case, decentralized control methods become implementable. In decentralized methods, either dynamic or a static neighborhood is defined and vehicles try to align themselves according to their neighbors using appropriate formation constraints. This idea is mainly based on the works of biomathematicians like [4] and [5] which provide insight on animal group behaviors. In these studies it is stated that animal swarms including bird flocks, fish schools, viral colonies and ant swarms move as a result of member to member interactions rather than a central leader's commands of what to do [6]. As a result of these behaviors, a "swarm intelligence" that exceeds the capabilities any individual member arises from these member-to-member interactions. A very

H. Bruyninckx et al. (Eds.): European Robotics Symposium 2008, STAR 44, pp. 93–101, 2008.

straightforward example to this phenomenon is highway traffic, where cars eventually form lines and preserve distances from each other.

Various methods for decentralized control are behavior based methods proposed in [7], artificial potential methods proposed in [8, 9] and graph based methods proposed in [10]. Formation graphs are also investigated in [11, 12], and [13] where various control systems using formation graphs are presented along with their stability analyses. Graph based systems are shown to be successful in terms of achieving a unique stable formation. However the major drawback of these systems is that they require highly connected rigid communication patterns in order to guarantee a unique formation that does not depend on initial conditions. An example to this dependency is shown in section 3. The other missing point in these studies, is that they generally consider the vehicles as a point mass in order to simplify their kinematics and dynamic models.

This work futher develops the graph based methods and aims to solve unique formation problem by using a different formation graph. The distance matrix of the graph is projected onto coordinate axes and resulting projected matrices are used in calculation of the control signal. Presented system is shown to reach unique desired formations independent of initial conditions even for loosely connected and non-rigid communication patterns. Additionally this work extends the kinematic models of vehicles considering non-holonomic constraints, which play important role in controlling and navigating them.

The article is organized as follows: In the following section Kinematic and Dynamic Models of Vehicles are presented. Next, basic definitions and methods are summarized. After that, illustrative simulations are given and conclusions are presented in the final section.

2   Kinematic and Dynamic Model of Vehicles

Each vehicle in the vehicle groups has the state vector $x_i = (q_{x_i} \ q_{y_i} \ \theta_i \ p_i \ \omega_i)^T$, $x \in \Re^5$, where q_{x_i} and q_{y_i} represent position vector, θ_i represents orientation angle and p_i, ω_i represents linear and angular velocities respectively for the vehicle i and this can be shown in figure 1. Dynamical equations the single vehicle can derived by the following nonlinear equation set:

$$
\begin{pmatrix} \dot{q}_{x_i} \\ \dot{q}_{y_i} \\ \dot{\theta}_i \\ \dot{p}_i \\ \dot{\omega}_i \end{pmatrix} = \begin{pmatrix} p_i cos(\theta_i) \\ p_i sin(\theta_i) \\ \omega_i \\ 0 \\ 0 \end{pmatrix} + \begin{pmatrix} 0 & 0 \\ 0 & 0 \\ 0 & 0 \\ 1/m_i & 0 \\ 0 & 1/J_i \end{pmatrix} \begin{pmatrix} F_i \\ \tau_i \end{pmatrix}
\tag{1}
$$

where F_i and τ_i are force and torque inputs affecting center of the vehicle i. As it is seen in the state equation, state transition terms are nonlinear while the input terms in linear relationship. Hence the state equation, in general, can be given as $x_i = f(x_i) + g_i u_i$. It is obvious that the input vector u for vehicle i is formed as $u_i = (F_i \ \tau_i)$. Due to constraints in the wheels of the vehicles, the input variable F may not cause a motion along some directions. In our general approach those forces are derived from gradient operator of the potential fields. Hence, lets assume that the external forces are applied a specific

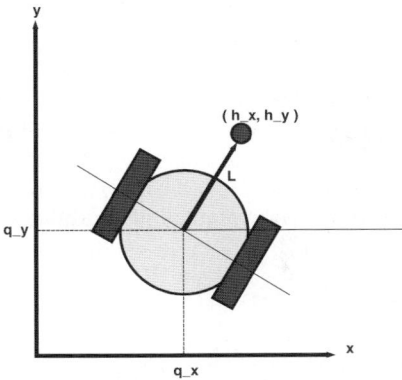

Fig. 1. A single vehicle

point instead of the center of the vehicle, namely, handle point, as it is shown in figure 1. The distance of the point from the center is given by L parameter. Then point h_i is defined by:

$$h_i = q_i + L_i \begin{pmatrix} cos(\theta_i) \\ sin(\theta_i) \end{pmatrix} \tag{2}$$

by taking time derivation of both sides of the above equation:

$$\dot{h}_i = \begin{pmatrix} cos(\theta_i) & -L_i sin(\theta_i) \\ sin(\theta_i) & L_i cos(\theta_i) \end{pmatrix} \begin{pmatrix} p_i \\ \omega_i \end{pmatrix} \tag{3}$$

It is possible to define a closed map such that $\Sigma(x_i) : \Re^5 \to \Re^5$ which maps the state vector x_i into a state vector assigned at handle point h_i:

$$\beta_i = \Sigma(x_i) = \begin{pmatrix} q_{x_i} + L_i cos(\theta_i) \\ q_{y_i} + L_i sin(\theta_i) \\ p_i cos(\theta_i) - L_i \omega_i sin(\theta_i) \\ p_i sin(\theta_i) - L_i \omega_i cos(\theta_i) \\ \theta_i \end{pmatrix} \tag{4}$$

$$\beta_i = \begin{pmatrix} \beta_{1_i} & \beta_{2_i} & \beta_{3_i} & \beta_{4_i} & \beta_{5_i} \end{pmatrix}$$

The mapping Σ between x_i and β_i is diffeomorphism [11, 12] and its inverse is given by:

$$x_i = \Sigma^{-1}(\beta_i) = \begin{pmatrix} \beta_{1_i} - L_i cos(\beta_{5_i}) \\ \beta_{2_i} - L_i sin(\beta_{5_i}) \\ \beta_{5_i} \\ \beta_{3_i} cos(\beta_{5_i}) + \beta_{4_i} sin(\beta_{5_i}) \\ (-1/L_i)\beta_{3_i} sin(\beta_{5_i}) + (1/L_i)\beta_{4_i} cos(\beta_{5_i}) \end{pmatrix}$$

Thus, the inverse mapping supply position and velocity vectors for the handling point. The orientation of a vehicle θ_i is uncontrollable as a result of the inverse mapping, however the orientation will always be aligned with the velocity vector in translational motion. For the sake of simplicity, the vehicle dynamic model will then be assumed as a double integrator in order to get rid of inertial parameters such as m_i and J_i. Thus, the dynamic of vehicle i is represented by:

$$\ddot{h} = \dot{p} \tag{5}$$

formation control of these vehicles is explained in the following chapter.

3   Formation Control with Graphs

In this section, several basic concepts and the control system presented in [11] are summarized. It is important to note that [11] provides the base and the proofs for the ideas and formulations presented here and interested reader is encouraged to read it also.

A directed graph $G = (V, E)$ consists of a vertice set $V = \{v_1, v_2, .., v_n\}$ and an edge set E. Here $E \subset V^2$ and each e_{ij} element of E is defined as $e_{ij} = (v_i, v_j)$ for vertices v_i and v_j that have a connection. First element, v_i is called "head" and the second element v_j is called "tail" of the edge. If the head and the tail of an edge are the same graph is said to contain a self-loop at that vertice. Throughout this article we will assume graphs that contain no self-loops. The connectivity number of vertice v_i is defined as the edges leaving v_i and denoted by $|v_i|$. Connectivity number of a vertice is the number of neighbors of that vertice. The degree of a graph is defined as $max(|v_i|)$ where $i \in [1, n]$ and denoted as $deg(G)$. Note that $deg(G)$ is the maximum number of neighbors that any vertice has in G.

Similarly a formation graph is a triplet $G = (V_e, C, D)$ that consits of an extended vertice set V_e, a connectivity set C and a distance set D. V_e is defined as $\{V, v_\infty\}$ where v_∞ is a virtual vertex at infinity and used only as a notational element in order to make connectivity numbers for all vertices equal. This is required in order to represent C and D sets as matrices. C and D are $[deg(G) \times n]$ matrices.

C matrix shows the neighboring relations and is defined as in equation 6.

$$c_{ij} = \begin{cases} j & ; \text{ if } v_i \text{ aligns itself according to } v_j \\ \infty & ; \text{ otherwise} \end{cases} \tag{6}$$

D matrix shows the distance constraints for each edge and is defined as in equation 7.

$$d_{ij} = \begin{cases} \|q_j - q_i\| & ; \text{ if } v_i \text{ aligns itself according to } v_j \\ b & ; \text{ otherwise} \end{cases} \tag{7}$$

Note that neighboring relations are one-way, meaning that if v_j is a neighbor to v_i, v_i may not be a neighbor to v_j, thus C and D matrices are not symmetrical.

Here q_j and q_i show the positions for ith and jth vehicles respectively, and d_{ij} is the desired euclid distance between the two. An example formation graph is shown in figure 2. Corresponding V_e, C and D matrices are given in equation 8.

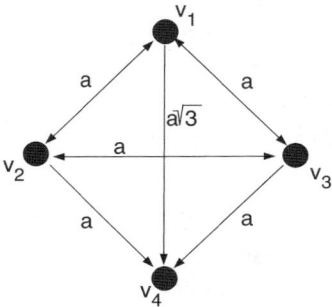

Fig. 2. A formation graph with 4 vehicles

$$V_e = \{v_1, v_2, v_3, v_4, v_\infty\} \qquad (8)$$

$$C = \begin{bmatrix} v_2 & v_3 & v_\infty \\ v_1 & v_3 & v_\infty \\ v_1 & v_2 & v_\infty \\ v_1 & v_2 & v_3 \end{bmatrix} \qquad (9)$$

$$D = \begin{bmatrix} a & a & \infty \\ a & a & \infty \\ a & a & \infty \\ \sqrt{3}a & a & a \end{bmatrix} \qquad (10)$$

Consider n identical vehicles with point mass dynamics described in equation 11 and a formation graph G. Here q is the state information of vehicles. In this work state information is 2D position information which implies that p and u are similar to velocity and acceleration respectively.

$$\dot{q}_i = p_i \qquad (11)$$
$$\dot{p}_i = u_i$$

In order to obtain the control signal for such a system, artificial structural potantial $V(q)$ is used. The definition of $V(q)$ is given in equation 12.

$$V(q) = < \Phi(q), \Phi(q) > \qquad (12)$$

Here $\Phi(q)$ is the structural constraint vector and $\Phi(q) = \{\phi_1, .., \phi_s\}$ where s is the edge count of graph G. ϕ_i is defined in equation 13 and may be interpreted as the distance to the desired d_{ij} value for the edge between ith and jth vehicles.

$$\phi_l(q_i, q_j) := ||q_i - q_j|| - dij ; \qquad (13)$$

With the help of Langrangian and Hamiltonian equations for the system, Murray et al. [13] states that the control signal in equation 14 achieves local stabilization of vehicles to the formation specified by graph G.

$$u_i = \frac{\bar{u}}{|J_i|} \sum_{j \in J_i} \lambda_1 \sigma(||q_i - q_j|| - d_{ij})\mathbf{u}_{ij} - \bar{u}\lambda_2 \sigma(p_i) \tag{14}$$

$$\mathbf{u}_{ij} = \frac{q_j - q_i}{||q_j - q_i||} \tag{15}$$

Here $\bar{u}$ is an upper bound for u and $\sigma(y)$ is defined as in equation 16. λ_1 and λ_2 are two real numbers such that $\lambda_1 + \lambda_2 = 1$.

$$\sigma(y) = \frac{y}{\sqrt{1 + ||y||^2}} \tag{16}$$

As stated and proved in [11] the control law stated above guarantees local formation stabilization with bounded feedback, which means vehicles starting from any initial conditions in the space forms and keeps the desired formation. However the direction of formation is not specified. A graph needs to be rigid and non-foldable in order to be able to represent a unique formation. Simulations of this system shows that different initial conditions may yield to different flocking points and formation orientations.

Figure 3-a and 3-b shows two sample runs of a system for a group of 7 vehicles starting from different initial conditions. In each setting, vehicles construct the formation in different locations and formations are rotated arbitrarily.

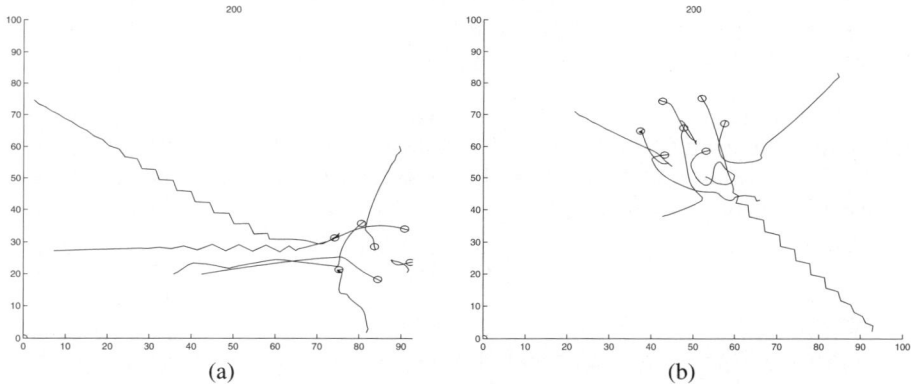

(a)   (b)

Fig. 3. Effect of initial conditions on flocking

In the following section, a modification to formation graph is presented as a solution to this problem.

4   Numerical Applications

In this section, some numerical simulations are presented in order to illustrate the behavior of various systems.

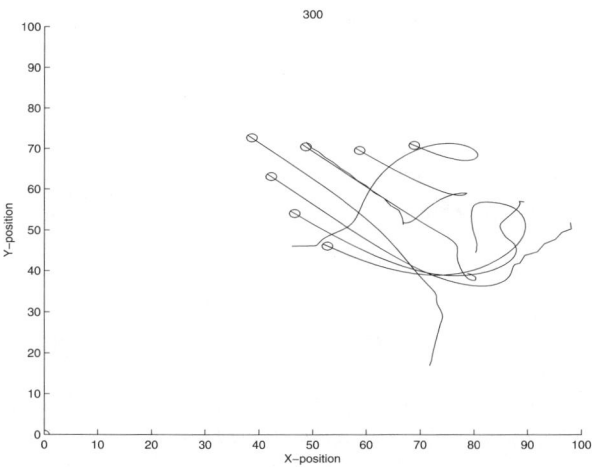

Fig. 4. 7 vehicles forming a V-shape

The first simulation shows 7 vehicles, whose initial positions are set randomly. The vehicles form a V-shape formation.

Figure 5 contains two columns of graphs where each column has six graphs that show the horizontal and vertical positions of all vehicles. Left column contains horizontal positions and right column contains vertical positions againts time.

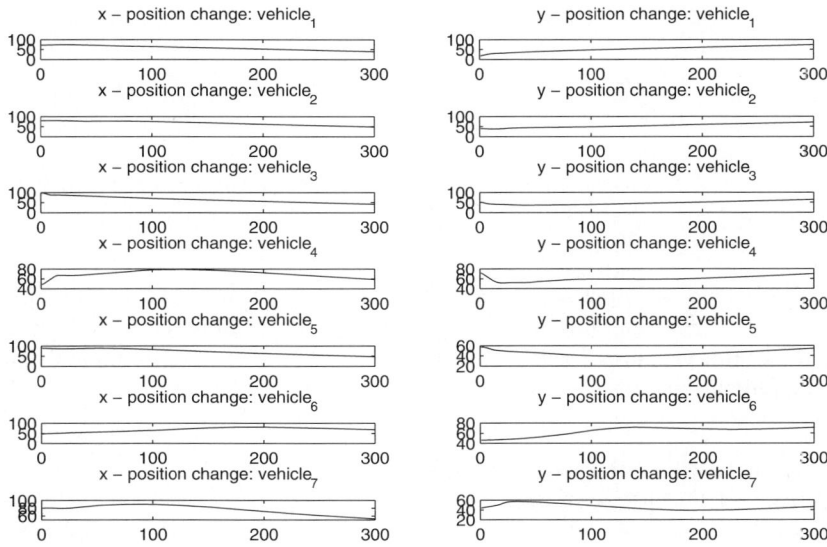

Fig. 5. Position changes of 7 vehicles in figure 4

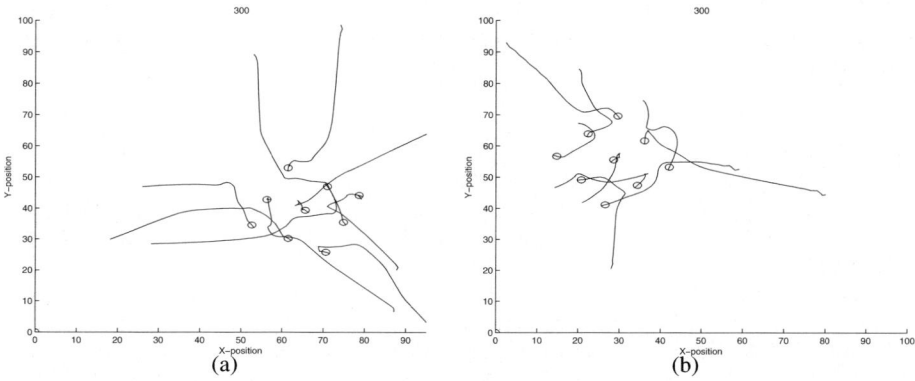

Fig. 6. Behavior of 9 vehicles forming a square shape formation graph under randomly chosen initial conditions

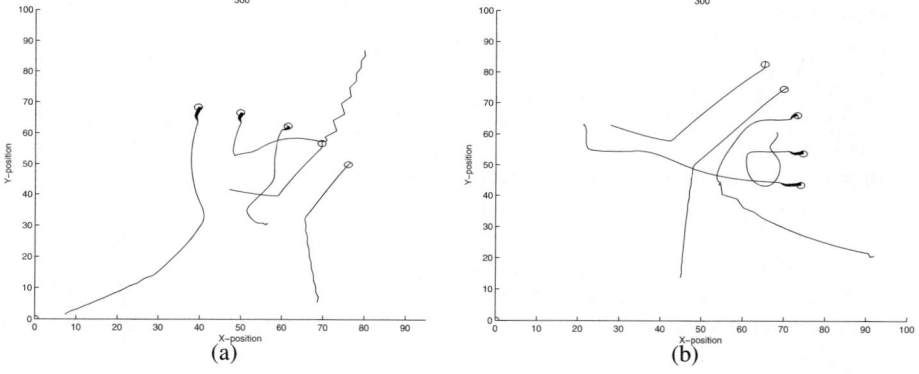

Fig. 7. Behavior of 5 vehicles forming a linear formation graph under randomly chosen initial conditions

Figures 6 and 7 illustrate the linear and square-shaped formation graphs. First one shows two runs of regular formation graphs with arbitrarily chosen initial conditions.

On the contrary, figures 7-a and 7-b shows linear formations under randomly chosen initial conditions.

5   Conclusions and Future Work

In this work, kinematic constraints have been successfully added to formation graph method for decentralized multi vehicle system. Various examples of formations are given as simulation results. Further study includes definition of a dynamical neighborhood in order to prevent collisions between non-neighboring vehicles in the final formation.

References

1. Zelinski, S., Koo, T.J., Sastry, S.: Optimization based formation reconfiguration planning for autonomus vehicles. In: Proceedings of the 2003 IEEE conf. robot. automat., pp. 3758–3763 (2003)
2. Schouwenaars, T., De Moor, B., Feron, E., How, J.: Mixed integer programming for multi-vehicle path planning. In: European Control Conference, pp. 2603–2608 (2001)
3. Chasparis, G.C., Shamma, J.S.: Linear-programming-based multi-vehicle path planning with adversaries. In: Proceedings of the 2005 American control conference, vol. 2, pp. 1072–1077 (2005)
4. Okubo, A.: Aspects of animal grouping. Adv. Biopysics 22, 1–94 (1986)
5. Warburton, K., Lazarus, J.: Tendency-distance models of social cohesion in animal groups. J. of Theoretic Biology 150, 473–488 (1991)
6. Resnick, M.: Turtles, termites, and traffic jams - explorations in massively parallel microworlds. MIT Press, London (1994)
7. Balch, T., Arkin, R.C.: Behavior-based formation control for multirobot teams. IEEE Transactions on Robotics and Automation 14, 926–939 (1998)
8. Leonard, N.E., Fiorelli, E.: Virtual leaders artificial potentials and coordinated control of groups. In: Proc. of the 40th IEEE Conference on Decision and Control, Orlando, FL, USA (2001)
9. Elkaim, H.G., Siegel, M.: A lightweight control methodology for formation control of vehicle swarms. In: IFAC World Congress, Prague, Czech Republic (2005)
10. Fax Alexander, J., Murray Richard, M.: Graph laplacians and stabilization of vehicle formations. In: IFAC World Congress, Barcelona, Spain (2002)
11. Olfati-Saber, R., Murray Richard, M.: Distributed cooperative control of multiple vehicle formations using structural potential functions. In: IFAC World Congress, Barcelona, Spain (2002)
12. Carlos, G., Morgansen, A.K.: Stabilization of dynamic vehicle formation configurations using graph laplacians. In: IFAC World Congress, Prague, Czech Republic (2005)
13. Olfati-Saber, R., Murray Richard, M.: Flocking with obstacle avoidance: cooperation with limited information in mobile networks. In: Olfati-Saber, R., Murray Richard, M. (eds.) Proceedings of IEEE Conference on Decision and Control, vol. 2, pp. 2022–2028 (2003)
14. Lawton, J.R.T., Beard, R.W.: A Decentralized Approach to Formation Maneuvers. IEEE Transactions on Robotics and Automation 19, 933–941 (2003)

Global Urban Localization of an Outdoor Mobile Robot with Genetic Algorithms

Can Ulas Dogruer[1], A. Bugra Koku[2], and Melik Dolen[2]

[1] Mechanical Engineering Department, Hacettepe University 06800, Ankara, Turkey
dogruer@hacettepe.edu.tr
[2] Mechanical Engineering Department, Middle East Technical University 06532, Ankara, Turkey
{kbugra,dolen}@metu.edu.tr

Summary. The localization of mobile robots has been studied rigorously in the past. However, only a few studies have focused on developing specific Genetic Algorithms (GAs) to address the localization problem effectively. In this study; the global urban localization of an outdoor mobile platform is considered with the utilization of the odometer, the laser-rangeq finder measurements and the digital maps created from the relevant satellite images on the Internet. The localization issue is formulated as a constrained optimization problem. The study proposes a GA-based technique to solve the problem at hand efficiently.

Keywords: Localization, Mobile Robotics, Genetic Algorithms, Remote Sensing Technology.

1 Introduction

Localization of mobile robots with limited sensory resources has been studied via various methods in the past including the Extended Kalman Filter (EKF) [1], the Monte Carlo Localization (MCL) [2] and the Grid Based Localization [3][4]. Despite the apparent success of these methods in controlled environments, they do have some well-known drawbacks such as numerical instability, divergence (or premature convergence), and computational cost. For instance, the EKF requires reliable measurement- and process noise models with good estimates on the covariance matrices beforehand. Ineffective noise models often times degrade the performance of the method. Note that in the long run, the estimates of the EKF may diverge from the optimal states (i.e. true location of the robot) since the system equations for the EKF is obtained by a first-order Taylor expansion of the nonlinear dynamic model of the robot. Likewise, the other popular method, MCL, is an iterative technique that may prematurely converge to an incorrect location if insufficient number of candidate locations (with poor statistical distribution) is taken into consideration. Similarly, the grid based localization is generally effective not only at the cost of huge computational burden, but these methods require significant amount of storage space as well.

Genetic algorithms (GAs), which may overcome the above-mentioned difficulties associated with the conventional techniques, have been lately employed as a novel tool in dealing with robot localization problem [5][6][7]. GAs, which were first introduced to the scientific community by Dr. John Holland in 1975, have emerged as powerful optimization tools. Since then, GAs have been applied to several engineering problems

H. Bruyninckx et al. (Eds.): European Robotics Symposium 2008, STAR 44, pp. 103–112, 2008.
springerlink.com
© Springer-Verlag Berlin Heidelberg 2008

successfully. GAs are utilized to optimize structural engineering systems [8][9], production planning [10], and control systems [11]. GAs have recently found their use in mobile robotics research. For instance, Duckett et al. [5] applied GAs to mapping problem while Moreno et al [6] employed Kalman Filter to enhance the search space of GAs. Kwork et al. [7] has successfully utilized evolutionary programming to localize a mobile robot.

Performance of GAs on the Global Urban Localization (GUL) problem is to be studied in this work. That is, a mobile robot with a Wi-Fi enabled device (such as a laptop, a PDA or a cell phone) wakes up in an unfamiliar urban setting. Luckily, a wireless Internet connection is detected in the area so that the IP address obtained through this connection can be used to identify this region. With the virtue of the wireless connection, the robot accesses a website (such as Google Earth [12]) where the satellite images of the earth can be freely downloaded. By simultaneously going over the satellite images and scanning the environment via short-range sensors, the robot can find common features that might pinpoint its location or at least prune the search space. If standing still does not help, robot may start wandering to acquire more useful local information. Consequently, the robot finds its whereabouts. For short, it performs 'global localization' limited to urban settings. This problem will be referred to as GUL problem. To the best of our knowledge, the work presented in this paper, is the first attempt to the solution of such a problem within a restricted scope.

In this study, an optimization problem for the GUL is first formulated for a mobile robot (platform) equipped with a laser range finder and odometer. For this purpose, the relevant sensory inputs as well as the digital maps created from the satellite images (to be downloaded from the Internet according to the scenario), are used together to express the localization problem at hand as a constrained optimization problem. In this scheme, the odometer data is to impose relation among successive moves on the path while the digital map restricts the plausible positions in the robot's domain. Once the optimization problem is formulated, the localization performance thru the GAs is then to be investigated. The organization of the paper is as follows: in section two, the background on the GAs are reviewed; in section three, the theory of the proposed solution together with definition of localization problem suitable for GAs are given. In section four, the simulation results are presented while the last section focuses on the key results of the paper.

2   Background

Genetic Algorithms (GAs), which are inspired by the laws of natural selection, are the search algorithms suitable for the solution of nonlinear optimization problems. By nature, GAs are different than the traditional optimization methods in terms of global convergence, parallelism, efficiency, and robustness. The basic principal behind GA is the survival of the fittest individual through stochastic yet structured information exchange. Note that GAs work with the coded representation of an optimization problem. As the basic object of GA, the genetic code (or chromosome), which is generally represented as finite-length character string, encodes the relevant parameters of a search domain. The representation of the domain along with the encoding of design vector plays a key role in the dynamics of the algorithms. The sequential GAs work with a collection of genetic codes called *population* at a particular instant. Since GAs do not employ auxiliary knowledge such as the gradient of the objective function, they exclusively rely on a properly formulated fitness (cost, or objective) function to evaluate the merit of each individual string in the population. After the fitness of a given population is evaluated, three stochastic operators (*reproduction, crossover,* and *mutation*) are systematically applied to transform one population to the next sequentially until an acceptable solution is found.

Even though GAs are frequently employed to solve unconstrained optimization problems in literature, they can be conveniently modified to handle constraints with the use of constraint-transformation techniques and penalty functions [13]. To be specific, let us consider a traditional optimization problem:

$$
\begin{aligned}
\text{Minimize:} \quad & f(\mathbf{x}) \\
\text{Subject to:} \quad & g_j(\mathbf{x}) \geq 0 && j = 1,......,J \\
& h_k(\mathbf{x}) = 0 && k = 1,......,K \\
& x_i^l \leq x_i \leq x_i^u && i = 1,......,n
\end{aligned}
\tag{1}
$$

where $f(\mathbf{x})$ is the objective function to be minimized; $g_j(\mathbf{x})$ are the inequality constraints (a total of J); $h_k(\mathbf{x})$ are the equality constraints (a total of K). Note that each element of the design vector $(\mathbf{x})$ has a specific range of $\left[x_i^l \ x_i^u \right]$. A solution to this objective function is said to be feasible when it also satisfies the imposed constraints. It is customary to convert a constrained optimization problem given in the form of (1) into an unconstrained one as

$$
P(\mathbf{x}, \mathbf{R}, \mathbf{r}) = f(\mathbf{x}) + \sum_{j=1}^{J} R_j \left\langle g_j(\mathbf{x}) \right\rangle^2 + \sum_{k=1}^{K} r_k \left[h_k(\mathbf{x}) \right]^2
\tag{2}
$$

where the objective function is augmented with constraints. Such a scheme usually works well with GAs that are originally designed to cope with constrained optimization problems. Note that the determination of penalty factors (like$[R_j, r_k]$) is critical in the solution of problem. That is, large penalty factors emphasize the constraints whereas small ones stress the objective function. Large penalty factors will force map and odometer data consistent solution, but put less emphasis on the search capability of objective function. The magnitude of the penalty factors issue will be revisited at the next chapter.

3   Problem Formulation

In the previous section, the theory of GAs is briefly reviewed. In this part, the localization problem is to be formulated as a constrained nonlinear programming problem. To study the performance of the proposed technique, a number of field tests are to be carried out first. During these tests, the odometer of the mobile platform is expected to yield its incremental position while the laser range finder is to give 180-degree polar scan of the surroundings. The preliminary studies show that when coupled, these two information sources may be efficiently employed to localize a mobile robot in the global sense with the use of a digital map. Such maps can be created using the satellite images freely available thru the Internet. The use of these maps enhance the implementation time greatly and is a vital parts of the localization technique with GAs.

As discussed in previous section, there are two important aspects of GA: the encoding of the parameters into chromosomes and finding a simple yet effective objective/fitness function. Via drawing analogy to the *matching algorithms,* the objective function can be simply formulated as

$$f(\mathbf{x}, \mathbf{y}) = -\sum_{i=1}^{N} P[O_i | M, \{x, y\}] \tag{3}$$

In Equation (3), $P[O_i|M, \{x,y\}]$ represents the probability of observing a polar scan data on the given the map M while $\{x,y\}$ stand for the robot position on this map. $P[O_i|M, \{x,y\}]$ is computed by computing the statistical similarity of experimental polar scan with virtual scans taken at various particles distributed on map. The similarity measure is assumed to be correlation coefficient, and negative correlation values are set to zero. The fitness value is then computed by using N previous measurements. The position close to the true location will have higher probability where as the position vector $\{x,y\}$ far away from the true location will yield low probability or fitness value.

Note that the odometer data is used as a constraint such that the possible locations in a sequence representing the true path are connected to each other. No single location is independent from the location computed before (or after). Since the odometer data is not precise; a range must be defined around a particular odometer datum. This range in odometer data also relaxes the constraints leaving a room for search. The inequality constraints employing the odometer data are designed as follows:

$$r_{i+1} < \theta_{i+1} - r_i < \theta_i \leq d_i^o < \theta_i^o + \delta^o \qquad (i = 1, \ldots, N) \tag{4}$$

where d_i^o (distance) and θ_i^o(angle) are the odometer readings expressed in polar coordinates. Similarly, r_i and θ_i denote the actual location of the mobile robot at the ith step. Since a search within a circular window centered around each odometer reading is to be performed, δ^o corresponds to a relaxation term for the ith reading. Hence, these constraints are coupled to form a chain of relevant data.

The equality constraint, which is defined as a double sum over the multiplication of a candidate path by a masking matrix representing plausible locations, can be given as

$$C_{eq} = \sum_{(i)} \sum_{(j)} A_{path} \otimes B_{mask} \tag{5}$$

where $\otimes$ stands for the element-wise matrix multiplication. The path (matrix) can be represented with ones and a viable location in the masking matrix could be characterized by zeros. Thus, the masking matrix is to be filled with ones elsewhere. In the ideal case, the element-wise multiplication of path- and masking matrices gives a zero-matrix. If the path coincides with implausible positions, the double sum yields non-zero terms pointing out to a violation in the equality constraint.

Consequently, the general objective function of the unconstrained optimization function that could serve as a fitness function for the GAs, can be expressed as

$$P(\mathbf{x}, \mathbf{R}, \mathbf{r}) = -\sum_{i=1}^{N} P[O_i | M, \{x, y\}] + r \sum_i \sum_j A_{path} \otimes B_{masking}$$

$$\sum_{i=1}^{N} R_i \{ r_{i+1} < \theta_{i+1} - r_i < \theta_i - d_i^o < \theta_i^o - \delta^o \}$$

(6)

As pointed out in previous section, the values of penalty factors play a critical role in the solution: $r$penalty factor puts emphasis on the map information while the penalty factor R accentuates the odometer data and the sequential constraints which determine the flexible shape of the rough path.

4   Simulation

In order to evaluate the performance of the proposed method, two experimental studies are conducted. A mobile platform equipped with a laser range finder, a notebook, and a battery pack is driven inside the campus of the Middle East Technical University. The laser range finder, which is aligned with the heading of the mobile platform, records the 180-degree polar scan of the environment in XY plane at a constant elevation. The incremental displacement of the mobile platform is recorded roughly at each meter of travel with the utilization of a tape meter that serves as a crude odometer. Fig. 1 shows the mobile platform used in the experimental study along with a typical laser-range finder polar scan.

Simulation is performed in the MATLAB^{TM} environment using the GA toolbox [14]. The experimental data consisting of odometer and laser range finder scans are recorded in the field and are later fed into the simulation program in an off-line fashion. The parameters of MATLAB GA toolbox used in the simulation are given in Table 1. The satellite image of the local environment, which is downloaded from the Google Earth [12], is shown in Fig. 2 and the segmented image is also shown in the same figure. Since the approach described here relies on the digital map of the environment, the segmentation of the satellite image, which is further elaborated in [15], is a vital part of the proposed paradigm.

For the first case, the path of the mobile platform is illustrated in Fig. 3a. Likewise, the final path and the corresponding masking matrix are shown in Fig. 3b. As can be seen, the path found by the GAs precisely matches with the original trajectory. In this simulation, the mobile platform is localized after it moves approximately 80 meters in a local area of 56000 square meters. The execution time of the algorithm was approximately 750 s.

To highlight the advantage of GAs over existing (iterative) methods; this particular case is also simulated using the MCL method in [2]. That is, In the following case, the Sampling Importance Resampling (SIR) particle filter algorithm is implemented. The results are illustrated in Fig. 4 where the MCL cannot localize the mobile platform until it reaches to the 12th step of the simulation and doesn't yield any useful information before that instance. It should be pointed out that the GAs also allow the user not only to compute the final location at a certain instant but also gives the flexibility to calculate the complete path right from the start. Furthermore, the MCL [2] and the Grid Based

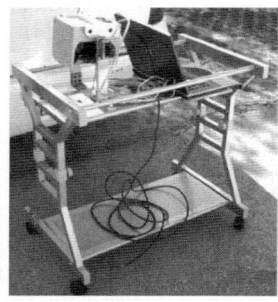 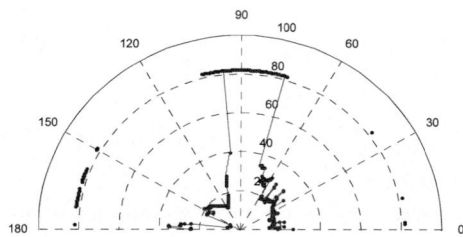

Fig. 1. Mobile platform used in the study (left) and typical laser range finder data (right)

 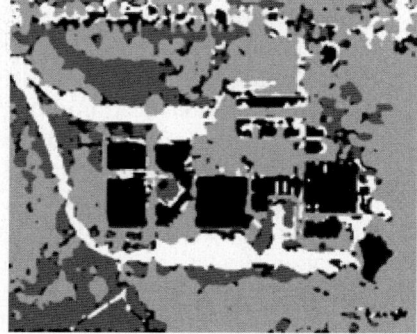

Fig. 2. Satellite image of part of Middle East Technical University nearby Mechanical Engineering Department (left figure), segmented image (right figure)

Approaches [3], which are iterative in nature, cannot compute the path backwards in time. Furthermore, these algorithms localize the robot at a certain time slice of the navigation for at least one step with full confidence but they cannot give any hint about what is going right before that instant in the simulation. This complete path gives us the opportunity to correlate the shape of the computed path with the one being constructed with the odometer data (plus a magnetic compass) roughly. This is an alternative checkpoint along the way to guarantee a successful localization.

A second simulation is conducted inside a larger environment. The simulation results are illustrated in Fig. 5. In this simulation, the mobile platform is localized after it has moved 100 meters within an area of 270,000 square meters. Although the computation time is longer (5780 seconds), the precision of localization with GAs is quite high since it takes into account the whole information collected up to the current position. As a baseline the convergence speed of MCL which runs with m particles is $O \propto m^{-\frac{1}{2}}$ [2].

Performances of GAs for both cases (i.e. statistical attributes) are illustrated in Fig. 6. In this figure, the current/final best individuals found by GA are demonstrated. Note that

Table 1. Parameters of Matlab GA toolbox

Crossover Function	Uniform
Crossover Fraction	0.8
Elite Count	5/20
Fitness Scaling Function	Fitness Scaling Rank
Mutation Function	Adaptive Feasible
Population Size	50;200
Population Type	Double Vector
Selection Function	Stochastic Uniform

a) Satellite image and path   b) Masking matrix and final path

Fig. 3. Actual robot path printed on satellite image and Possible path printed on the masking image

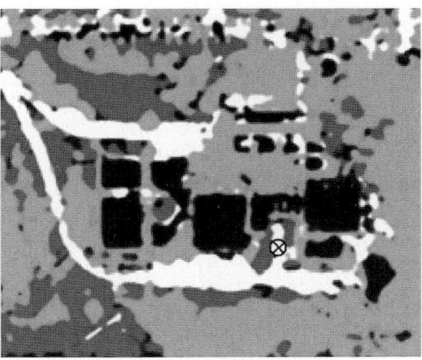

Fig. 4. Localization of mobile platform with the MCL at the 12th simulation step

the order of the variables is $[\{x_1,y_1\}, \{x_2,y_2\}, \ldots \{x_i,y_i\} \ldots, \{x_N,y_N\}]$. The optimization is terminated in three generations when the change in fitness value drops below a fixed tolerance.

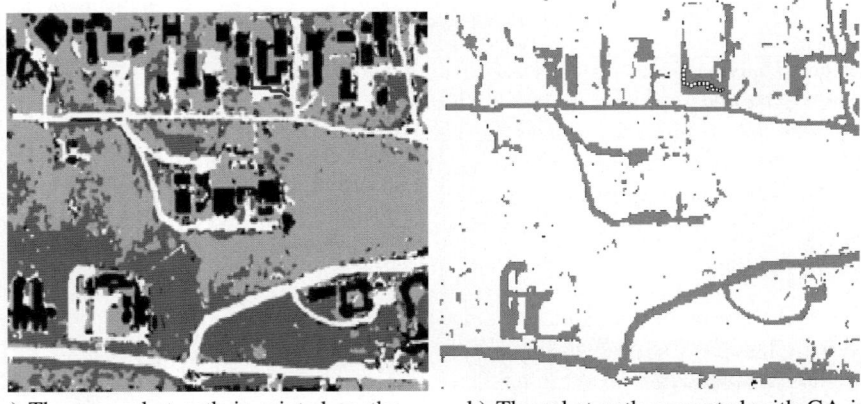

a) The true robot path is printed on the map with red circles

b) The robot path computed with GA is printed on the map with red circles

Fig. 5. Actual robot path and localization of mobile platform with GAs

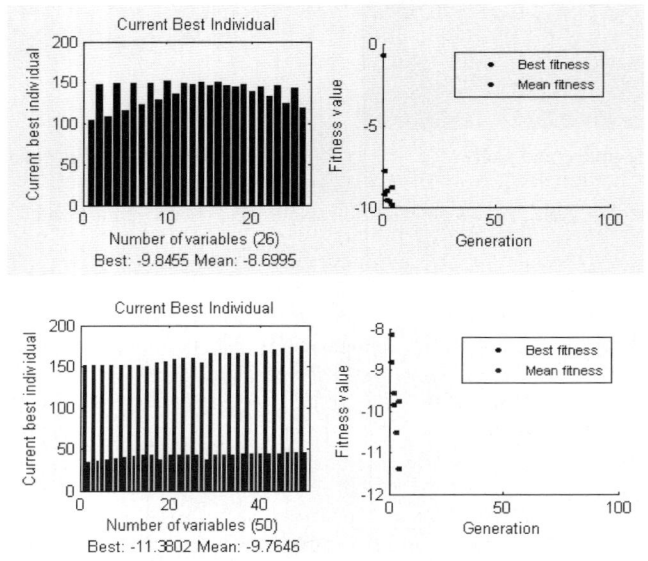

Fig. 6. Performance of GA for results published in Figure 3 & 5 (top & bottom figure)

5   Conclusions and Discussion

In this study, the localization of a mobile robot was studied with the utilization of GAs. First, the localization problem was expressed as a constrained optimization problem. Using odometer data and a digital map, the necessary constraints of the problem were formulated. As discussed earlier, the use of satellite images along with a suitable localization technique allows the solution of GUL problem. Digital map used in this study was obtained by segmenting the (freely available) satellite images on the Internet. Maps, which were created with a special segmentation technique, have simplified the outdoor localization significantly. Though it is described elsewhere [15], the segmentation scheme is a vital part of the localization approach proposed here.

To assess the merit of the proposed approach, two case studies have been conducted. The preliminary results indicate that this approach is viable and quite precise within the context of mobile robot localization. Consequently, the performance of the GAs on GUL problem was found to be satisfactory. That is, localization with GAs seems significantly better than its counterparts since the objective function is optimized globally. When the localization problem is solved with GAs globally, the complete path right from the start of navigation up to the final location can be found. Thus, a mobile robotic system with limited resources can locate itself in an urban setting as it moves around the environment. Here, GAs serve as viable alternatives to existing localization techniques (like well known techniques like Extended Kalman Filter or Monte Carlo Localization).

Acknowledgements. This work was supported in part by the State Planning Organization of Turkey (DPT) under Grant BAP-08-04-DPT. 2003(06) K120920-24.

References

1. Leonard, J.J., Durrant-Whyte, H.F.: Mobile Robot Localization by Tracking Geometric Beacons. IEEE Transactions on Robotics and Automation 7, 376–382 (1993)
2. Thrun, S., Fox, D., Burgard, D., Dellaert, F.: Robust Monte Carlo localization for mobile robots. Artificial Intelligence 128, 99–141 (2001)
3. Fox, D.: Markov Localization: A Probabilistic Framework for Mobile Robot Localization and Navigation. PhD Dissertation, Institute of Computer Science III, University of Bonn, Germany (1998)
4. Thrun, S., Burgard, W., Fox, D.: A Probabilistic Approach to Concurrent Mapping and Localization for Mobile Robots. Autonomous Robot 5, 253–271 (1998)
5. Duckett, T.: A Genetic Algorithm for Simultaneous Localization and Mapping. In: Duckett, T. (ed.) IEEE International Conference on Robotics and Automation, Proceedings. ICRA 2003, pp. 14–19, 434–439 (2003)
6. Moreno, L., et al.: A Genetic Algorithm for Mobil Robot Localization Using Ultrasonic Sensors. Journal of Intelligent and Robotics Systems 34, 135–154 (2002)
7. Kwok, N.M., Liu, D.K., Dissanayake, G.: Evolutionary Computing Based Mobile Robot Localization. Engineering Applications of Artificial Intelligence 19, 857–868 (2006)
8. Sheng-Li, Z., et al.: Genetic Algorithms for Optimal Design of Underground Reinforced Concrete Tube Structure. Machine Learning and Cybernetics 4, 2307–2311 (2004)
9. Srinivas, V., Ramanjaneyulu, K.: An Integrated Approach for Optimum Design of Bridge Decks Using Genetic Algorithms and Artificial Neural Networks. Advances in Engineering Software 38, 475–487 (2007)
10. Ip, W.H., Li, Y., Man, K.F., Tang, K.S.: Multi-product Planning and Scheduling Using Genetic Algorithm Approach. Computers and Industrial Engineering 38, 283–296 (2000)
11. Krishnakumar, K., Goldberg, D.E.: Control System Optimization Using Genetics Algorithm. Journal of Guidance, Control and Dynamics 15 (1992)
12. Google Earth Software,@ Google, http://earth.google.com

13. Carlos, A.C.C.: Theoretical and Numerical Constraint-handling Techniques Used with Evolutionary Algorithms: A Survey of The State of The Art. Comput. Methods Appl. Mech. Eng. 191, 1245–1287 (2002)
14. MATLAB software, Mathworks, http://www.mathworks.com/
15. Dogruer, C.U., Koku, B., Dolen, M.: A Novel Soft Computing Algorithm to Segment Satellite Images for Mobile Robot Localization and Navigation. In: Dogruer, C.U., Koku, B., Dolen, M. (eds.) the IEEE/RSJ International Conference on Intelligent Robots (IROS 2007), San Diego, California, pp. 2077–2082 (2007)

Grip Force Control Using Vision-Based Tactile Sensor for Dexterous Handling

Norinao Watanabe[1] and Goro Obinata[2]

[1] Graduate School of Engineering, Nagoya University,Furo-cho, Chikusa-ku, Nagoya, 464-8603, Japan
    n.watanabe@dynamics.mech.nagoya-u.ac.jp
[2] EcoTopia Science Institute, Nagoya University, Furo-cho, Chikusa-ku, Nagoya, 464-8603, Japan
    obinata@mech.nagoya-u.ac.jp

Summary. In this paper, we propose a force control method for robot grippers based on the degree of slippage using vision-based tactile sensor. Our tactile sensor consists of a CCD camera, LED lights, an acrylic plate and a spherical elastic body. The feature of this sensor is to measure "stick ratio", which indicates the degree of slippage between a robot gripper and a grasped object. The stick ratio is used to control the grip strength. The proposed method doesn't require any other information such as the weight of object and the friction coefficient. The gripper with the control can hold the grasped object while the robot gives a certain movement to the object. It is shown that the method makes it possible to achieve dexterous handling only using signals from the tactile sensor.

1 Introduction

Humans have the ability to sense the degree of slippage between the grasped object and the contact surface of fingertip with their distributed tactile receptors. The ability makes it possible for humans to prevent from the slippage of grasped object under various circumstances. Such dexterous handlings are achieved by feeding back the signals from the tactile receptors to muscle control system through neural networks. Therefore, the tactile sensing must be a key point for establishing dexterous handlings by robots when we want to mimic skilled human functions. So as to establish tactile sensing, many methods and sensors have been proposed during these three decades, which are based on electrical resistance, capacitance, electromagnetic component, piezoelectric component, ultrasonic component, optical component, and strain gauge [1], [2]. There exist many problems with these sensors to be solved for the practical usages. For an example, the sensor which consists of elastic body and strain gauges requires an array with many gauges and the wiring to signal processing devices. Moreover, the process for obtaining the values of contact forces and the friction coefficients requires a certain amount of computational resources; thus, it is difficult to achieve the real time processing [3]. On the other hand, vision sensors have been introduced because the wiring is not required in the contact part to the object [4], [5], [6]. The introduction of vision sensor makes the size small and the wiring simple. However, the sensing of friction coefficient is not considered in those papers. Piezoelectric sensors have a certain potential

H. Bruyninckx et al. (Eds.): European Robotics Symposium 2008, STAR 44, pp. 113–122, 2008.
springerlink.com

to solve the problems of size and wiring but there has not been a practical solution yet for measuring friction coefficient [2].

Another aspect for achieving dexterous handling like human is the integrative usage of sensing information. For preventing from the slippage of grasped object, we have to take the degree of slippage between the object and the gripper into the consideration when we determine the grip force. More concretely we have to determine the grip force based not only on the contact forces/moments but on the degree of slippage such as friction coefficient. Thus, this point requires multiple usage of force/moment sensors and extra sensors for the friction coefficient if the coefficient is unknown. However, it is hard in most cases to implement a number of sensors into the small finger tips of robot grippers. So as to avoid multiple usage of tactile sensors, we have proposed a new design of tactile sensors for multiple measuring of contact information including friction coefficient [8], [9].

In this paper, we discuss the real time estimation of degree of slippage on the contact surface with a vision based tactile sensor, which has been proposed by the authors [7], [9]. In addition, we propose a new real time method of image processing to obtain the degree of slippage, and apply the method to control the grip force. The experimental results show that the proposed method can achieve a dexterous handling of the object with the prototype sensor and the gripper of two fingers.

2   Vision Based Tactile Sensor

We proposed a vision-based sensor for multiple measuring of contact information including friction coefficient [7]. The configuration is illustrated in Fig. 1. This shows that the sensor consists of a CCD camera, LED lights, a transparent acrylic plate, and an elastic body. The elastic body, which is made of transparent silicon rubber and has grid pattern or dotted pattern on the spherical surface as shown in Fig. 2, is to contact the object. The CCD camera is to take pictures of the spherical surface from the flat surface side of the elastic body.

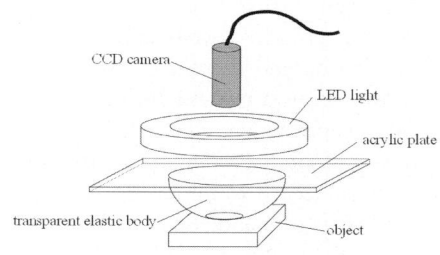

Fig. 1. Structure of vision-based tactile sensor

We assume in this study that the elastic body touches a flat surface of rigid object and the pictures taken by the CCD camera will be used to estimate the contact information while touching.

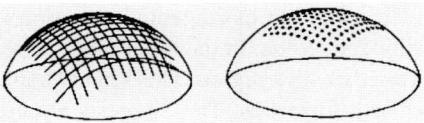

Fig. 2. Examples of shape and pattern on the surface of the elastic body

3   Measurement of Contact Force and Moment

We can estimate the contact area from the brightness of each pixel in a picture of contact surface which is taken by the CCD camera. The relation between the contact area and the applied normal force is known and can be determined for the specified elastic material; that is, we can obtain the estimation of the normal force from the contact area. We have experimentally confirmed this relation [7]. Next, we can also measure the displacements of the dots or the intersection points of the elastic body when both normal and tangential forces are applied. We found experimentally that the displacements of dots/intersection-points in the contact plane are a function of the applied normal and tangential forces. The normalization with the contact area on the displacements cancels out the effect of the applied normal force. This means that the applied tangential forces can be estimated from the normalized displacements of the central dot/intersection-point in the contact area [8].

We can measure the movements of the dots or the intersection points on the body surface while a moment is applied. The difference between the pre-contact and post-contact pictures includes the information for the displacements of the dots or the intersection points on the surface. Using a technique of signal processing, we can extract the rotation angles around the center. We recognized that the estimated angles depend on the applied normal forces. However, the normalization to the applied normal force makes it possible to estimate the applied moment from the displacements of the dots or the points.

The obtained relations for a prototype sensor to estimate the applied normal force, tangential force, and moment are given in [7], [9] with the process of the image processing on the pictures.

4   Estimation of Degree of Slippage

4.1   Incipient Slippage and Definition on Degree of Slippage

So as to prevent from the slippage of the grasped object, we need to obtain the conditions of contact surface between the gripper and the object. The coefficient of static friction is a key for handling the object without slipping. When the contact occurs between a curved surface and a flat surface, the pressure between the two surfaces distributes in the contact area. If the pressure of contact surface takes a lower value than the constant value which is determined by both the surface conditions and the materials, the relative motion in tangential direction is possible in the area. The pressure distribution between the gripper and the object divides the contact area into two parts in general. In one part of contact surface, the relative motion in tangential direction is possible. We call the part of area as incipient slippage region. In the other part, the relative motion is impossible. This part of area is called as stick region. This kind of contact always occurs when human touches the object with fingertips. This suggests a potential of human sensing that some groups of receptors in cutaneous sensory system catch the degree of slippage on the basis of this fact without macroscopic slippage. The schematic view of finger contact and the definition of the two regions are illustrated in Fig. 3. If we distinguish the two parts of area from the picture of CCD camera with our sensor, we can estimate the degree of slippage from the ratio of the stick area to the whole contact area. The ratio is defined by

$$\phi = \frac{S_s}{S_c} = \frac{r_s^2}{r_c^2} \tag{1}$$

where S_s is stick area, S_c is contact area, r_s is radius of stick area, and r_c is radius of contact area. We call the ratio as stick ratio, and it relates directly to the friction coefficient. In the cases of contact between spherical and flat surfaces, the incipient slippage occurs in peripheral part of contact.

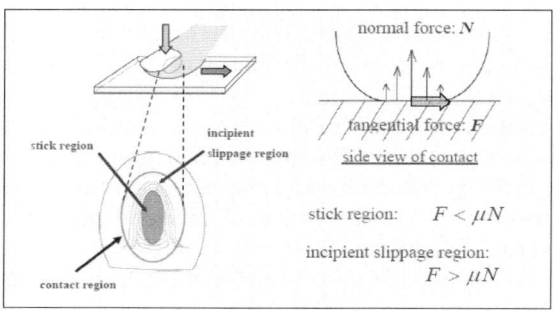

Fig. 3. Schematic view of finger contact and definition of incipient slippage region

4.2   Experimental Result of Behavior on Incipient Slippage

So as to confirm the phenomena of incipient slippage, we conducted an experiment using a prototype sensor. First, we identified the positions of all dots when only a normal force was applied. The distance of each dot from the central one in the contact area is called as initial distance. Next, we applied small additional force in tangential direction and increased the magnitude gradually. The dots in stick region moved in the same direction as the additional force. The displacements were almost equivalent to the relative displacement between the sensor and the object because the surface in the stick region clung to the object while moving. The dots in incipient slippage region moved shorter distances on the one hand because slippage occurred in the region. It is noted that macroscopic slippage did not occur at the moment while the surface in stick region moved with the object. This kind of movement is possible because the sensor body for contact is elastic. The experimental result is summarized in Fig.4. In the figure, the abscissa axis is initial distances from the central dot of contact and the vertical axis is the displacements of the dots after three levels of tangential forces were applied. The three kinked lines are corresponding to three levels of the tangential forces. Each relative displacement between the sensor and the object is also shown. It can be seen that the dots near the central dot moved the almost same distance as the relative displacement between the sensor and the object and that the dots far from the central dot did not. Moreover, it is noted that the number of dots moved with the object decreased as the applied tangential force increased. We can estimate the radius of stick region based on the displacements of dots. The estimation of radius is also shown with the distance from the central dot in Fig. 4 Macroscopic slippage will occur when the radius reaches to zero. This result leads to the possibility for estimating the degree of slippage from the displacements of the central and peripheral dots.

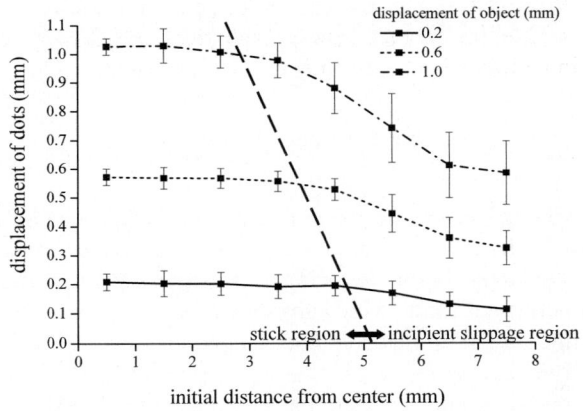

Fig. 4. Identifying incipient slippage region

4.3   A New Method for Estimating the Stick Ratio

We propose a new method for estimating the stick ratio only from measurements of the sensor. The method uses the relative displacements of peripheral dots to the central dot. The radius of stick region can be determined by comparing the relative displacements with the threshold.

Fig. 5 shows images captured by a CCD camera before and after the tangential force is applied. Four dots around the center are colored with red. Those red dots indicate a reference frame of coordinate. Each dot in every images can be identified the correspondence to the original one with respect to the reference frame. Then the displacement of each dot d_k is measured.

As mentioned in 4.2, the incipient slippage region spreads from the periphery to the center in the contact area. We can assume that the central dot is always in the stick region unless macroscopic slippage occurs because the maximum normal pressure is given around the center of the contact area. This fact was confirmed experimentally as shown in Fig.4. In this study, we define the stick region as a smallest circle which includes dots satisfying the following relation:

$$|d_k - d_0| < \delta \qquad (2)$$

where d_0 and d_k are the displacements of the central dot and the dot in stick region respectively, and where δ is a certain minute threshold.

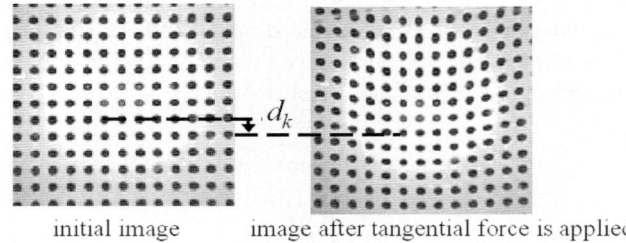

initial image   image after tangential force is applied

Fig. 5. Displacement of dots

Now we describe the proposed image processing method to find the dots of stick region which satisfy the relation (2). First, we take two images and convert them into gray scale . One is a image before applying a tangential force and the other is a image after applying a tangential force, which are shown in Fig. 5. Second, the initial image is moved along the direction of the applied tangential force. The moving distance is taken as that of the central dot. Third, the moved initial image is subtracted wholly from the image captured after a tangential force is applied. This process makes the dotted pattern disappear in the stick region because the displacement in the stick region is almost equivalent to that of central dot. The examples of images are shown in Fig. 6. In these examples, the normal force keeps at a constant magnitude, and the images are in order of the tangential force magnitudes. The white circles show the stick regions. It is confirmed with these images that the proposed method can extract the stick region and the stick region decreases from peripheral part of the contact area as the tangential force increase.

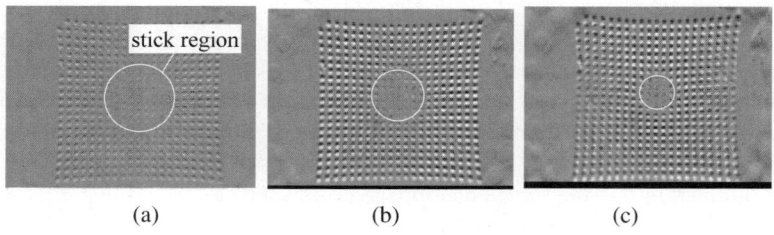

(a)   (b)   (c)

Fig. 6. Transition of stick region decreasing as the applied tangential force increase

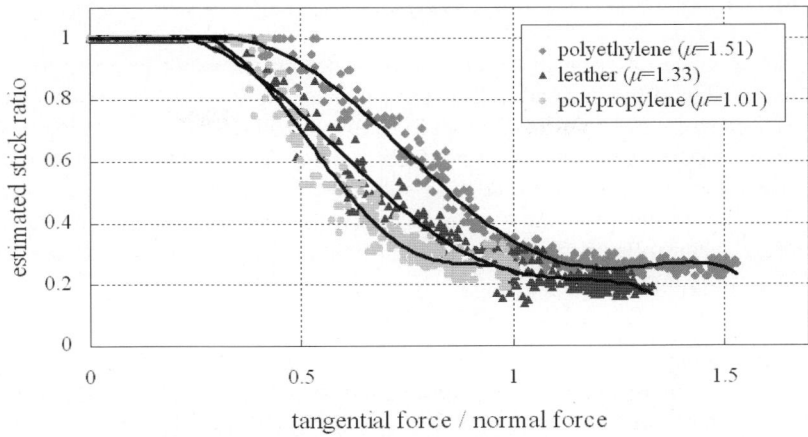

Fig. 7. Estimated stick ratio: μ means friction coefficient

We carried out experiments under the conditions of three levels of friction coefficient in order to show the effectiveness of the proposed method. The result is summarized in Fig. 7. We used three materials with a certain punctured surface to have different levels of friction coefficient. The materials are polyethylene, leather, and polypropylene. We determined the values of friction coefficient with the ratio of tangential force to the normal force at occurrence of the macroscopic slippage. The values of friction coefficient are shown at the explanatory note in Fig. 7. The abscissa indicates the ratio of the applied tangential forces to the applied normal forces, and the vertical axis is the stick ratio which is estimated by our proposed method. The three types of the markers for each plot are corresponding to the materials, and the three curved lines are the fitted curves which express the averages for the three different materials. We can see in Fig. 7 that the variance for the three different materials took a certain magnitude. This is not so curious since it is well known that the phenomena of friction are sensitive for perturbations of the several factors. In the extreme regions near one and zero of the friction coefficient, the estimated stick ratios are not necessarily reasonable to distinguish the different friction coefficients. However, in the middle range we can see the capability of our method to distinguish different friction coefficients; that is, the lower stick ratio means the lower friction coefficient in the middle range. Note that the middle range is most important to prevent from macroscopic slippage for robot grippers because the prevention can be achieved by keeping the stick ratio at a certain value in the middle range and that the grip force determined from the stick ratio should be suboptimal for most of the practical usages. Although our proposed method cannot estimate smaller stick ratios under 0.15, we can still use the estimated value at a larger stick ratio to prevent from slippage because we can keep the value by feedback control of contact force of robot grippers.

5 Control System for Robot Grippers Using Tactile Sensor

In this section, we describe the control system design for robot grippers to prevent from the slippage of the grasped object. The feedback signal from the proposed tactile sensor is used to control the grip strength for stable handling of the object. Fig. 8 shows overview of experimental setup for grip force control. The control system consists of the tactile sensor with image processing software, a voice coil motor, and a simple proportional controller with gain K, which is shown as the block diagram in Fig. 9. The grip force f_n is controlled by feeding back the stick ratio ϕ as the following equation:

$$f_n = K(\phi^* - \phi) \tag{3}$$

where K is a feedback gain and ϕ^* is the target stick ratio. The controller amplifies the deviation $\phi^* - \phi$ by K, and transmits the calculated value to the amplifier of voice coil motor. The signal f_n is converted to the electrical current by the amplifier. The voice coil motor generates the grip force under the control and the generated force is proportional to the provided current. This feedback mechanism keeps the stick ratio around the target point.

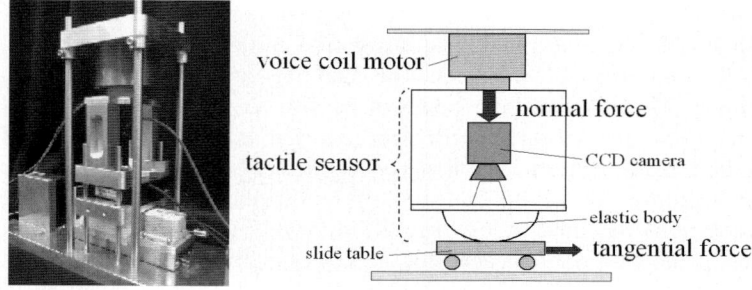

Fig. 8. Overview of the experimental setup

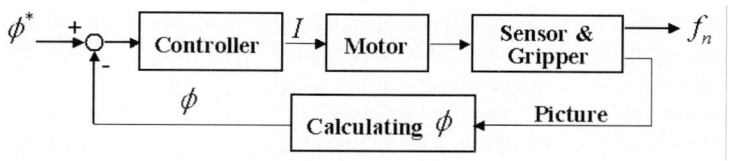

Fig. 9. Block diagram of grip force control system

In the control experiment, CCD camera (TOSHIBA: IK-SM43H) captured the images in 30 frames per second. We used image processing tool HALCON 7.1 (MVTec) in this control system. The processor of the PC machine as the digital controller was Celeron 2.8 GHz. The tangential force was measured by a load cell and it was used only for analysis not for the control. In this experiment, the torque around the vertical axis to the contact surface was not considered, and the direction of the tangential force was fixed. However, these conditions of the experiment doesn't lose generality and can be expanded to the case where the direction of the tangential force is unknown. The experimental results of this control system are given by Fig. 10, in which the tangential force was applied repeatedly by increasing and decreasing. The feedback gain was set as K=5.0, and the target stick ratio was set as ϕ^* =0.7. The tangential forces were applied manually; so, the curves in Fig. 10 are not typical functions of time. The manipulated variables of control system correspond to curves in "current" of the second row which are proportional to the generated normal forces. The current of one ampere is corresponding to the force of 3.5 N. Before the tangential force was applied, the empty weight of tactile sensor was applied as the contact force of 4.7N. The estimated stick ratios show that a macroscopic slippage occurred while the control did not work. Moreover, it is shown in the figure that the control resulted in keeping values of estimated stick ratio over 0.1. The values around 0.1 are for the macroscopic slippage occurring. These experimental results prove prevention from slippage by the control. It should be noted that the control system works only using signals from the tactile sensor.

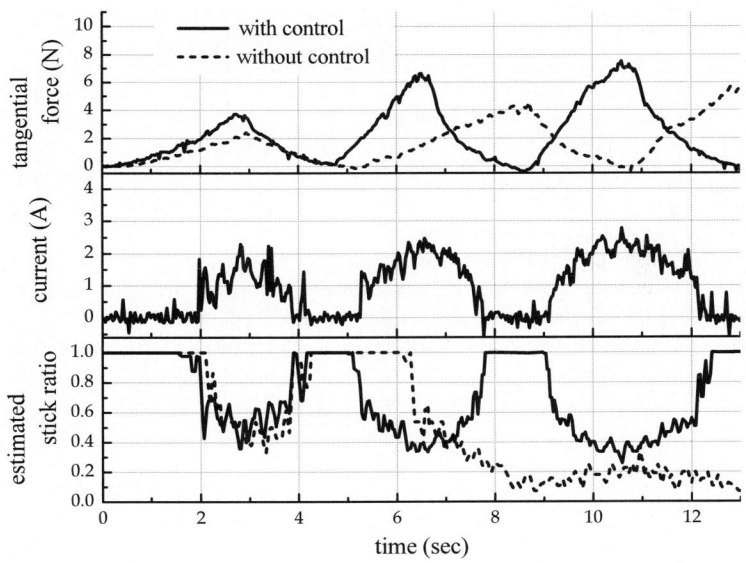

Fig. 10. Result of stick area control: Comparison between with and without control

6   Conclusion

We have investigated on a vision-based tactile sensor for measuring multi-dimensional force and moment of contact. The structure of sensor is simple because the method is vision-based. We defined the stick ratio as an index for indicating the degree of slippage. We have also proposed a new method to estimate the stick ratio for preventing from slippage of grasped object. It is shown with the experiments that the sick ratio is a key index for non-slip handling of robot grippers. We demonstrated the control system for keeping the estimated stick ratio around a set point. It is shown that the proposed sensor has the potential for dexterous handling like human.

The purpose-built integrated circuit for the image processing of this vision-based sensor may be required to achieve high speed control against disturbances in high frequency band. The exact relation of the defined stick ratio or the estimated stick ratio to the exact friction coefficient is an important problem to be solved.

References

1. Shinoda, H.: Contact Sensing, a State of the Art. Trans. of Japanese Robotic Society 20(4), 385–388 (2002)(in Japanese)
2. Lee, M.H., Nicholls, H.R.: Tactile sensing for mechatronics -a state of the art survey-. Mechatronics 9, 1–31 (1999)
3. Maeno, T.: Grip force control by detecting the internal strain distribution inside the elastic finger having curved surface. Trans. of Japanese Society of Mechanical Engineers 64(620), 1258–1265 (1998)(in Japanese)
4. Ohka, M.: Sensing characteristics of an optical three-axis tactile sensor under combined loading. Robotica 22, 213–221 (2004)
5. Ferrier, N.J., Brockett, R.W.: Reconstructing the shape of a Deformable Membrane from Image Data. The International Journal of Robotics Research 19(9), 795–816 (2000)
6. Kamiyama, K.: Development of a vision-based tactile sensor. Trans. of Society of Electrical Engineers of Japan, Series E 123(1), 16–22 (2003)(in Japanese)
7. Obinata, G., Kurashima, T., Moriyama, N.: Vision-based tactile sensor using transparent elastic fingertip for dexterous handling. In: Proceedings of 36th International Symposium on Robotics, TU41(CDROM), Tokyo, Japan (2005)
8. Obinata, G., Watanabe, N.: A new control method for robot gripper using vision-based tactile sensor. In: Proceedings of the Third International Conference on Autonomous Robots and Agents, Palmerston North, New Zealand, pp. 527–532 (2006)
9. Obinata, G., Ashish, D., Watanabe, N., Moriyama, N.: Vision Based Tactile Sensor Using Transparent Elastic Fingertip for Dexterous Handling. In: Kolski, S. (ed.) Mobile Robots: Perception & Navigation, pp. 137–148 (2007)

HNG: A Robust Architecture for Mobile Robots Systems

Robin Jaulmes and Eric Moliné

Délégation Générale pour l'Armement (French Defence Procurement Agency), Technical
Expertise Centre, 16 bis avenue Prieur de la Cote d'Or, 94114 Arcueil Cedex
{robin.jaulmes,eric.moline}@dga.defense.gouv.fr

Summary. Researchers in robotics agree that control architectures must be reactive, modular, standardized, reliable, and should allow the use of multiple functions. In the last few years, the open challenges organized by the DARPA have shown that robustness can be obtained and that real autonomous systems can be built. We conducted an analysis of several control architectures, and especially of those who had good performances in terms of reliability. The purpose of this analysis was to select the best mechanisms in the available architectures and to build an architecture that would allow us to proceed to benchmarking and prototyping for perception, planning and control algorithms on real systems in difficult environments. As a result, the Hybrid and Network-based Generic architecture HNG uses multiple processes, publish/subscribe communication mechanisms, and its control is based on contract net mechanisms. Our preliminary tests show the potential of this approach.

Keywords: robotics, control architecture, real applications, robustness, reliability, modularity, IPC, multi-agent, contract-net protocol.

1 Introduction

The current challenge in mobile robotics architectures is to build robust control architectures that can use not only resources of one robot, but resources of multiple robots inside a network. These architectures must integrate the best algorithms that are available "off the shelf" but must also adapt quickly to emerging technologies and principles: as the control of a robot can be achieved through the combination of algorithms from many different domains (image processing, automatics, data fusion, data filtering, artificial intelligence, or machine learning), bringing the contributions from these domains into the architecture need to be as easy as possible.

There are many existing control architectures [5] and most of the time they are not compatible, which means that an algorithm developed in a framework is not always easily adapted to another. As more and more projects address the standardization issue [1][4][8], libraries of algorithms are developed, and increase in size and quality. However they are different and it is necessary, in order to proceed to performance assessment [7], to use a common framework.

The HARPIC [2] architecture was designed in order to study two principles: the adjustable autonomy, which allows the user to give the robot commands of different levels, and the attention function, which allows a comparison between similar perception algorithms. The architecture was adapted so that multiple robots can be controlled from one PC [9]; algorithmic chains changing automatically through a plan execution

H. Bruyninckx et al. (Eds.): European Robotics Symposium 2008, STAR 44, pp. 123–131, 2008.

were also studied [6]. The Hybrid Network-based Generic architecture we build (HNG) synthesizes the best aspects of HARPIC while adding successful concepts developed in other architectures of the literature.

2   Functional Analysis

The first step of our study is a functional analysis: we identify the main functions we want HNG to achieve. These functions are:

1. A communication tool allowing modules to exchange messages among them, and allowing modules to run on different computers or robots.
2. Maintenance tools allowing as most fault-tolerance as possible and an easy detection and analysis of causes of malfunction. Tools to test modules without using the whole architecture need to be available, as well as tools for parameters centralization.
3. Control mechanisms allowing automatic module chains reconfiguration when the robot operator changes the control mode or when some modules are not efficient anymore. These should offer reliability.
4. A framework allowing to easily build modules and encapsulate algorithms from different domains (image processing, filtering, control, AI,...) into modules.
5. An initial set of modules, allowing access to sensor data, control over robot actuators, an interface with the operator, and including off-the-shelf algorithms.

These functions are all intertwined but we took great care to identify their interfaces. So, while conceiving the software architecture and taking technical choices, this decomposition allowed us to put each function in separate classes, so that if for one of them a better technical solution became available, the function could easily be updated without any need to change the other functions. In the following paragraphs, we will describe the technical solutions we chose in V1.0 for these functions.

3   Communication and Maintenance Mechanisms

We conducted a survey of several control architectures [5] in order to find solutions that we could use in HNG. It appeared that the simulators Stage and Gazebo from the Player/Stage project answered our needs for simulation, [4], that the watchdog and heartbeat mechanisms in latest Carmen [8] answered our maintenance needs, as did, because it wraps efficiently the UNIX sockets and Task Control Protocol and includes access control, the IPC library it uses [10]. We also used Carmen's parameter server concept, with a single centralized server in order to have all our initialization and update parameters of the system in the same file.

In our architecture, modules are processes, and not threads. Therefore, modules do not share memory. Communications use the publish/subscribe paradigm: the only permanent interlocutors of modules in a given sub-network is its Central. It is informed by the modules in its sub-network of their interests in messages with a given name and a given format (they send subscribe messages). Modules can also define new message names and formats and inform Central that they wish to send them (they send publish messages). Whenever connections appear between publisher and subscriber in the sub-network, Central routes the messages[1], and the sender never has to know who its receivers are.

We encapsulated the IPC library into a C++ class and adapted its interfaces. We made easier and fool-proof the manipulation of multiple centrals with multiple incoming and outgoing types of messages. The use of multiple centrals allows the coexistence of different sub-networks. Moreover all the system is robust to wireless communication failure, because even if it is not linked to the global system, a robot is still able to act on its own in order to achieve its mission as long as its own central is still running.

Watchdogs can be used as they are in Carmen. The purpose of the watchdog process is to intercept error signals concerning modules and re-launch them when they receive one of them. We extended this mechanism by using for each module context saves, in a file, at regular intervals. When a module crashes and restarts, it can load its context from the file, so that the most important information is not lost. Furthermore, information about crashes and restarts are logged in a unique file.

We believe that these mechanisms allow HNG to be a globally stable architecture, even though some of its modules may be unstable. As we focus more on the prototyping aspect than on industrial robustness, we often handle hazardous code and it is necessary that we have as few architecture crashes as possible. It is also important that we understand quickly where software malfunctions come from, which is guaranteed by the Operating System since we have processes and not threads. We can use its tools to determine the amount of memory and CPU taken by each module.

Using multiple processes for our modules also allows us to expand the architecture on multiple computers for the control unit, and for each of the robot, and even to build a whole sensor network. The solution we chose is the most convenient off-the-shelf solution we could find. It might be preferable to change the communication mechanisms later on. Our first identified improvement will be to use non robust but quicker protocols like UDP for the less important data transfers.

4   Control Mechanisms: Obtaining Robustness

The control mechanisms are very important in HNG. The architecture should be able to incorporate short control loops as well as long-term planning algorithms. It should be able to incorporate verified and robust mechanisms able to guarantee the integrity of the platform as well as experimental and potentially unstable mechanisms that are being tested. Furthermore, the control should be able to self-adapt in case of technical malfunction of some of the sensors, algorithms, or actuators inside the system. We also want the system to easily manage the use of its resources in case the mission changes. We describe the problematic the control mechanisms have to solve in 4.1.

To answer to this problematic, we base the control mostly on the following principles. An introspection of the modules is made through a query/reply communication protocol that is described in 4.2: queries express needs and answers express capabilities. It allows *functional links* to be built between modules. The links themselves may cause

---

[1] Direct connexions are also possible once publisher and subscribers are known.

communication channels to be open, which allows data exchanges when it is necessary. As the architecture allows several modules to reply to the same need, we introduce an evaluation mechanism that allows comparison between replies. Objective comparison of the costs of the modules is made by the resource managers described in 4.3.

4.1   Definitions

The problematic for the controller is the following. The architecture has N modules $\{m_1,\ldots,m_N\}$. Each of these modules can be activated or deactivated, has a set of input (the data they can subscribe to) and a set of outputs (the data they can publish). Each input or output has a type, and an "output to input" connection can be made only if their type is the same. Let the matrix **C**, whose values are on $\{0;1\}$ indicate what the connection are between inputs and outputs; $\mathbf{C}(i, j) = 1$ if and only if input i is connected to output j. What we have to define is a way to decide who is active and to obtain matrix C at any given time. Multiple inputs can be connected to a single output, which means that a published message has several subscribers. Multiple outputs can also be connected on a single input, but then a priority needs to be defined as inputs may contradict themselves.

The modules m_i can evaluate their aptitude to fulfill its purpose at a given time t, which defines the function $f_{mi}(t)$ whose values are on $\{0;1\}$. Its value is 1 if the module "works" and 0 otherwise. We also suppose that some modules can evaluate the quality of their achievements at a given time t, which defines $Q_{mi}(t)$ whose values are on $[-1;+1]$. This value may be biased. Let the vectors $\mathbf{R}_{mi}(t)$ be the instant load of the modules on the resources. The vectors are of size Z, where Z is the total number of resources available in the system (the CPU and RAM of every computer, and the communication load for every central).

One module is called "the manager". The purpose of the control mechanisms is to make this module able to work and to maximize its quality. However, the sum of the resources consumed by the systems need to remain under the threshold. The following inequality (see Eq. 1) must be verified at all times ($\mathbf{R}_{max}$ is a vector characterizing the total amount of resources the system is able to give).

$$\forall t, i \quad \sum R_{m_i}(t) \leq R_{\max} \tag{1}$$

4.2   The Query/Answer Protocol

Since at first no information is available about the factors that have an influence on the performance of the manager module and that many combinations is available, a blind search of possible combinations is impossible. So we make the following assumption.

Every module is associated to a vector of needs **N** and a vector of capabilities **K**. Services are the needs to which the module can answer. Each of these needs/services may be associated to an input/output type T. We assume that these lists of all the modules verify the following properties:

-    If, for every need n_i of the vector **N** of a given module m, another module m', such that $f_{m'}(t) = 1$, has in its vector **K** n_i, and if T is either undefined or correspond to an input/output connection of type T that is made between m and m' (then we say that m' answers this need of m), $f_m(t) = 1$.

- $Q_m(t)$ increases if any of the $Q_{m'}(t)$ increases (m' answering a need of m).
- Every need of a given module has a priority. The need with the smallest probability may be unnecessary to obtain $f_m(t) = 1$ and have less effect on the overall quality.

Using these assumptions, we can define a protocol that guarantees having, when it is possible, the best quality under the resource constraint. This protocol is inspired from Contract-Net [11]. In a first phase, which is illustrated on an example in Figure 3, the manager module makes queries that are answered by modules, whose needs are answered by other modules, recursively, until every need is answered. When queries are answered, there might be multiple providers. In that case, the value of each provider $V_{m_i}(t)$ is computed, by taking into account their estimated quality, the "cost of the resources" vector $C_R(t)$ given by the resource manager described in 4.3, and their effective cost, also computed by the resource manager. The past evaluations E_n, given by the clients to the providers at times $T(E_n)$ also have an influence (see Eq. 2). These evaluations allow the system to be robust in cases where the qualities given by modules are different from the real qualities; τ_{m_i} is a characteristic of the module m_i.

$$V_{r1_i}(t) = Q_{m_i}(t) - R_{m_i}^{Eff}(t).C_R(t) + \sum_{E_n \in E(m_i,t)} E_n e^{-\frac{t-T(E_n)}{\tau_{m_i}}} \qquad (2)$$

In the second phase, contracts are notified between clients and providers: the client uses the query message to inform its potential providers of its choice and of the corresponding value. As soon as a provider knows it has been chosen, it signs the contract with all its sub-provider. Each time a contract is notified, the corresponding provider is activated and the necessary publish/subscribe connections are made, according to the mechanism illustrated in Figure 4 on an example.

At regular intervals, values are updated. Clients can also send rewards or penalties to their providers if they can evaluate it (in particular variations in their estimated quality can be sent as penalties to their providers). Each concurrent of a given provider verify if its value becomes higher than the value of the current contractor. If it is the case, it can send to the client its own value and the client can then chose it as its new provider, which cancels its previous contract.

This mechanism allows dynamic reconfiguration of the architecture, and guarantees that in case a module goes down another one will take its place. Also, the query/answer protocol allows to easily define priorities between queries and therefore priorities when several outputs go to the same input. The order of priority is defined by doing a depth first search of the query tree shown on Figure 4. Therefore, the priority between the three inputs shown on Figure 2 is the following: highest priority commands are those sent by the security module commands; second priority is those sent by the user interface; third priority is those sent by the servo-controller.

4.3   The Resource Management

The resource management system is made of several resource managers (one by computer): each resource manager has the interfaces described in Figure 1. The resource manager uses the information it receives to compute the cost of each resource. This cost is updated at regular intervals. If it becomes rare (available resource is below a given threshold) then the cost increases. Otherwise, it decreases.

$$R_{m_i}^{Eff}(t) = R_{m_i}(t) + \sum_{j \in F(m_i)} \frac{R_{m_j}^{Eff}(t)}{N_R(m_j)} \tag{3}$$

The resource manager also computes effective costs of the modules according to the following equation (see Eq. 3), where $F(m_i)$ is the set of modules answering the needs of module m_i and $N_R(m_j)$ is the total number of queries m_j is currently answering. These costs are computed with a recursive function.

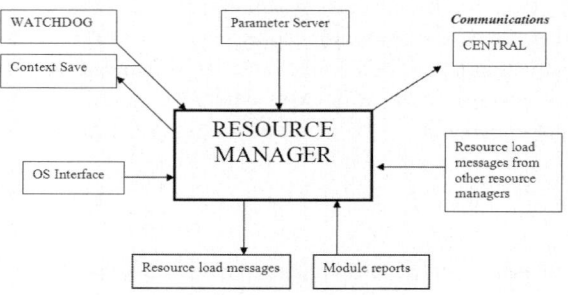

Fig. 1. Interfaces of the resource manager

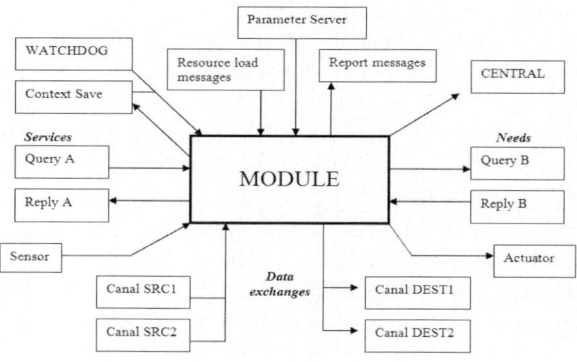

Fig. 2. Interface of the modules

5   The Modules

Modules are the "bricks" of HNG. They offer to the architecture algorithmic capabilities as well as sensor or actuator interfaces. They are interfaced all the mechanisms we

Table 1. Modules available in HNG V1.0

Module	Services	Needs	Data in	Data out
Manager		Robot integrity Human control		
Security	Robot integrity	Robot control LIDAR data	LIDAR data	Commands
Interface	Human control Beacon designation	Robot control Follow wall Reach beacon Reach goal Localisation Mapping Video Beacon tracking Wall detection	Map Position Images	Commands Goal Beacon
Robot	Robot control Odometry		Commands	Odometry
Simulator	Robot control Odometry LIDAR data Video		Commands	Odometry LIDAR data Images
Laser	LIDAR data			LIDAR data
Camera	Video			Images
Planification	Reach goal	Reach waypoint Mapping	Goal Map Position	Waypoint
Servo- controller	Reach waypoint	Robot control Localisation	Waypoint Position	Commands
SLAM	Localisation Mapping	Odometry LIDAR data	LIDAR data Odometry	Map Position
Beacon reacher	Reach beacon	Robot control Track beacon	Beacon	Commands
Visual tracker	Track beacon	Beacon Designation Video	Images Beacon	Beacon
Wall follower	Follow wall Detect wall	Robot control LIDAR data	LIDAR data	Commands
GPS/IMU	Localisation			Position
Maps	Mapping			Map

have describe and add their own mechanisms. Figure 2 describes the interfaces modules of HNG can have. All the interfaces in the upper part of Figure 2 are common to every module in the architecture. However, the interfaces from Sensors to Actuators are specific to other modules. From a software point of view, the upper part is coded in a generic Module class, and the lower part in a specific ModuleX class that inherits from Module. This mechanism guarantees that all the communication, control and maintenance interfaces are perfectly identical for every module in HNG. In HNG V1.0 we

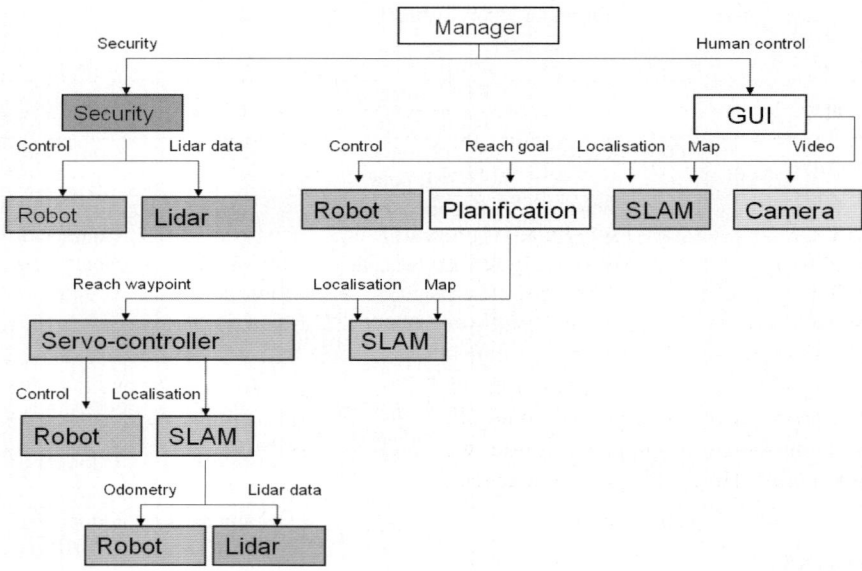

Fig. 3. The "query tree" on the example scenario

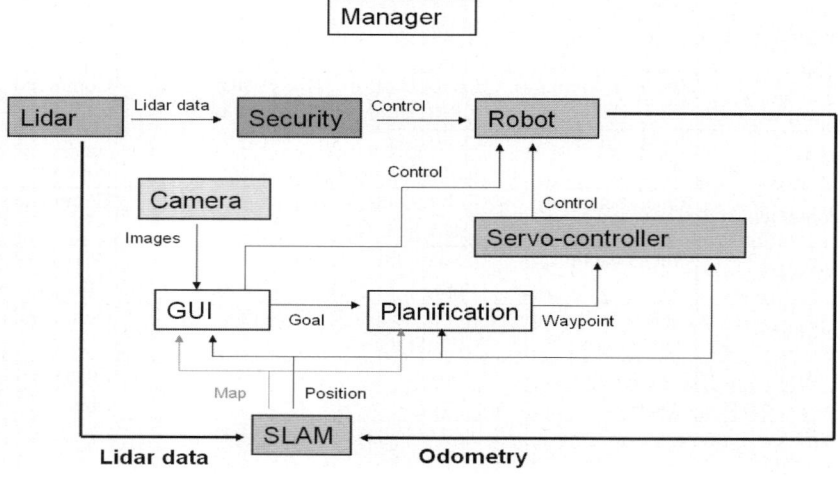

Fig. 4. Data exchanges between Modules on the example scenario

will have the set of modules described in Table 1. A scenario of the queries and data exchanged is illustrated with some of these modules on Figures 3 and 4.

6 Conclusion

HNG is an architecture built to be reliable and robust and to allow an easy integration and testing of innovating mechanisms. The query/answer mechanisms allow an easy reconfiguration in case of malfunctions, the watchdog and context saves allow robustness to execution errors. HNG should be able to design very quickly robust solutions for robot control even if it is based on C++ modules that are not perfectly verified and in particular open source or off-the-shelf modules. Therefore, HNG should become a very useful tool for the evaluation of algorithms, sensors or concepts. Many mechanisms of HNG have already been developed and tested. However, we still need to finish the integration and testing of the combined mechanisms. Our next objective is to then extend HNG to networks of combined ground sensors and unmanned vehicles. We also want to study how the security aspect of the architecture can be formalized and how we can build proofs of reliability that would allow the architecture to be used with UAVs: a change in the communication system would then be probably necessary: the way we implemented HNG allows such changes.

References

1. Baillie, J.-C.: URBI: A Universal Language for Robotic Control. International Journal of Humanoids Robotics (2004)
2. Dalgalarrondo, A.: Intégration de la fonction perception dans une architecture de contrôle de robot mobile autonome. Thèse de doctorat (2001)
3. Dufourd, D., Dalgalarrondo, A.: Integrating human/robot interaction into robot control architectures for defense applications. In: CAR 2006 (2006)
4. Gerkey, B., et al.: Most valuable Player: A robot device server for distributed control. In: IEEE/RSJ Intl. Conf. on Intelligent Robots and Systems (2001)
5. Kramer, J., Scheutz, M.: Robotic Development Environments for Autonomous Mobile Robots: A Survey. Autonomous Robots 22(2), 101–132 (2007)
6. Lacroix, S., Joyeux, S., Lemaire, T., Bosch, S., Fabiani, P., Tessier, C., Bonnet, O., Dufourd, D., Moliné, E.: Projet Acrobate, Algorithmes pour la coopération entre robots terrestres et aériens (2006)
7. Lambert, M., Jaulmes, R., Godin, A., Moliné, E., Dufourd, D.: A methodology for assessing robot autonomous functionalities. In: Intelligent Auton. Vehicles (2007)
8. Montemerlo, M., Roy, N., Thrun, S.: Perspectives on Standardization in Mobile Robot Programming: The Carnegie Mellon Navigation (CARMEN) Toolkit. In: IEEE/RSJ International Conf. on Intelligent Robots and Systems (2003)
9. Sellem, P.: Navigation coopérative par échange de représentations de l'environ-nement (2000)
10. Simmons, R. Inter Process Communication library, www.cs.cmu.edu/~IPC/
11. Smith, R.: The Contract Net Protocol: High-Level Communication and Control in a Distributed Problem Solver. IEEE Transactions on Computers, C-29 12, 1104–1113 (1980)
12. Thrun, S., et al.: Stanley: the robot that won the DARPA Grand Challenge. Journal of Field Robotics 23(9), 661–692 (2006)

Information Relative Map Going Toward Constant Time SLAM

Viet Nguyen and Roland Siegwart

Autonomous Systems Laboratory, Swiss Federal Institute of Technology (ETH Zurich), Rämistrasse 101, CH-8092 Zürich, Switzerland
{viet.nguyen,roland.siegwart}@mavt.ethz.ch

Summary. The paper presents the Information Relative Map algorithm for solving SLAM. Instead of estimating directly the relative quantities as in other relative mapping approaches, the proposed algorithm estimates the canonical quantities, the information vector and information matrix, using the Information filter. The estimation algorithm has constant time complexity without any approximation or linearization. The correlation between observed quantities are fully taken into the estimation. Furthermore, only independent relative quantities from observations are mapped so that the required computation is significantly reduced. The algorithm is empirically evaluated by testing on more than 100 simulated problem instances and the real world Victoria park dataset. The comparison with an existing implementation of the FastSLAM and EKF algorithms clearly shows a better performance in map precision and speed.

Keywords: Relative Mapping, Information Filter, SLAM.

1 Introduction

In the Simultaneous Localization and Mapping (SLAM) problem, a mobile robot has to be able to autonomously explore an unknown environment with its on-board sensors, incrementally build a map of the environment while simultaneously localize itself relative to this map.

Most of the up-to-date proposed SLAM solutions are inspired from a seminal paper by Smith et. al. [1] which use the extended Kalman filter (EKF) for solving SLAM. In [2, 3], it has been proved that EKF converges for linear SLAM problems where the robot motion model and observation model are linear functions. For general nonlinear SLAM problems, the authors of [4, 5] have pointed out that conventional EKF based SLAM yields an inconsistent map. The inconsistency comes from the linearization introduced by the EKF.

The second challenge of the standard EKF SLAM is the quadratic complexity for updating the covariance matrix. It has been recognized as the main obstacle in scaling EKF SLAM algorithms to large real world problems. Many techniques have been suggested to reduce the computational requirements of SLAM. Most of them are mainly constructed based on the creation of a hierarchy in the stochastic map framework. By using this idea, it will be possible to postpone some global updates and only process local maps in regular steps. Another related approach is to neglect the correlation between elements of different submaps. To preserve consistency, usually ad-hoc strategies

H. Bruyninckx et al. (Eds.): European Robotics Symposium 2008, STAR 44, pp. 133–144, 2008.
springerlink.com

(like inflation of the uncertainty within sub-maps) are taken. Few examples of applying some of these ideas are Decoupled Stochastic Mapping [6], Compressed EKF Filter [7], Constrained Local Sub-map Filter [8], Network Coupled Feature Maps [9] and Divide and Conquer [10].

One approach for solving SLAM is to use relative mapping. The first mathematical formulation was introduced in [11]. Later [12, 13] proposed relative mapping algorithms based on quantities invariant to the robot pose, i.e. to shift and rotation. Both algorithms estimate the relative distances between landmarks pairwise. However, they do not take into account any correlation between the distances and thus the estimation is local or suboptimal.

Another algorithm has been proposed in [3] in which a relative map filter is coupled with the Geometric Projection Filter (GPF) in the estimation. The second filter provides a means to produce a geometrically consistent map from the relative map estimate by solving a set of linear constraints. The constraints are extracted from the dependency among the map quantities. Both filters are optimal because the dynamics/observation equations are both linear and they are based on the Kalman Filter. However, the elements used in this algorithm are invariant to shift only, not rotation.

In another research, [14] introduced an algorithm in which the correlation between the observed quantities are also considered in the estimation. Following the line, [15] described a technique in order to enhance the consistency of the relative map estimate by using a geometric filter. Both algorithms approximate the covariance matrix to be block diagonal and thus enable to obtain constant update time. More recently, [16] proposes two methods for enhancing the relative map precision by exploiting the geometrical structure of the map. Again, they also make an approximation for the covariance matrix being block diagonal. The proposed methods are shown to obtain very good performance. However only simple simulation results are given.

Much of recent research has been focused on the Information matrix, the inverse of the covariance matrix, where the sparseness of the matrix can be exploited for SLAM updates [17, 18]. The Sparse Extended Information Filter introduced in [19] approximates the information matrix by a sparse matrix that allows update in nearly constant time. However, the inaccuracies introduced through the sparsification can generate overconfident estimations of the state [20]. The latter work also proposed a modified version, the Exactly Sparse Extended Information Filter, shown that the map consistency theoretically preserved. However, the filter is no longer constant time complexity.

In this paper, we propose the Information Relative Map (IFRM) algorithm which is based on the relative mapping concept. However, instead of estimating directly the relative map in covariance form, the canonical quantities in the information form are estimated by using the Information filter. The key idea is inspired from the simplicity of the update step equations and the complexity of the prediction step where the latter can be ignored from the concept of relative mapping. The proposed IFRM algorithm has constant time complexity. More importantly, the estimation is exact: there is no approximation or linearization in the estimation. For the purposes of data association and other navigation tasks which are practically required, we propose also an approximation method to recover the absolute local map from the information vector and matrix. The method is independent from the map estimation and has constant time complexity. It is adaptive so that the precision of the local map and the required computation can be dynamically adjusted based on the measurement noise level.

In the framework of relative maping, an independent work of [16] has pointed out the observation that *"the inverse of the covariance matrix (the information matrix) is block diagonal and therefore its computational complexity is independent of the number of features"*. As we will see in the paper, the proposed algorithm inherits from the information filter that the update step has constant time complexity. The block diagonal property of the information matrix is just a consequence. In fact, the block diagonal property is not required in the estimation. In other words, the complexity is independent of the number of features even in the case that the information matrix is fully populated.

2   Brief of Relative Mapping

In relative mapping, the map state contains only relative quantities between landmarks. These quantities are invariant to the robot motion error, i.e. shift and rotation. In this paper, for simplicity we use the distances between point landmarks as the relative quantities, similarly to the ones used in [12, 13, 14]. The techniques described here can be also applied to different types of relative quantities, e.g. relative distances and angles as in [3, 21]. Figure 1 depicts an example of map construction using a relative map of distances between landmarks.

Usually, the existing approaches use the Kalman filter to update the state vector and covariance matrix. In general, the covariance matrix is fully correlated if one considers the correlation between the observed quantities. The update involves a full size matrix multiplication and thus has a quadratic complexity. In order to reduce the complexity to a constant time, one needs to approximate the covariance matrix as being block diagonal [14, 15] or not consider the correlation between the quantities in an observation and the covariance matrix becomes diagonal [3, 16]. The map consistency is enhanced when the dependency between the map elements are enforced at the end by means of a geometric filter [3, 15]. However, it is well known that ignoring the correlation would lead to a suboptimal solution. It is also a high risk that the algorithm diverges before the robot completes the whole trajectory. During this work, we have implemented the method described in [14] and tested on the Victoria park dataset. When the correlations between elements are ignored, the algorithm diverges very soon before reaching the end.

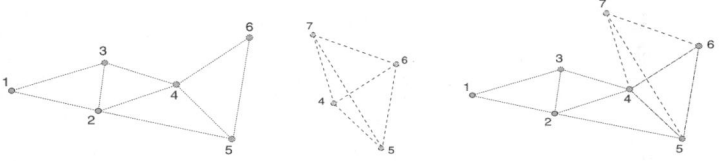

Fig. 1. *Left:* the relative map before the observation. *Middle:* the observation. *Right:* the relative map obtained by fusing the information coming from the old map and the observation. In all the three figures the map state only contains the indicated distances between the landmarks.

3   The IFRM Algorithm

Our IFRM algorithm is inspired from the properties of the Information Filter equations
[17, 18]: In absolute mapping, the prediction equations have cubic complexity and the
update equations are additive with constant complexity (usually the number of quantities in an observation is much smaller than the map size). If we apply the information
vector/matrix into a relative mapping algorithm, we will see that the prediction equations are empty and the update equations are constant time. In order to describe the
IFRM algorithm in detail, we adopt the equations used in [18] as follows.

Prediction step:

$$\bar{\Omega}_t = (A\,\Omega_{t-1}^{-1}A^T + R)^{-1} \tag{1}$$

$$\bar{\xi}_t = \bar{\Omega}_t\,(A\,\Omega_{t-1}^{-1}\xi_{t-1} + B\,u_t) \tag{2}$$

Update step:

$$\Omega_t = C^T Q^{-1} C + \bar{\Omega}_t \tag{3}$$

$$\xi_t = C^T Q^{-1} z_t + \bar{\xi}_t \tag{4}$$

where ξ and Ω are the information vector and matrix, respectively. We call (ξ, Ω)
the information map. These quantities relate to the gaussian variable vector $x(\mu, \Sigma)$ by
the following equations:

$$\xi = \Sigma^{-1}\mu \qquad\qquad \Omega = \Sigma^{-1} \tag{5}$$

$$or \qquad \mu = \Omega^{-1}\xi \qquad\qquad \Sigma = \Omega^{-1} \tag{6}$$

The system dynamics matrices A, B and the observation matrix C are defined in the
gaussian equations:

$$x_t = A\,x_{t-1} + B\,u_t + \varepsilon_t$$
$$z_t = C\,x_t + \delta_t$$

The square matrices R and Q are the covariances of the zero-mean noise variables ε
and δ, respectively. For clarity the time index t of the matrices is not shown.

In relative mapping, we assume that the features are fixed and thus the relative quantities between the features do not change over time. Therefore, the matrix A is identity, the
matrices B and R are zero. The prediction step is empty, meaning $\bar{\Omega}_t = \Omega_{t-1}$, $\bar{\xi}_t = \xi_{t-1}$.
In the update step, the matrix Q has constant size m, the number of quantities in an ob-
servation, which is usually much smaller than the map size n. The observation matrix C
(dimension $m \times n$) has the property that there is only one nonzero element of value 1 in
each row. Thus, the two equations in the update step can be performed in constant time.
Overall, the estimation of IFRM has constant time complexity. Notice again that for the
case of absolute mapping the prediction step requires an inversion of a full size matrix.

The IFRM algorithm has one important advantage that the estimation is exact: the
estimation is fully described by the information filter. There is no approximation or
linearization in the estimation. Furthermore, the correlation between quantities in observations are taken into account (i.e. matrix Q is fully correlated) so that the map
consistency is theoretically preserved.

It is easy to see from (3) that the information matrix Ω is block diagonal (the observation matrix C has the structure similarly to that of a permutation matrix), each update

with a new observation is local. However, the block diagonal property is not required for the constant time complexity of the update step. This property holds even after the loop closing. However, it is not true when the covariance form is used because the co-variance matrix Σ is fully correlated after closing the loop. Once the covariance matrix is fully correlated, a direct update to (μ, Σ) (i.e. Kalman filter equations) can not be performed in constant time without any approximation. However, nothing comes for free. The cost of working with the information map is that whenever μ or Σ is required, one would need to perform one matrix inversion. Particularly in this case, one would need μ, i.e. most updated estimate of the relative map, to construct the absolute map. The absolute map (or a submap) is needed for the data association or other navigation tasks. In order to avoid this cost and keep the total algorithm complexity constant, the following subsection will describe a strategy to compute approximately one small part of μ which is sufficient to construct the local submap.

The IFRM algorithm uses independent quantities in each observation. It has been shown in [22] that this strategy improves the map consistency and also significantly reduces computational expenses by having a smaller map size. However, maintaining only geometrically independent quantities in the map estimate requires a relative map geometric filter [3, 15] which is not considered in this paper.

3.1   The Estimation

The IFRM algorithm implements the two update equations (3) and (4) of the Information Filter. The pseudo-code is shown in Algorithm. 1 where *NumObs* is the number of relative quantities in the current observation, *ObsID(i)* is the map element *ID* of the quantity i. Notice that the function *UpdateIFRM* has constant time complexity.

Algorithm 1. *UpdateIFRM*

---

1 $q = Q_t^{-1}$
2 **for** $i = 1{:}NumObs$ **do**

3   **for** $j = 1{:}NumObs$ **do**

4         $\Omega(ObsID(i), ObsID(j)) \mathrel{+}= q(i,j)$

5 $p = q * z_t$
6 **for** $i = 1{:}NumObs$ **do**

7      $\xi(ObsID(i)) \mathrel{+}= p(i)$

3.2   Local Submap Construction

The absolute map can be constructed from the estimated relative map and given lo-cations of the two known landmarks. The absolute map is usually required for data association and other navigation tasks. For recovering the relative map from the infor-mation map, one needs to compute μ using (6) in which Ω is a full size matrix. This

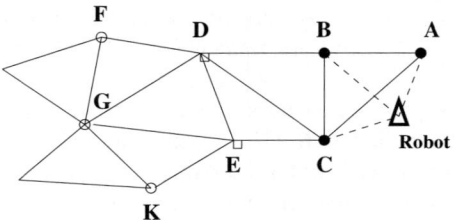

Fig. 2. An example showing the submaps when $d_1 = 1$, $d_2 = 1$. The local submap (LM): $\Phi_0 = \{A,B,C\}$. The extended LM level 1 (ELM1): $\Phi_1 = \{D,E\}$. The extended LM level 2 (ELM2): $\Phi_2 = \{F,G,K\}$.

calculation is expensive and should be avoided for online robot navigation except when the final map is generated at the end.

In this subsection, we will describe a strategy to generate *a local submap*. A local submap is a small area around the robot and contains the observed landmarks in the current observation. Note that this local submap is not the same as the local map in the robot coordinate frame. The landmark locations in a local submap are in the global coordinate. In order to keep the procedure constant time complexity, we compute approximately the locations of the landmarks in the local submap. The precision of the approximation is dynamically adjustable based on the measurement noise level. Before going into detail, let's define the notations as following:

- Φ_0: set containing landmarks currently observed. It's called the local submap or LM.
- Φ_1: set of landmarks not in Φ_0, connecting to Φ_0 by at most d_1 steps and $d_1 \geq 0$. It's called the extended local submap level 1 or ELM1.
- Φ_2: set of landmarks not in Φ_0 and Φ_1, connecting to Φ_1 by at most d_2 steps and $d_2 \geq 0$. It's called the extended local submap level 2 or ELM2.
- Φ_∞: set of all landmarks not in Φ_0, Φ_1, Φ_2.
- $\Gamma_0, \Gamma_1, \Gamma_2, \Gamma_\infty$ are the sets of observed distances between landmarks in $\Phi_0, \Phi_1, \Phi_2, \Phi_\infty$, respectively.

d_1 and d_2 are called the diameters of ELM1 and ELM2, respectively. Note that $\Phi_0 \cup \Phi_1 \cup \Phi_2 \cup \Phi_\infty = \Phi$ is a set of all landmarks; $\Gamma_0 \cup \Gamma_1 \cup \Gamma_2 \cup \Gamma_\infty = \Gamma$ is set of all observed distances. An example is shown in Fig. 2 when $d_1 = 1$ and $d_2 = 1$ and the robot currently observes 3 landmarks $\Phi_0 = \{A,B,C\}$. The observed distances $\Gamma_0 = \{AB, BC, CA\}$ are used to update the information map as described above.

From (6) we have:

$$\Gamma = \Omega^{-1}\xi$$

Decompose Γ, Ω and ξ into components we have:

$$\begin{bmatrix} \Gamma_\infty \\ \Gamma_2 \\ \Gamma_1 \\ \Gamma_0 \end{bmatrix} = \begin{bmatrix} \Omega_\infty & \Omega_{\infty,2} & \Omega_{\infty,1} & \Omega_{\infty,0} \\ \Omega_{\infty,2}^T & \Omega_{2,2} & \Omega_{2,1} & \Omega_{2,0} \\ \Omega_{\infty,1}^T & \Omega_{2,1}^T & \Omega_{1,1} & \Omega_{1,0} \\ \Omega_{\infty,0}^T & \Omega_{2,0}^T & \Omega_{1,0}^T & \Omega_{0,0} \end{bmatrix}^{-1} \begin{bmatrix} \xi_\infty \\ \xi_2 \\ \xi_1 \\ \xi_0 \end{bmatrix} \tag{7}$$

We want to compute approximately Φ_0 (the local submap) or equivalently Γ_0 in the Eq. 7. Instead, we will compute an approximation of $\Phi_{0+1} = \Phi_0 \cup \Phi_1$, or equivalently $\Gamma_{0+1} = \Gamma_0 \cup \Gamma_1$.

As already mentioned, the information matrix Ω remains the property of block diagonal even after the loop closing. In (7), the submatrix $\Omega_{\infty,0}$ is a zero matrix because the distances Γ_0 are not observed simultaneously with any distance of Γ_∞. Similarly, the matrices $\Omega_{\infty,0}$ and $\Omega_{2,0}$ are zeros. Since we are only interested in computing Γ_0 and Γ_1, we can approximate $\Omega_{\infty,2} = 0$. The approximation allows us to compute Γ_{0+1}, thus Φ_{0+1}, in constant time. The intuition is that the correlation between the further away distances Γ_2 and Γ_∞ (i.e. $\Omega_{\infty,2}$) has very small influence to the local distances Γ_1, Γ_0. Equation (7) is then reduced to:

$$
\begin{bmatrix} \Gamma_2 \\ \Gamma_1 \\ \Gamma_0 \end{bmatrix} = \begin{bmatrix} \Omega_{2,2} & \Omega_{2,1} & 0 \\ \Omega_{2,1}^T & \Omega_{1,1} & \Omega_{1,0} \\ 0 & \Omega_{1,0}^T & \Omega_{0,0} \end{bmatrix}^{-1} \begin{bmatrix} \xi_2 \\ \xi_1 \\ \xi_0 \end{bmatrix}
\tag{8}
$$

where Γ_∞ is not included. The square matrix in (8) has the size equal to the number of distances of $\Gamma_0 \cup \Gamma_1 \cup \Gamma_2$ which is usually much smaller than the global map size.

It is important to emphasize that the approximation does not affect the estimation of the information map. The local submap is used in this case only for the data association. Furthermore, the precision of the approximation is dynamically adjustable. If d_2 increases, Γ_2 is approaching Γ_∞ and the computed values of Γ_1 and Γ_0 using (8) are more accurate to their exact values. Similarly, if d_1 increases, ELM1 is approaching the global map and thus LM is more accurate. In fact, our simulation shows that when $d_2 \geq 2$ the differences between the computed values of Γ_1 and Γ_0 using the approximation (using (8)) and their exact values (using (7)) are less than 1%.

4 Experimental Results

For the evaluation, the IFRM algorithm is tested on simulated datasets and the real world Victoria park dataset. The results are compared with ground truth (for the simulation) and the results obtained by using the FastSLAM algorithm [23] and a standard EKF algorithm. The IFRM is implemented in Matlab. We use the Matlab implementation of FastSLAM and EKF algorithms written by Tim Bailey [24] for the comparison. Special thanks to the author.

Figure 3(a) shows an instance of the simulated map and robot trajectory. The map consists of 38 point landmarks and has a size of $250m \times 200m$. The robot travels two times the loop and makes 1744 observation steps. The data association is known.

The simulated laser scanner has a view angle of 180° and looks toward the front direction. The scanner provides the bearings and ranges of the landmarks within a range of $60m$ from the robot. The range and bearing measurements are generated as gaussian quantities with the variances (σ_R^2, σ_B^2). The scanner frequency is $4Hz$ and the robot speed is $3m/s$. The simulated odometry provides the measurements of speed and steering angle as gaussian quantities with the variances (σ_V^2, σ_G^2) where $\sigma_V = 0.4m/s$ and $\sigma_G = 4^o$.

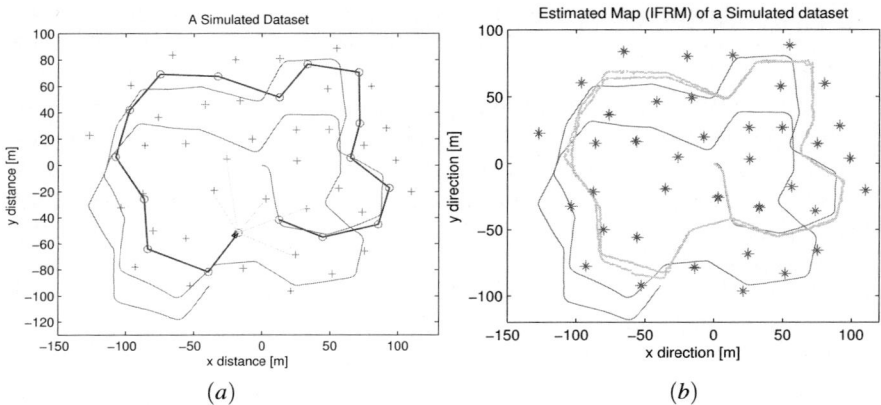

Fig. 3. (*a*) An instance of the simulated map and robot trajectory. Red crosses represent the actual locations of the landmarks. Circles are the true way points, blue line approximately represents the true robot trajectory. Red line is the raw odometry trajectory. The robot starts from location (0,0) and travels the loop twice.

(*b*) An estimated map by using the IFRM algorithm. Blue stars are the estimated landmarks whose locations are almost the same as those of the true landmarks (red crosses). Green line is the estimated robot trajectory. This simulated instance is generated when $\sigma_R^2 = 0.5m^2$ and $\sigma_B^2 = 1.5deg^2$.

Figure 3(*b*) displays the results obtained by using the IFRM algorithm when $\sigma_R^2 = 0.5m^2$ and $\sigma_\theta = 1.5^o$. The mean location error on the estimated landmark locations, defined as the distance between the estimated location of a landmark and its true location, is $E_m = 0.25m$. In this case, the diameter values $d_1 = 1$ and $d_2 = 2$ are used.

We test the algorithms on two sets of simulated data. The first one includes data problems generated with $\sigma_R = 0.5m$ and σ_B varies in the range $[0.25^o : 3.5^o]$. The second one includes data problems generated with $\sigma_B = 2deg$ and σ_R varies in the range $[0.25m : 2.0m]$. For each value pair of (σ_R, σ_B) we generate 10 problem instances. In total there are 130 problem instances. Each problem instance is input to the algorithms. This is to ensure the algorithms are tested with the same data. The mean location errors of the 10 runs are computed.

The results are plotted on Fig. 4(*a*) and Fig. 4(*b*). The notations FSv1-20p and FSv2-20p stand for FastSLAM version 1 and FastSLAM version 2, respectively, with 20 particles. As one can see, the IFRM algorithm produces more precise maps than other algorithms for all the cases. The map location error is almost increasing linearly with the measurement errors. During the experiments, we notice that the map errors obtained by using the FSv1 and FSv2 algorithms are different between the 10 runs by a large amount. The nondeterministic behavior is also observed when we repeat several times the two algorithm on the same problem instance.

Figure 5 displays the average running time of the algorithms on the simulated datasets. Given the same map structure and the same laser sensor range of 60m, the running time of an algorithm is approximately constant over different problem instances.

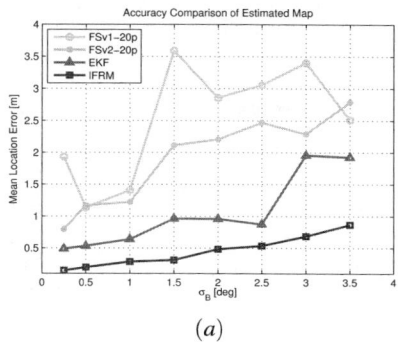
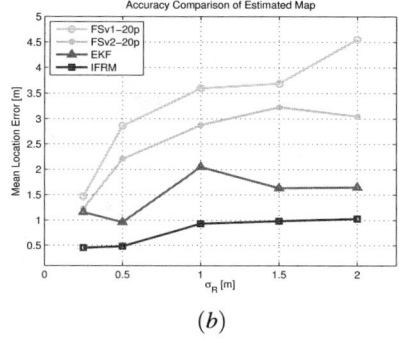

$$(a) \qquad\qquad (b)$$

Fig. 4. Comparison of map precision obtained by using four algorithms IFRM, EKF, FastSLAM v1/v2 both with 20 particles at (a) Different bearing noise levels and the range variance σ_R^2 is fixed at $0.5m^2$ and (b) Different range noise levels and the bearing variance σ_B^2 is fixed at $2deg^2$

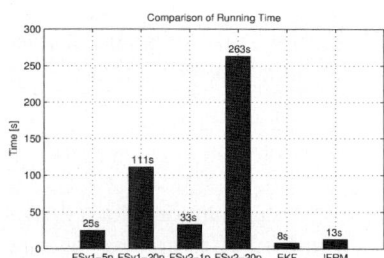

Fig. 5. Comparison of average running time on the simulated datasets

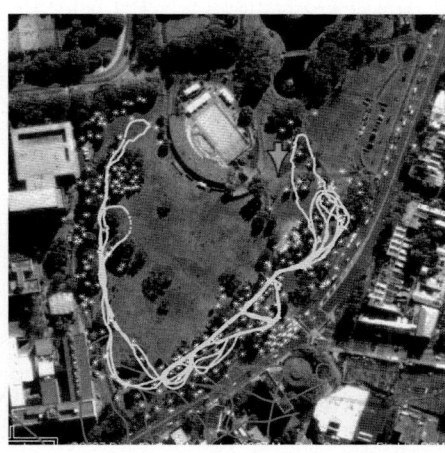

Fig. 6. The built map on the Victoria Park dataset by using the IRFM algorithm. The red line is the raw odometry trajectory. The green line is the estimated robot trajectory. The yellow stars are the estimated locations of observed objects (e.g. trees, human)

The running time of FSv1 and FSv2 using 5 and 1 particles, respectively, are also included. The EKF algorithm is the fastest one in this simulation. However, when the map size increases the EKF gets slower because its complexity is approximately cubic.

Finally, the algorithms are tested with the well known Victoria Park dataset. The dataset includes 6774 observation steps of a map having 236 point landmarks (trees). The data association is known. Figure 6 shows the estimated landmark locations and

Table 1. Running time on the Victoria Park dataset

Algorithm	Time
IFRM	87s
Func. UpdateIFRM	3s
EKF	201s
Func. KF_Cholesky_update	105s
FSv1-50p (*)	~120s

robot trajectory obtained by using the IFRM algorithm. Unfortunately we do not have the ground truth of the map for performing a qualitative evaluation. For this dataset, periodically every 200 observation steps, the IFRM algorithm performs a big update of local maps with $d_1 = 8$ and $d_2 = 2$. This is mainly because in this map there are parts which are very sparse. There are moments that the robot observes only 1 landmark in more than 10 successive observation steps. These observations are not used since the IFRM algorithm requires at least 2 landmarks to be able to compute one relative quantity.

The running time of the algorithms are shown in Table. 1. The algorithms IFRM and EKF complete the whole dataset. The algorithm FSv1 using 50 particles diverges before reaching the step $1000th$. For some reason we are not able to get the implementation of FSv2 to work with this dataset. Notice from Table. 1 that the main function `UpdateIFRM` of the IFRM algorithm uses only $3s$ where the main function `KF_Cholesky_update` of the EKF algorithm uses $105s$. This is because IFRM has constant time complexity and EKF has approximately cubic time complexity.

5   Conclusions and Discussions

The paper presents the Information Relative Map IFRM algorithm in which the Information filter is applied to the relative map concept. Instead of estimating the relative quantities, the algorithm estimates the Information map, the information form of the relative map. The correlation between the quantities are taken into account and thus there is no information loss. The estimation has constant time complexity without any approximation and linearization. Only independent quantities in each observation are considered so that the required computation is significantly reduced. In order to recover the absolute map from the Information map estimate, a strategy is described in which different levels of local submaps can be computed. The strategy is dynamically adjustable based on the measurement noise level.

The IFRM algorithm is empirically tested and compared with an existing implementation of the FastSLAM and EKF algorithms using more than 100 simulated problem instances and the Victoria park dataset. The results clearly show that the IFRM algorithm obtains a much better performance in map precision and speed.

We notice a problem that when the measurements are too noisy (e.g. setting $\sigma_B >= 4^o$ or $\sigma_R >= 4m$) the IFRM becomes unstable in constructing the absolute map or local submaps. The problem is not because of the approximation used in constructing local submaps but from the common method on how an absolute map is computed from a relative map [3, 14, 15, 16]. This problem has never been addressed in the relative mapping literature. Finding a robust method to recover the absolute map from a relative map will be one of our objectives.

In this paper, the diameter values of the extended local submaps are pre-selected by hand. However, one can see that the diameters can also be chosen based on the map topology. For example, if the environment consists of many connected places, d_1 can be selected so that ELM1 would span to cover approximately the place containing the robot. Inside the place, the observations (quantities) are correlated but much of them are not correlated to the quantities in other places.

In the IFRM algorithm, only independent quantities are considered in each observation. However, the final map in general contains dependent elements. The concept of geometric filter [3, 15] can also be applied in order to enhance the map consistency. If a constraint enforcement was performed in the information space, the information matrix would be fully populated in general. A strategy to construct constraints of only nearby landmarks so that the information matrix remains block diagonal is possible. In both cases, the proposed estimation algorithm still works since it does not use the block diagonal property. It would affect the local submap construction procedure, however. One could apply the same approximation procedure without using the block diagonal property and increase the diameters. This will be our future work.

Acknowledgment

This work has been supported by the Swiss National Science Foundation No 200021-101886 and the EU project Cogniron FP6-IST-002020. We would like to thank Ahad Harati for very helpful discussions.

References

1. Smith, R., Cheeseman, P.: On the Representation and Estimation of Spatial Uncertainty. International Journal of Robotics Research 5(4) (1987)
2. Dissanayake, M., Newman, P., Clark, S., Durrant-Whyte, H., Csorba, M.: A Solution to the Simultaneous Localisation and Map Building (SLAM) Problem. IEEE Transactions on Robotic and Automation 17(3), 229–241 (2001)
3. Newman, P.: On the Structure and Solution of the Simultaneous Localization and Mapping Problem. Ph.D. dissertation, Australian Center for Field Robotics, Uni. of Sydney (1999)
4. Julier, S., Uhlmann, J.: A counter example to the theory of simultaneous localization and map building. In: Proceedings of ICRA 2005 (2005)
5. Castellanos, J.A., Neira, J., Tardós, J.D.: Limits to the Consistency of EKF-Based SLAM. In: Proceedings of the 5th IFAC Symposium on IAV, Lisbon (2004)
6. Leonard, J.J., Feder, H.J.S.: Decoupled Stochastic Mapping. IEEE Journal of Oceanic Engineering, 561–571 (2001)
7. Guivant, J., Nebot, E.: Optimization of the Simultaneous Localization and Map-Building Algorithm for Real-Time Implementation. IEEE Trans. on Robotics and Automation 17(3), 242–257 (2001)
8. Williams, S.: Efficient Solutions to Autonomous Mapping and Navigation Problems. Ph.D. dissertation, Uni. of Sydney, Australia (2001)

9. Bailey, T.: Mobile Robot Localisation and Mapping in Extensive Outdoor Environments. Ph.D. dissertation, University of Sydney (2002)
10. Paz, L., Jensfelt, P., Tardos, J., Neira, J.: EKF SLAM Updates in O(n) with Divide and Conquer SLAM. In: Proceedings of ICRA 2007 (2007)
11. Smith, R., Self, M., Cheeseman, P.: Estimating Uncertain Spatial Relationships in Robotics. In: Autonomous Robot Vehicles, Springer, Heidelberg (1990)
12. Csorba, M., Uhlmann, J.K., Durrant-Whyte, H.F.: A Sub Optimal Algorithm For Automatic Map Building. In: Proceedings of ACC 1997 (1997)
13. Deans, M.C., Hebert, M.: Invariant Filtering for Simultaneous Localization and Mapping. In: Proceedings of ICRA 2000 (2000)
14. Martinelli, A., Nguyen, V., Tomatis, N., Siegwart, R.: A Relative Map Approach to SLAM based on shift and rotation invariants. Robotics and Autonomous Systems 55(1), 50–61 (2007)
15. Nguyen, V., Martinelli, A., Siegwart, R.: Handling the Inconsistency of Relative Map Filter. In: Proceedings of ICRA 2005 (2005)
16. Martinelli, A., Siegwart, R.: Exploiting the Information at the Loop Closure in SLAM. In: Proceedings of the International Conference on Robotics and Automation - ICRA (2007)
17. Maybeck, P.: Stochastic Models, Estimation and Control. Academic Press, London (1979)
18. Thrun, S., Burgard, W., Fox, D.: Probabilistic Robotics. MIT Press, Cambridge (2005)
19. Thrun, S., Liu, Y., Koller, D., Ng, A., Ghahramani, Z., Durrent-Whyte, H.: Simultaneous Localization and Mapping With Sparse Extended Information Filters. International Journal of Robotics Research (2004)
20. Eustice, R., Walter, M., Leonard, J.: Sparse Extended Information Filters: Insights into Sparsification. In: Proceedings of IROS 2005 (2005)
21. Wang, Z., Huang, S., Dissanayake, G.: Decoupling Localization and Mapping in SLAM Using Compact Relative Maps. In: Proceedings of IROS 2005 (2005)
22. Nguyen, V., Martinelli, A., Siegwart, R.: Improving the Consistency of Relative Map. In: Proceedings of the IROS 2006, Beijing, China (2006)
23. Montemerlo, M., Thrun, S., Koller, D., Wegbreit, B.: FastSLAM: A factored solution to the simultaneous localization and mapping problem. In: AAAI (2002)
24. Tim bailey's website, http://www-personal.acfr.usyd.edu.au/tbailey/

Measuring Motion Expressiveness in Wheeled Mobile Robots

João Sequeira

Instituto Superior Técnico / Institute for Systems and Robotics, Lisboa, Portugal
jseq@isr.ist.utl.pt

Summary. This paper addresses the measurement of motion expressiveness in wheeled mobile robots.

A neural network based supervised learning strategy is proposed as a method to fuse information obtained from the measurement of selected features. The choice of these features is made to reflect the visual quality of the trajectory and hence carries semantic ambiguities that are filtered out through the ability to generalize knowledge by the neural network.

The paper presents results with two features that might be significant in what concerns motion expressiveness, namely, how confident/hesitant is the motion and whether or not contains local loops that might indicate, for example, a call for attention by the robot towards a group of humans.

1 Introduction

Motion expressiveness (ME) is a concept for which most humans can provide acceptable definitions, all of them with close semantics. Roughly, an expressive motion triggers some kind of emotion in people observing it.

In everyday life, human and animal societies use individual motion to express multiple behaviors and emotions, e.g., aggressiveness, anxiety, attention, autistic, curiosity, dominance, egotistic, love, neutral, submissive, etc. In mobile robotics contexts a natural ME definition can be extrapolated directly from the human locomotion context, that is, ME is an index that expresses the ability of a trajectory executed by a robot to convey a meaning to a human observer. Some locomotion behaviors by humans have clear and socially accepted meanings and hence are readily classified by people as expressive.

Expressiveness is a common concept in Information Sciences and software engineering, namely in interface analysis and design. Usability metrics often include expressiveness measures along with measures of other concepts such as concision, simplicity, transparency, and scriptability (see for instance [14]). Some authors also argue that there is a tradeoff between usability and expressiveness, [3, 15], which could then provide alternative approaches to ME measuring.

Image processing techniques have been used to identify dynamic models of humans and animals using motion capture from video images (see for instance [5] for an application related to the generation of realistic movements by animation creatures). Sequences of expressive motions, chosen according to some classification criteria, can thus be processed to identify the relevant features of the models and the corresponding

H. Bruyninckx et al. (Eds.): European Robotics Symposium 2008, STAR 44, pp. 145–154, 2008.
springerlink.com

control requirements such that expressive motion can be replicated by robotic models. As an example, motion capture combined with frequency decomposition techniques and discriminant analysis is used in [1] to obtain Laban parameters for motion and classify the intensity of dance movements.

Nevertheless, most people can identify an expressive motion when seeing it just by observing and estimating a number of features, often unconsciously. Roughly, a motion qualifies as expressive when it conveys a meaning to an external observer. Some loco-motion behaviors by humans have clear, socially accepted, meanings and hence qualify as expressive motion.

Assessing the ME of an anthropomorphic robot tends naturally to be easier as the motion is immediately compared with the corresponding human capabilities. In robotic heads the combined motion of neck, eyes, eyebrows and mouth easily induces an emotional response in external observers (see for instance, [7, 9]). For legged locomotion the current state of the art in biped robotics has still to improve to reach the same degree of expressiveness of human walking. Similarly for manipulation through anthropomorphic robot arms.

Psychology identifies motion related features and the emotions they trigger in external observers. The Arouse-Valence-Stance space is used to represent EM in face movements, [9], also can also be used to map the basic emotions into real motion of robotic heads.

In wheeled mobile robots, ME has been mostly related to behavioral control of teams of robots, e.g., flocking and foraging (see for instance [12]). Expressive motion by single robots has been addressed also with the help of anthropomorphic features, [17, 11], hence masking the effect of pure motion with the motion of the body of the robot itself.

Some results refer that people tend to prefer "machine-like appearance, serious personality and round shape", [7]. This argument supports ME analysis directly based in the estimation of relevant features in the trajectories performed by the robots.

In information sciences well defined models for expressiveness have been proposed. The weighted linear combination,

$$E = c_1 e_1 + c_2 e_2 + c_3 e_3 + c_4 e_4 \tag{1}$$

has been proposed in [4]. The c_i are constant weights, e_1 is estimated as the number of data elements in an information system, e_2 supports high-resolution concepts by allowing the user to distinguish between entities when the differences are very small and it is estimated from the fan-out of the entities at various levels in a data model, e_3 is estimated as the count of synonyms (multiple synonyms for the same entity increase the probability that the system can support users from different backgrounds where different terminology is used to express the same concept), and e_4 is estimated as the number of query types supported by the information system.

The e_i components in (1) are of course specific to information systems but the linear structure suggests a linear independence/decoupling assumption between them that is worth to explore in the robotics context.

A meaningful ME measure, that is a measure that can provide coherent information on ME, can be used to influence motion control. For instance, if the goal behavior is to socialize with a group of people, approaching them with a confident motion, and, due to obstacles in the way, the real time evolution of the ME measure shows a hesitant motion, then the motion parameters might need an adjustment to better reflect the goal behavior. This is the ultimate goal for this study.

2   A Neural Network Based Classifier

Within the human activities motion expressiveness is an ill-defined concept. Each person has a personal idea on what an expressive motion is, relying on the observation of a number of features chosen according some personal, often unconscious, criteria. The estimates for the values of these features are combined, often using ill-defined rules, to yield an estimate of a personal expressiveness index. Still, most people can identify an expressive motion when seeing it just by observing and estimating those features and associating this to a meaning.

Given the ambiguous nature of an ME index it seems appropriate to perform its estimation using supervised learning. This means that the values obtained are correct up to the human reference knowledge and training data used during the learning stage.

The architecture of the proposed ME estimator is shown in Figure 1. Feature extractors map the observations into a real number that expresses the strength of the feature. Features can be grouped according specific contexts to yield ME measures valid in these contexts Each context might be interpreted as corresponding to some specific perspective to estimate ME. For example, in an active surveillance application, using robots, it may be useful to distinguish between ME in normal and threat situations. The ME measures at the output of each context can then be further combined to obtain a more general index.

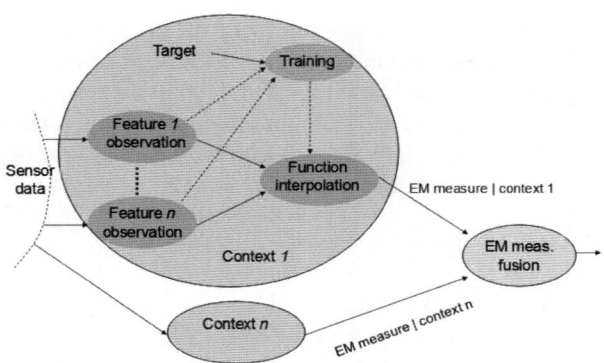

Fig. 1. The classifier architecture

The supervised learning approach in Figure 1 is a classical architecture. The function interpolation combines the feature values at the input such that the output yields a priori known target values for the specific input values that form the training set. Features can be grouped such that they are identified with particular situations, or contexts. ME indexes obtained for each of these contexts can then be fused to yield a global index.

As aforementioned, the selection of adequate features is quite an arbitrary process as it depends on personal feelings that induce subjective decision criteria. In what concerns wheeled mobile robot trajectories, there are a number of features that humans most likely use to qualify a trajectory as expressive. Among these, one can identify,

- the low and high frequency components of the trajectory, expressing the amount of smoothness or sharp movements (low frequency components in angular velocity measures a tendency for a trajectory to contain loops);
- the distance to objects or people, expressing the comfort or distress in the interaction;
- the maximum and minimum linear and angular velocities, expressing anxiety;
- the number of changes in the signs of the linear and angular velocities, expressing hesitation or confidence in the movement;
- the direction of the motion, expressing an intention;
- the space spanned by each of the coordinates, expressing how active a robot is (the larger the span the bigger the activity of the robot; it may be interpreted as having a robot deeply committed to complete survey the space);
- the entropy, expressing a measure of the organization in the trajectory (an entropic trajectory might induce feelings of low confidence or high anxiety).

The problem of combining the values of all the features in a context can be set using multiple approaches. If each feature observed yields a random variable with some probability distribution then the fusion of multiple features can be seen as the solution of combining these variables to minimize the variance of the estimation error between pre-specified target outputs and the outputs obtained for a priori known inputs. Under assumptions of gaussian input disturbances and knowledge on the feature dynamics, the Extended Kalman Filter (EKF) can be used to fuse all the feature measurements. However, features will often behave according nonlinear and nonsmooth models, difficult to identify, and hence such techniques might not be directly applicable to this problem.

A supervised learning technique based on neural networks is used in this paper. The motivation for this approach is drawn mainly from the fact that (i) standard data fusion structures, namely the EKF, also contain a linear combination of error inputs, and (ii) the model (1) also suggests that a linear structure be used to combine the values of the features.

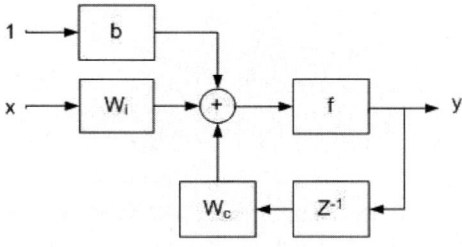

Fig. 2. An Elman network cell

The typical structure of a neural network includes a linear combination stage and, depending on the specific type of network, some additional elements that can shape the dynamics to arbitrary nonlinear models. Elman neural networks, [6], are characterized by the inclusion in the linear combination stage of past output values. Figure 2 shows the building blocks that form one cell of an Elman network. At each cell, the inputs x are weighted by W_i and added a bias, b, and the output, y, weighted by W_c. At the output stage , f, multiple functions can be used, e.g., linear and tansig. When using linear output f blocks, each cell is a first order dynamic system and hence an Elman network is a mesh of first order systems. By establishing adequate interconnections among the cells and shaping the weights W_i and W_c it is possible to create arbitrary linear systems and approximations of nonlinear systems. Elman networks have been used in speech recognition, [16], plant monitoring, [18], and nonlinear dynamics identification, [8]. Multiple learning algorithms can be used to compute the W weights. In this work a standard backpropagation rule is used.

Defining the proper network topology for a given problem is in general a difficult issue. Some theoretical results point to two hidden layers when the network is to model a generic boolean function, [2]. Evolutionary programming has been used to determine the number of hidden cells in neural networks, [10, 13].

3 Experiments

In this section two experiments are presented. The features used in these experiments were chosen aiming at recognizing hesitating-confident motion, in the first experiment. In the second one, the features chosen aim at measuring the curiosity by the robot towards a person or a group of persons. In both experiments the a Pioneer robot is teleoperated in a number of independent runs. A subset of these runs is used to train the neural network. The remaining runs are used to assess the performance of the classifier and its ability to generalize the knowledge acquired during the training stage.

During the training stage, an ME value is assigned to each of the trajectories by a human based on its visual qualities. An adequate visual quality might trigger the desired emotion.

In the first experiment the network has size $[30, 50, 10, 1]$. The second experiment was carried out in two versions, (i) the network size is $[5, 10, 10, 1]$ in the first version and $[70, 150, 100, 1]$ in the second one. In both cases the number of cells in each layer was empirically chosen (relatively easy to find in the first case, it is a single variable monotonic function; more difficult in the second case).

The robot used is of unicycle type. The input data for the ME measure is formed by the linear and angular velocities used to control the robot. For the first experiment the feature chosen was the number of changes in the sign of the linear velocity. In the second experiment the feature to measure the ME is the existence of localized loops in the trajectory. In the first version of the second experiment, the feature is measured by the low frequency components in the angular velocity that generates the trajectory. If a number of the low frequency components present in the angular velocity are higher than those corresponding to high frequencies then the trajectory tends to loop. The relative weight among these components can be used to separate between localized loops and loops spanning large areas. In this case, a 512 point FFT is used to get the components corresponding to the three lowest frequencies, followed by a normalizing transformation. For the second version, the low frequency components of the linear velocity are also used. Figures 3 and 4 show the complete data set (of trajectories) used in the experiments.

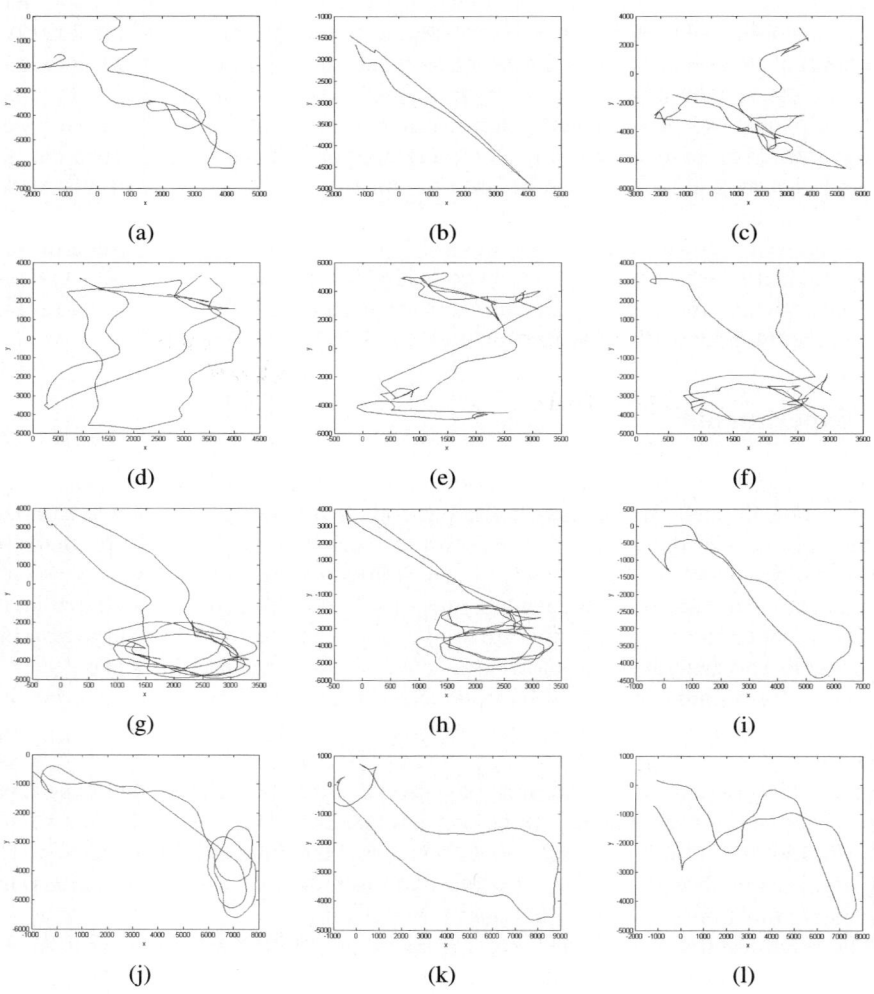

Fig. 3. Complete set of trajectories generated by the experiments

Table 1 shows the data for both experiments, the sets used to train the network, the target output for each training set, and the obtained output for each test set. The ME index values, i.e., the output of the network, during the training stage are chosen from the visual observation of the trajectories exclusively.

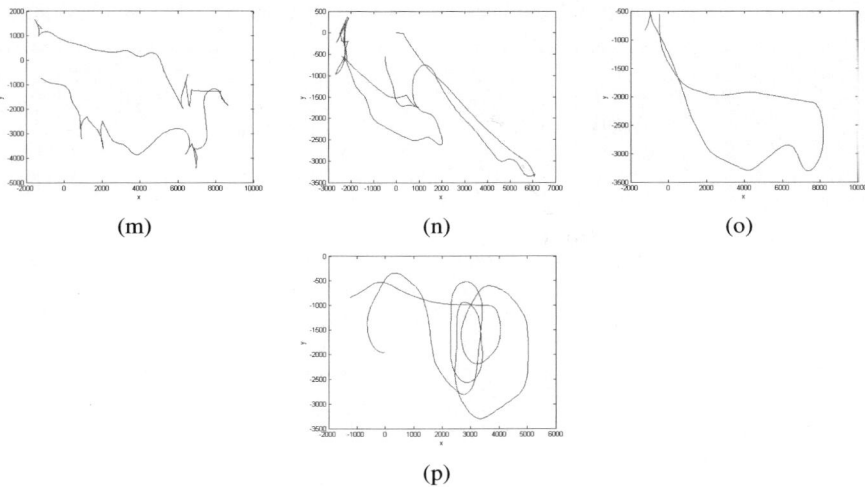

Fig. 4. Complete set of trajectories generated by the experiments (cont.)

Table 1. Data and results for experiments 1 and 2

Set	Experiment 1			Experiment 2		
	Feature f_v	Target	Output	Feature f_ω	Target	Output
(a)	8	0.20	0.1946	8.38419, 9.18992, 10	0.1	0.0976
(b)	17	0.15	0.2029	1.26535, 3.95569, 10	0.1	0.0753
(c)	55	0.90	0.8732	2.35201, 7.09803, 10	0.1	0.1207
(d)	20	0.35	0.2873	10, 0.52747, 0.739253	0.1	0.2212
(e)	54	0.85	0.8619	10, 0.874886, 0.197938	-	1.0610
(f)	54	0.85	0.8619	10, 2.75835, 0.40712	-	1.0634
(g)	35	0.60	0.6016	10, 0.0973174, 0.217039	1	1.0209
(h)	30	0.50	0.5155	10, 0.241414, 0.37905	1	0.8405
(i)	6	-	0.2744	10, 8.12918, 4.15008	-	−0.1817
(j)	13	-	0.1181	10, 3.7375, 0.0230103	1	1.1500
(k)	16	-	0.1762	10, 7.51246, 2.19928	-	−0.2317
(l)	15	-	0.1521	10, 7.51036, 2.24441	-	−0.2342
(m)	34	-	0.5853	10, 1.64281, 0.0744162	-	1.1585
(n)	52	-	0.8386	0.9487, 10, 0.263107	-	0.0853
(o)	15	-	0.1521	10, 8.077, 3.38752	-	−0.2012
(p)	10	-	0.1315	10, 0.0152748, 0.168055	1	0.8770

In experiment 1, the network output (the ME index) increases with the "hesitation" of the robot. The target ME values were assigned based uniquely on the visual observation of the smoothness of the trajectories. This means that the plots hide those changes that are not associated to changes in motion direction. Given the target ME index values, this experiment amounts to identify an almost monotonic function of a single variable.

Table 2. Experiment 2 with additional inputs

Set	Feature f_ω	Feature f_v	Target	Output
(a)	8.38419, 9.18992, 10	10, 8.14684, 3.71772	0.1	0.1063
(b)	1.26535, 3.95569, 10	0.101767, 2.24074, 10	0.1	0.0895
(c)	2.35201, 7.09803, 10	10, 5.58878, 4.30554	0.1	0.1061
(d)	10, 0.52747, 0.739253	10, 1.24861, 0.105222	0.1	0.2138
(e)	10, 0.874886, 0.197938	10, 6.32206, 1.29682	-	0.4349
(f)	10, 2.75835, 0.40712	6.00649, 10, 1.35242	-	1.0180
(g)	10, 0.0973174, 0.217039	0.0598811, 10, 1.63062	1	0.9394
(h)	10, 0.241414, 0.37905	2.04612, 10, 3.77873	1	1.0432
(i)	10, 8.12918, 4.15008	0.770519, 4.64428, 10	-	0.3285
(j)	10, 3.7375, 0.0230103	2.04532, 10, 0.499195	1	1.1237
(k)	10, 7.51246, 2.19928	10, 5.39222, 0.405971	-	0.3682
(l)	10, 7.51036, 2.24441	10, 7.44558, 0.939264	-	0.5930
(m)	10, 1.64281, 0.0744162	2.74215, 3.30769, 10	-	0.4964
(n)	0.9487, 10, 0.263107	8.11451, 0.0382724, 10	-	0.4113
(o)	10, 8.077, 3.38752	7.89761, 10, 3.39284	-	0.7089
(p)	10, 0.0152748, 0.168055	10, 1.96887, 4.92446	1	0.7861

This is a simple task for an Elman network. The results show that the network is able to properly identify such function despite the disturbance introduced by sets (a) and (b).

In what concerns experiment 2, the network is able to identify a function that returns values close to 1 when there are localized looping characteristics in the trajectory. Localized loops such as those shown in Figure 4f-g might have been generated by a robot moving around a group of people trying to catch their attention. Loops covering large areas, such as those shown in Figure 4d, are more likely to be related to exploratory other than curiosity behaviors.

The net output indicates that sets (e), (f), (g), (h), (j), (m), and (p) contain localized loops. However, the features selected are not enough to define a one-to-one map between the observations space and the ME index space. The result obtained with the set (m) clearly does not correspond to the expectations created by the visual observation of the trajectory. This happens because the corresponding feature values are close to those of sets (f) and (j) which were taught as returning a high ME value. Also, the network fails on set (o) as its feature values match some of those used for training that do not contain any local loops, such as (k) and (l). Set (e) clearly contains localized loops and, though it is not clear from the visual observation what was the intention of the robot, the network classified it has having loops. Eventually, for some data sets these problems can be solved by carefully tuning the network topology and/or using additional frequency components of the angular velocity at the input of the network.

More generally, components from both the linear and angular velocity are required (as they define completely the characteristics of a trajectory for a given robot model). Table 2 shows results using the 3 lowest frequency components of both the linear and angular velocity with network topology $[70, 150, 100, 1]$. Clearly, there is a better discrimination in the sense that higher values are associated with localized loops better defined. Set (m) was now dropped and sets (o) and (p) are classified as close (as they are given close ME index values) which makes some sense as they contain loops that span wider areas than those shown in the taught sets.

4 Conclusions

The paper addressed the measurement of the expressiveness of trajectories described by wheeled mobile robots. Encouraging results on mapping robot movements into a space that can be identified with human idea of motion expressiveness were obtained with the supervised learning approach described in the paper.

The paper addressed exclusively the direct problem, that is, computing a ME index from the observation of trajectories. Still to address is the inverse problem of generating a trajectory with a given ME index and accounting for the robot and environment motion constraints. Also, including ME measures in standard motion control strategies will require the selection of the adequate temporal portions of a signal from where to extract features.

This study addressed only the direct problem of measuring a ME index. Future work will address the inverse problem, that is generating motion that corresponds to some ME index. Also, different ME values obtained from different features will exhibit some overlapping, i.e., comparable ME values will be obtained for identical data sets, whereas very different ME values can obtained for close data sets. By combining multiple such ME measures it might be possible to further improve the discrimination among data sets. The technique can be applied to all sorts of robots and hence naturally leads to challenging problems related to the fusion of ME measures taken from multiple robots, e.g., mobile manipulators and teams of robots.

Acknowledgments

This work was supported by the European Project FP6-2005-IST-6-045062-URUS, and ISR/IST plurianual funding through the POS_Conhecimento Program that includes FEDER funds.

References

1. Lees, A.: Expressive Motion. PhD thesis, New York University (2006)
2. Brightwell, G., Kenyon, C., Paugam-Moisy, H.: Multilayer neural networks: one or two hidden layers? In: Procs. NIPS 1996 (1996)
3. Buchanan, G.: Expression and Interpretation in Spatial Hypertext. In: Procs of the 4th Workshop on Spatial Hypertext, ACM Hypertext Conference, Santa Cruz, USA (2004)
4. Ceruti, M., Robin, S.: Infodynamics: Analogical analysis of states of matter and information. Int. Journal on Information Sciences 177, 969–987 (2007)
5. Liu, C.-Y.K.: Towards a Generative Model of Natural Motion. PhD thesis, University of Washington (2005)
6. Elman, J.L.: Finding structure in time. Cognitive Science 14, 179–211 (1990)
7. Fong, T., Nourbakhsh, I., Dauntenhahn, K.: A survey of socially interactive robots. Journal of Robotics and Autonomous Systems 42, 143–166 (2003)

8. Gao, X.Z., Gao, X.M., Ovaska, S.J.: A Modified Elman Neural Network Model With Application to Dynamical Systems Identification. In: Procs. IEEE Int. Conf. on Systems, Man, and Cybernetics, Beijing, China, vol. 2, pp. 1376–1381 (1996)
9. Mayor, L., Jansen, B., Lorotte, A., Siegwart, R.: Improving the Expressiveness of Mobile Robots. In: Procs. of ROMAN 2002 (2002)
10. McDonnell, J.R., Waagen, D.: Determining Neural Network Hidden Layer Size Using Evolutionary Programming. In: Proc. World Congress on Neural Networks, vol. 3, pp. 564–567 (1998)
11. Mizoguchi, H., Sato, T., Takagi, K., Nakao, M., Hatamura, Y.: Realization of Expressive Mobile Robot. In: Procs. IEEE Int. Conf. Robotics and Automation (1997)
12. Ngo, T.D., Schioler, H.: An Approach to Sociable Robots through Self-distributed Energy. In: Procs. of the 2006 IEEE/RSJ International Conference on Intelligent Robots and Systems, October 2006, pp. 2192–2199 (2006)
13. Pengt, K., Ge, S., Went, C.: An Algorithm to determine the neural network hidden layer size and weight coefficients. In: Procs. 15th IEEE Int. Symp. on Intelligent Control, ISIC 2000, Rio, Patras, Greece, July 17-19 (2000)
14. Raymond, E.: The Art of Unix Programming. Addison-Wesley, Reading (2003)
15. Repenning, A., Ambach, J.: Tactile Programming: A Unified Manipulation Paradigm Supporting Program Comprehension, Composition and Sharing. In: Procs. 1996 IEEE Symp. of Visual Languages, Boulder, CO, Computer Society, pp. 102–109 (1996)
16. Rothkrantz, L.J.M., Nollen, D.: Speech Recognition Using Elman Neural Networks. In: Matoušek, V., Mautner, P., Ocelíková, J., Sojka, P. (eds.) TSD 1999. LNCS (LNAI), vol. 1692, Springer, Heidelberg (1999)
17. Scheef, M., Pinto, J., Rahardja, K., Snibbe, S., Tow, R.: Experiences with Sparky, a social robot. In: Procs. of the Workshop on Interactive Robot Entertainment (2000)
18. Seker, S., Ayaz, E., Türcan, E.: Elman's recurrent neural network applications to condition monitoring in nuclear power plant and rotating machinery. Engineering Applications of Artificial Intelligence 16(7-8), 647–656 (2003)

Modeling, Simulation and Control of Pneumatic Jumping Robot[*]

Grzegorz Granosik, Edward Jezierski, and Marcin Kaczmarski

Technical University of Lodz, Stefanowskiego 18/22, 90-924 Lodz, Poland
{granosik,edward.jezierski,marcin.kaczmarski}@p.lodz.pl

Summary. The paper presents some results in development of a jumping robot. Our research is conducted into the problem of changing mechanical impedance of driving system and its influence on features of a jump. After theoretical analysis and Matlab based simulations we have built a virtual model of a planar jumping robot with pneumatic driving system. The Working Model 2D (WM2D) software was employed for this task. Our investigations are focused on the stabilization of a robot's body during flight. We present some stable jumps simulated in WM2D and controlled from Matlab. We show robustness of controller at various jumping conditions.

Keywords: jumping robot, pneumatic drives, impedance control.

1 Introduction

Pneumatic actuators originally limited to simple motion between two hard stops, are now becoming more and more popular, and substituting in many cases electric drives. Especially robots intended to contact with ground, obstacles, co-operating with other robots and humans require drives strong, compliant and safe. In [1] we have shown a few projects of quite different machines all taking benefits from various pneumatic drives. There is a number of walking robots built in Robotics Laboratory of the Technical University of Lodz. Most of them are statically stable – mostly quadruped [2], worm-like [3] and bipeds. On the other hand we have conducted a lot of research into pneumatic drives: starting from the position control of a light weight manipulator [4], through position-force control of pneumatic cylinders [5], to the design of novel pneumatic drives [6].

We have also made some theoretical investigations in the field of impedance control of robots in contact with objects and surroundings [7], [8]. Having this experience we went back to the fundamentals of legged locomotion i.e. to control of jumping robots.

From biology we know that all animals, try to move in an energy efficient way. The important mechanisms available are muscles and tendons which make energy recuperation possible. For example, the Achilles tendon in a human leg can store up to one third of the motion's energy during running. The muscular system is capable of adapting stiffness characteristics in order to move in a wide range of different walking and

---
[*] This work was supported by Ministry of Science and Higher Education under grant No 3 T11A 023 30.

running patterns and still exploit the passive behavior of the actuation system. This is why in the very first jumping robot thrust was generated by pneumatic actuator [9], and why the idea of using naturally compliant actuators for driving walking robots is still and extensively explored [10].

A new concept of jumping robot with pneumatic driving system is presented in this paper. We start with theoretical background of the research; go through analysis of dynamics of 1DOF leg and simulation of 2DOF planar robot. In this area we focus our investigations on the aspect of changing mechanical impedance of driving system and its influence on features of a jump. For the modelling phase we have employed Matlab/Simulink software while for simulation Working Model 2D connected with Matlab. We have also built special experimental stand and prototype of planar robot. We are currently verifying proposed control algorithms on real robot.

2 Analysis of 1DOF Robot

2.1 Dynamical Model of the Robot

The structure of 1DOF robot is presented in Fig. 1. It consists of a platform and a pneumatic cylinder. A piston rod equipped with a rubber ball form the leg of the robot.

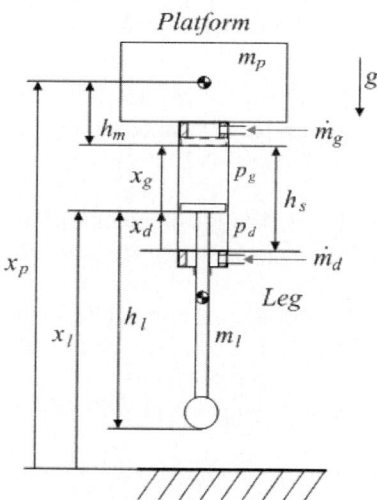

Fig. 1. The structure of 1DOF jumping robot

In the transient state, a pneumatic force produced by the cylinder plays an important role. This force is given by the formulae

$$f_p(t) = A_g p_g(t) - A_d p_d(t)$$ (1)

where A_g and A_d are cross-sections of upper and lower part of the piston, respectively. Volumes of upper and lower chambers are equal to

$$V_g(t) = V_{go} + A_g x_g(t) = A_g \left(l_{go} + x_g(t) \right),$$
$$V_d(t) = V_{do} + A_d x_d(t) = A_d \left(l_{do} + x_d(t) \right) \tag{2}$$

where V_{go} and V_{do} denote the minimum volume of chambers that follow from boundaries of the piston's movement. Scalar quantities l_{go} and l_{do} are much less than a piston stroke h_s. The values $x_g(t)$ and $x_d(t)$ are direct functions of a platform position $x_p(t)$ and a leg position $x_l(t)$, namely

$$x_g(t) = x_p(t) - x_l(t) - h_m, x_d(t) = x_l(t) - x_p(t) + h_s + h_m \tag{3}$$

The dynamical features of pressured gas in the upper chamber is described by the equation

$$\dot{p}_g = \frac{\kappa}{l_{go} - h_m + x_p(t) - x_l(t)} \times \left(-p_g(t) \left(\dot{x}_p(t) - \dot{x}_l(t) \right) + \frac{RT}{A_g} \dot{m}_g(t) \right) \tag{4}$$

The first term in brackets represents the negative internal feedback in the cylinder, while the last term describes the effect of increasing pressure in the chamber when the mass rate of flowing air is positive. R is an universal gas constant, and κ is a gas specific heat ratio. Similarly the time derivate of pressure in the lower chamber may be presented as follows

$$\dot{p}_d(t) = \frac{\kappa}{l_{do} + h_s + h_m - x_p(t) + x_l(t)}$$
$$\times \left(p_d(t) \left(\dot{x}_p(t) - \dot{x}_l(t) \right) + \frac{RT}{A_d} \dot{m}_d(t) \right) \tag{5}$$

Dynamics of the platform in a free vertical movement, as well as in the phase of contact of the leg with the base, is described by the equation

$$\frac{d^2 x_p(t)}{dt^2} = \frac{f_p(t) - f_t(t)}{m_p} - g \tag{6}$$

where $f_t(t)$ denotes the friction force in the cylinder, and g is a gravity acceleration.

Dynamics of the leg in a free vertical movement, i.e. when $x_l(t) > h_l$, is described by

$$\frac{d^2 x_l(t)}{dt^2} = -\frac{f_p(t) - f_t(t)}{m_l} - g \tag{7}$$

In the phase of contact of the leg with the base the impedance features have to be taken into account. The situation in presented in Fig. 2b. The compliance of the rubber foot could be presented as a spring characterised by the coefficient c_{sf}. On the other hand the compliance of the ground in a general case is described by c_{sg}.

Additionally, the effects of damping play also an important rule. The springs that model the rubber foot and the compliant base are linked in serial. Thus the resultant compliant coefficient is equal to $c_s = c_{sf} + c_{sg}$[11]. The spring force is described by

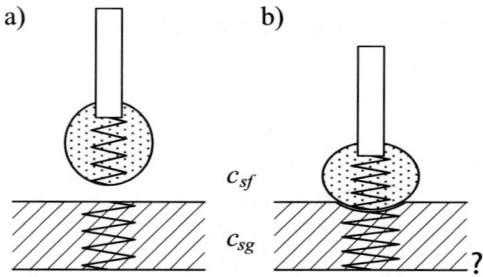

Fig. 2. Modelling of contact situation between the leg and a base

$f_s(t) = (x_l(t) - h_l)/c_s - f_d$, where f_d represents damping force in the system. Finally the dynamics of the leg in this phase is given by the equation

$$\frac{d^2x_l(t)}{dt^2} = -\frac{f_p(t) - f_t(t) - f_s(t)}{m_l} - g \qquad (8)$$

It is useful to introduce a new variable $x_t(t)$ that represents the position of the piston related to mid-point of the cylinder. This value is defined as follows

$$x_t(t) = x_l(t) - x_p(t) + h_m + \tfrac{1}{2}h_s \qquad (9)$$

2.2   Simulation Results

The dynamic features of the whole system were tested using Matlab/Simulink software package. The low-friction Clippard cylinder with a diameter of 1 inch and stroke 10 inches was used. The following parameters were taken in calculations $m_p = 2.45$kg, $m_l = 0.4$kg, $l_l = 0.300$m, $h_s = 0.254$m, $A_g \cong A_d = 7.9$cm^2. The parameters for modelling Coulomb and viscous friction effects in the cylinder were measured in a simple laboratory test.

Detailed simulation results for free-falling with verification on out test-bed are presented in [12]. Tests were performed for different initial positions of the piston in the cylinder and different initial levels of pressures in both chambers. Here we focus on the latter aspect. It is worth to underline that the mechanical impedance of a pneumatic cylinder depends on pressures in both chambers. In particular, the stiffness of the drive is nearly proportional to the sum of both pressures $(p_g + p_d)$[7].

The second series of simulations were performed to control jumps of the robot. The previous experience of authors proved that the simple models of air flow for mid-size valves given by Bobrow and McDonell [13] may be precise enough. For a charging mode of a valve the model is of the form

$$\frac{dm_s}{dt} = c_s\sqrt{p_s - p} \qquad (10)$$

and for venting of the valve the model is slightly different

$$\frac{dm_o}{dt} = c_o(p - p_o)$$ (11)

where p denotes pressure in the chamber, p_s is supplying pressure, $p_o = 1$bar, and c_s, c_o are constants taken from experiments.

An experiment of initialising of jumps was performed in which the system was working in an open loop. A simple algorithm to control three-way valves was based on the information S from a two-state sensor that detects contact of the leg with the base. When $S = 1$ the upper chamber is charged from the source $p_s = 8$bar, and in the case of $S = 0$ the upper chamber is vented. The altitude of jumps is increasing and tends to a certain stable level.

The altitude depends also on features of the ground. The results of an experiment are presented in Fig. 3. After initialising of jumps on a stiff ground a soft mat was put under the feet at the time t=10s. It caused that the ground stiffness was reduced and finally the altitude of jumps was tending to a lower level.

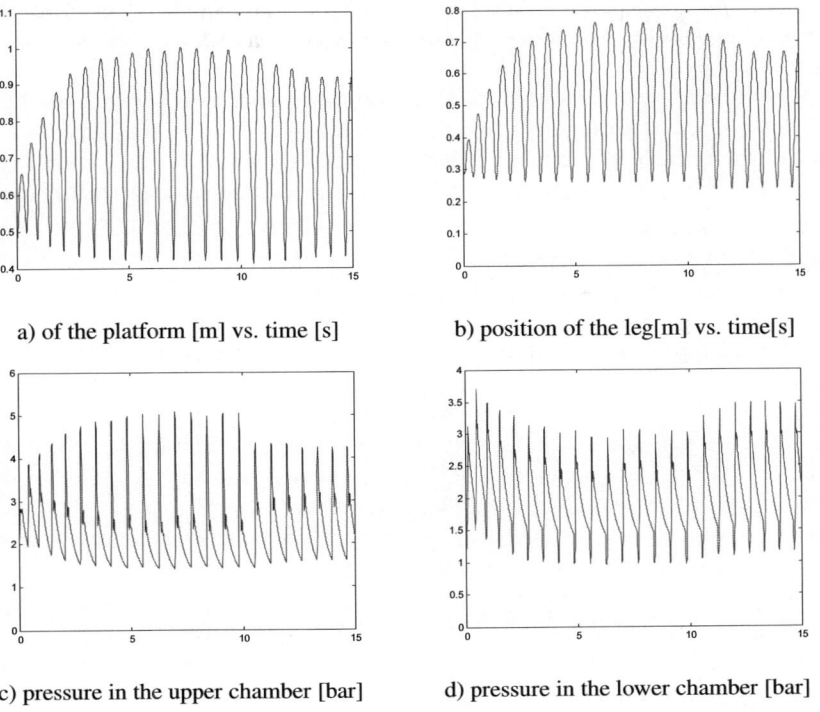

a) of the platform [m] vs. time [s]   b) position of the leg[m] vs. time[s]

c) pressure in the upper chamber [bar]   d) pressure in the lower chamber [bar]

Fig. 3. Positions of platform, leg, and pressures in chambers versus time in initialising jumps, and then after changing the ground at t=10s a) $x_p(t)$, b) $x_t(t)$, c) $p_g(t)$, d) $p_d(t)$

In this experiment the system was working in an open loop. However it is possible to introduce a feedback from the system measuring the vertical position of the robot.

3   Simulation of a Planar Jumping Robot

In the reported project we consider two constructions of a mechanical system: a biologically inspired one, with legs mimicking animal limbs as in the Kenken robot [14] and a simplified one, with thrust generated directly by a pneumatic cylinder (see Fig. 8). In both cases we can start by modeling of dynamics using a virtual leg model [9].

The detailed analysis and Matlab simulations of the swing during stance phase of the jump and stabilization of robot's body during flight were reported in [15]. Here we show verification of control algorithms in simulations of virtual model of planar jumping robot for different ground conditions.

3.1   Simulations in Working Model

Model of the planar hopping robot, created in the Working Model 2D program is shown in Fig. 4. For the simulations we took parameters of the real robot being designed. Total robot mass equals 4kg. Maximum shortness in length in stance phase is 0.3m. The spring stiffness equals 1000N/m.

To provide better stabilization of the robot body we used a cross bar with two masses placed on its ends. Both masses are equal to 2 kg. This kind of mass displacement increases total moment of inertia of robot's body and therefore decreases acceleration acting on the robot during flight phase, when the foot position is being set for landing.

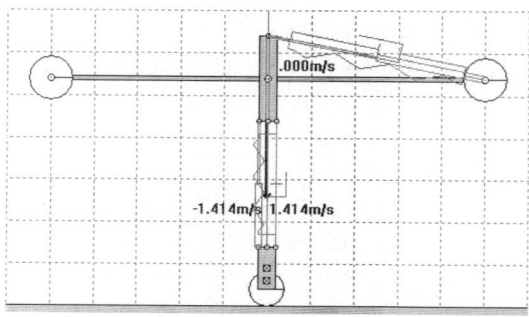

Fig. 4. Hopping Robot created in Working Model 2D

3.2   Algorithm of Robot Stabilization

Stabilization algorithm of our robot is schematically shown in Fig. 5. All calculations are made in Maltab. Regulator consists of four feedback loops. The main one is the circuit providing the requested angle of take off and assuming constant value of velocity. The calculated value is compared with the simulation and transformed into a correction in angle of the robot leg displacement β_{leg}. This correction is added to the value calculated with the inverse model of the hopping robot. The created function determines the value of leg's orientation $-\beta$. This calculation takes into account the values of the landing speed and the requested jump length. Landing velocity can be calculated in every step of calculation or may be set as a constant. That is why the velocity line is marked as dashed. In the second case the inverse model is outside the main feedback loop. We used a simple proportional controller in this loop and results were satisfactory.

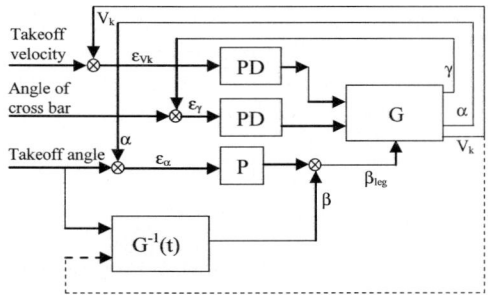

Fig. 5. Robot stabilization feedback loop

Another feedback value is the take-off velocity loop. The proportional-derivative element is used. It is characterized by a fast response time. It is a non-static element, but the difference between requested and steady state values is small and can be neglected.

During the flight phase, actuator trying to change orientation of robot's leg simultaneously acts on the cross bar with the same force. Therefore regulator for the cross bar stabilization is also required. It contains a PD element and it is used only in the stance phase while the foot is in contact with the ground. We chose the proportional-derivative element, because it has a fast response in transient phase.

3.3   Simulation Results

Simulation has been done in the environment created in WorkingModel 2D. Fig. 6 shows the virtual robot, obstacles and some settings while the following plot shows the displacement of the robot during the forward movement with changing of jump length. First range is the longest and it equals 0.4m. After 15 seconds it changes 0.2m, and at the end the displacement equals 0m. That means the robot 'stops' and jump vertically without any forward dislocation. Different inclination in the characteristic corresponds to the ranges of different jump length. It can be seen that the changes in the length of jumps also change the forward velocity of the hopping robot.

Plot in Fig. 7 shows that also uneven terrain almost does not change the forward speed. Inclination of the characteristic is constant in corresponding jump length. Height of the steps is 0.2m. We can assume that the inverted model is working correctly, because simple proportional regulator is enough to stabilize jump length and forward velocity at presence of disturbances. Some other experiments can be found in [15].

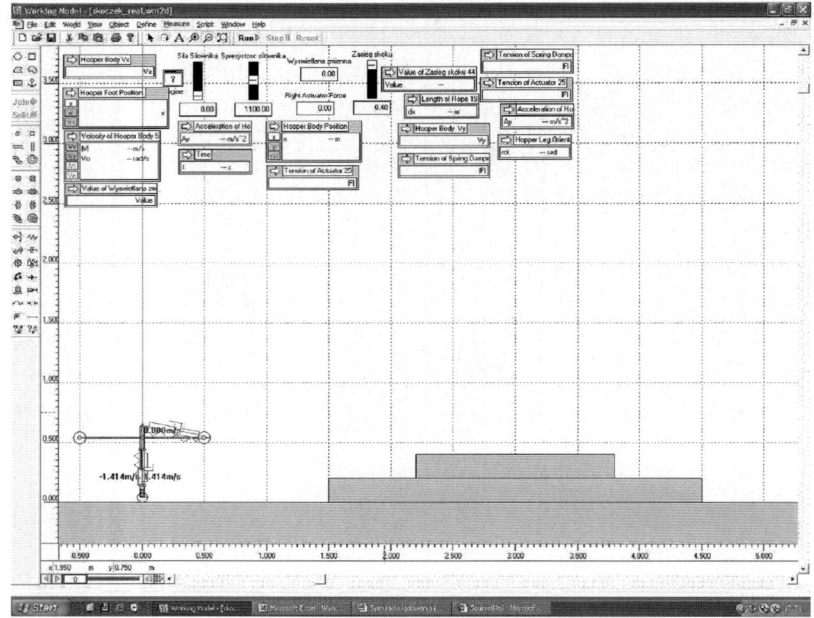

Fig. 6. Simulation environment of Working Model 2D

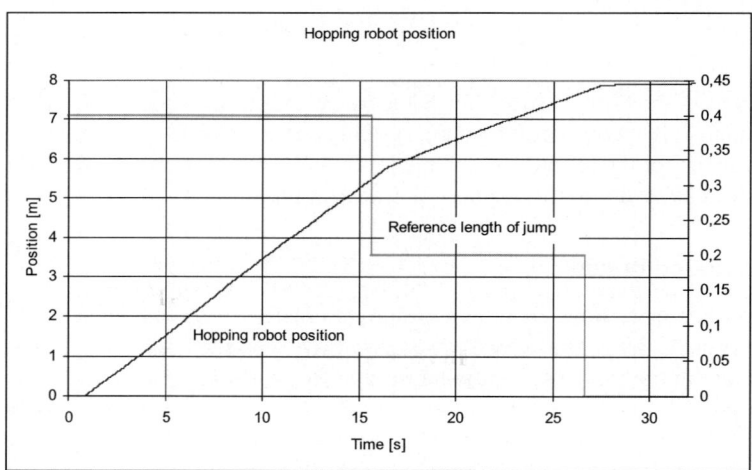

Fig. 7. Simulations of jumps in uneven terrain plot of hopping robot's position

Currently, we are implementing proposed algorithms on the simplified prototype with 2DOF (see Fig. 8). In the near future we expect to extend the idea of virtual leg to the robot with more complex kinematic structure.

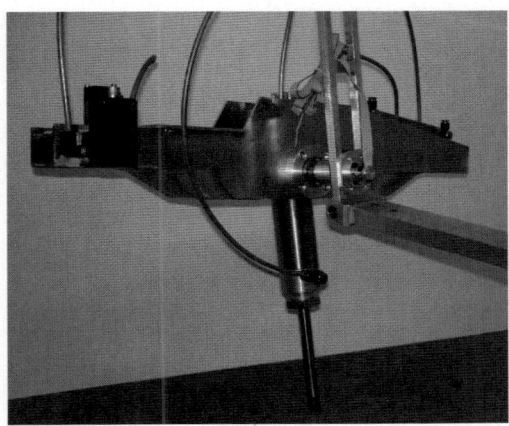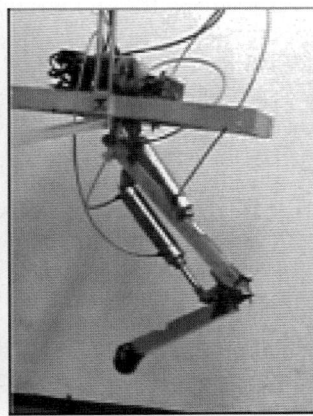

Fig. 8. Two prototypes of jumping robots: 2DOF simplified approach (on left) and 2DOF leg model (on right)

4 Conclusions

A simple jumping robot with a pneumatic cylinder as a main source of thrust is being developed. It allows to easily change the impedance of the leg and in this way to control the contact phase of jumps. After detailed mathematical analysis of the dynamics of 1DOF jumping system we have presented the simulation study of a one leg jumping robot. We used commercial simulation software Working Model 2D for modeling and Matlab package to realize control algorithms. We have obtained stable jumps for various ground conditions.

References

1. Granosik, G.: Pneumatic Actuators for Climbing, Walking and Serpentine Robots. In: Habib, M. (ed.) Bioinspiration and robotics: Walking and Climbing Robots, pp. 483–510 (2007), ISBN 978-3-902613-15-8
2. Dąbrowski, T., Feja, K., Granosik, G.: Biologically inspired control strategy of a pneumatically driven walking robot. In: CLAWAR conf., pp. 687–694 (2001)
3. Granosik, G., Kaczmarski, M.: Bellows driven, muscle steered caterpillar robot. In: Proc. CLAWAR Conf., pp. 743–750 (2005)
4. Jezierski, E., Mianowski, K., Collie, A.A.: ZarychtaD Design and control of a manipulator arm for a walking robot. In: Proc. of IEE Colloquium on Inf. Technology for Climbing and Walking Robots Digest, vol. 96/167, pp. 2/1–2/3 (1996)
5. Granosik, G.: An adaptive position/force control of a pneumatically driven manipulator, Ph.D. dissertation (in Polish), Inst. of Automatic Control, Tech. University of Lodz (2000)
6. Granosik, G., Borenstein, J.: Integrated joint actuator for serpentine robots. IEEE/ASME Trans. on Mechatronics 10, 473–481 (2005)
7. Granosik, G., Jezierski, E.: Application of a maximum stiffness rule for pneumatically driven legs of walking robot. In: Proc. CLAWAR Conf., pp. 213–218 (1999)

8. Jezierski, E.: From positional control of robots to impedance control (in Polish). In: Tchoń, K. (ed.) Postępy robotyki – sterowanie, percepcja, komunikacja, Warszawa, WKiŁ, pp. 13–36 (2006)
9. Raibert, M.H.: Legged robots that balance. MIT Press, Cambridge (1986)
10. Vanderborght, B., Verrelst, B., Van Ham, R., Van Damme, M., Lefeber, D., Meira, Y., Duran, B., Beyl, P.: Exploiting natural dynamics to reduce energy consumption by controlling the compliance of soft actuators. The Int. Journal of Robotics Research 25(4), 343–358 (2006)
11. Jezierski, E.: On electrical analogues of mechanical systems and their using in analysis of robot dynamics. In: Kozłowski, K.R. (ed.) Robot Motion and Control – Recent Developments, Berlin, pp. 391–404. Springer, Heidelberg (2006)
12. Jezierski, E., Granosik, G.: Modeling and Control of Jumping Robot with Pneumatic Drive. In: 13th IEEE IFAC Int. Conf. on Methods and Models in Automation and Robotics, Szczecin, Poland, pp. 941–945 (2007)
13. Bobrow, J.E., McDonell, B.W.: Modelling, identification, and control of a pneumatically actuated, force controllable robot. IEEE Trans. on Robotics and Automation 14, 732–742 (1998)
14. Sang-Ho, H., Mita, T.: Development of a Biologically Inspired Hopping Robot - "Kenken". In: Proc. of the 2002 IEEE Int. Conference on Robotics 8 Automation, Washington, DC (2002)
15. Granosik, G., Kaczmarski, M.: Simulation of jumping robot. In: 13th IEEE/IFAC Int. Conf. on Methods and Models in Automation and Robotics, Szczecin, Poland, pp. 1143–1148 (2007)

Multilayer Perceptron Adaptive Dynamic Control of Mobile Robots: Experimental Validation

Marvin K. Bugeja* and Simon G. Fabri

Department of Systems and Control Engineering, University of Malta, Msida, MSD 2080, Malta
{mkbuge,sgfabr}@eng.um.edu.mt

Summary. This paper presents experimental results acquired from the implementation of an adaptive control scheme for nonholonomic mobile robots, which was recently proposed by the same authors and tested only by simulations. The control system comprises a trajectory tracking kinematic controller, which generates the reference wheel velocities, and a cascade dynamic controller, which estimates the robot's uncertain nonlinear dynamic functions in real-time via a multilayer perceptron neural network. In this manner precise velocity tracking is attained, even in the presence of unknown and/or time-varying dynamics. The experimental mobile robot, designed and built for the purpose of this research, is also presented in this paper.

Keywords: Mobile robots, adaptive control, neural networks.

1 Introduction

The vast majority of wheeled vehicles suffer from some kind of inherent mobility restriction. These restrictions manifest themselves as nonholonomic constraints in the kinematic model, and often arise due to the underactuated nature of the drive mechanism. Consequently, the linearized kinematic models of these vehicles lack controllability, full-state feedback linearization is out of reach, and smooth time-invariant feedback stabilization is unattainable [2]. These characteristics render the motion control of nonholonomic mobile robots not only practically relevant but also theoretically interesting and challenging.

A vast number of past contributions on the control of nonholonomic wheeled mobile robots (WMRs) [9, 2] completely ignore the robot dynamics and rely on the assumption that the control inputs, usually motor voltages, instantaneously establish the desired wheel velocities. As expected, controllers which explicitly account for the robot dynamics due to its mass, friction and inertia [7, 4] lead to better control performance. However, as argued in [7] *perfect* knowledge of the robot dynamics is unattainable in practice. Moreover, these parameters can even vary over time due to loading, wear, and ground conditions. In response, a number of more advanced controllers have been proposed including: pre-trained neuro-controllers [10], parametric adaptive schemes [11], and robust sliding-mode methods [3]. A more powerful approach is that of online

---

* This work was supported by the National Grant, RTDI-2004-026.

H. Bruyninckx et al. (Eds.): European Robotics Symposium 2008, STAR 44, pp. 165–174, 2008.
springerlink.com

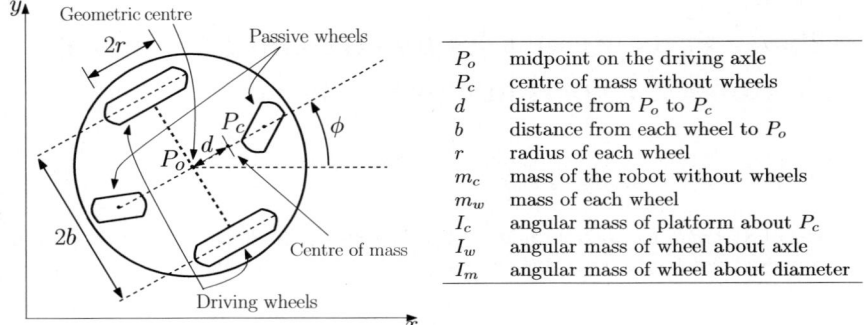

Fig. 1. Differentially driven wheeled mobile robot

functional-adaptive control, where the uncertainty is not restricted to parametric terms, but covers completely the dynamic functions themselves [8, 1].

Of all the proposed adaptive controllers, only a few have ever been implemented on a physical robot, among which one finds [5, 11, 6]. In [5] D'Amico et. al. propose radial basis function (RBF) artificial neural networks (ANNs) for the adaptive control of WMRs. However, this work disregards the robot dynamics since ANNs are used solely to approximate the inverse *kinematic* model of the vehicle. On the other hand, Wang et. al. [11] and Dixon et. al. [6] do consider the robot dynamics in their adaptive control methods, but address only *parametric* uncertainty in the dynamic model. In contrast, in this paper we employ *functional-adaptive* neuro-control to handle better the uncertainty in the *dynamic functions* of the WMR, and not just its parameters.

The main contributions in this paper are the presentation and analysis of a set of experimental results which validate and compare the employed multilayer perceptron (MLP) adaptive control scheme for the first time, after it was originally proposed in our previous publication [1], and tested by simulations only. In addition this paper outlines the design and implementation of the mobile robot designed and built for the purpose of this research. The rest of the paper is organized as follows. Section 2 develops the discrete-time dynamic model of the WMR. This is then used in the neuro-adaptive dynamic controller revisited in Sect. 3. Section 4 outlines the experimental setup and the related design and implementation issues. Experimental results are presented and compared in Sect. 5, which is followed by a brief conclusion in Sect. 6.

2   Modelling

The differentially driven WMR considered in this paper is depicted in Fig. 1. The passive wheels are ignored and the driving wheels are assumed to roll without slipping. The robot state vector is given by $q \triangleq [x \ y \ \phi \ \theta_r \ \theta_l]^T$, where (x, y) is the Cartesian co-ordinate of P_o, ϕ is the robot's orientation with reference to the xy frame, and θ_r, θ_l are the angular displacements of the right and left driving wheels respectively. The *pose* of the robot refers to $\boldsymbol{p} \triangleq [x \;\; y \;\; \phi]$.

2.1   Kinematics

The kinematic model of this WMR, detailed in [1], is given by:

$$\dot{\boldsymbol{q}} = \boldsymbol{S}(\boldsymbol{q})\boldsymbol{v} \,, \tag{1}$$

where the velocity vector $\boldsymbol{v} \triangleq [v_r \;\; v_l]^T \triangleq [\dot{\theta}_r \;\; \dot{\theta}_l]^T$, and $\boldsymbol{S} = \begin{bmatrix} \frac{r}{2}\cos\phi & \frac{r}{2}\cos\phi \\ \frac{r}{2}\sin\phi & \frac{r}{2}\sin\phi \\ \frac{r}{2b} & -\frac{r}{2b} \\ 1 & 0 \\ 0 & 1 \end{bmatrix}$.

2.2   Dynamics

The equations of motion of this WMR are given by:

$$\boldsymbol{M}(\boldsymbol{q})\ddot{\boldsymbol{q}} + \boldsymbol{V}(\dot{\boldsymbol{q}},\boldsymbol{q})\dot{\boldsymbol{q}} + \boldsymbol{F}(\dot{\boldsymbol{q}}) = \boldsymbol{E}(\boldsymbol{q})\boldsymbol{\tau} - \boldsymbol{A}^T(\boldsymbol{q})\boldsymbol{\lambda} \,, \tag{2}$$

where $\boldsymbol{M}(\boldsymbol{q})$ is the inertia matrix, $\boldsymbol{V}(\dot{\boldsymbol{q}},\boldsymbol{q})$ is the centripetal and Coriolis matrix, $\boldsymbol{F}(\dot{\boldsymbol{q}})$ is a vector of frictional forces, $\boldsymbol{E}(\boldsymbol{q})$ is the input transformation matrix, $\boldsymbol{\tau}$ is the torque vector, and $\boldsymbol{A}^T(\boldsymbol{q})\boldsymbol{\lambda}$ is the vector of constraint forces.

Deriving the WMR dynamics, requires the elimination of the kinematic constraints $\boldsymbol{A}^T(\boldsymbol{q})\boldsymbol{\lambda}$ from (2). This is detailed in [7, 1], and yields

$$\bar{\boldsymbol{M}}\dot{\boldsymbol{v}} + \bar{\boldsymbol{V}}(\dot{\boldsymbol{q}})\boldsymbol{v} + \bar{\boldsymbol{F}}(\dot{\boldsymbol{q}}) = \boldsymbol{\tau} \,, \tag{3}$$

where:

$$\bar{\boldsymbol{M}} = \begin{bmatrix} \frac{r^2}{4b^2}(mb^2 + I) + I_w & \frac{r^2}{4b^2}(mb^2 - I) \\ \frac{r^2}{4b^2}(mb^2 - I) & \frac{r^2}{4b^2}(mb^2 + I) + I_w \end{bmatrix} \,, \quad \bar{\boldsymbol{V}}(\dot{\boldsymbol{q}}) = \begin{bmatrix} 0 & \frac{m_c r^2 d\dot{\phi}}{2b} \\ \frac{m_c r^2 d\dot{\phi}}{2b} & 0 \end{bmatrix} \,,$$

$\bar{\boldsymbol{F}}(\dot{\boldsymbol{q}}) = \boldsymbol{S}^T(\boldsymbol{q})\boldsymbol{F}(\dot{\boldsymbol{q}})$, $I = (I_c + m_c d^2) + 2(I_m + m_w b^2)$, and $m = m_c + 2m_w$. To account for the fact that the controller is finally implemented on a digital computer, the continuous-time dynamics (3) are discretized through a first order forward Euler approximation with a sampling interval of T seconds, resulting in

$$\boldsymbol{v}_k - \boldsymbol{v}_{k-1} = \boldsymbol{f}_{k-1} + \boldsymbol{G}_{k-1}\boldsymbol{\tau}_{k-1} \,, \tag{4}$$

where the subscript integer k denotes that the corresponding variable is evaluated at time kT seconds, and vector $\boldsymbol{f}_{k-1}$ and matrix $\boldsymbol{G}_{k-1}$, which together encapsulate the WMR dynamics, are given by

$$\boldsymbol{f}_{k-1} = -T\bar{\boldsymbol{M}}_{k-1}^{-1}\left(\bar{\boldsymbol{V}}_{k-1}\boldsymbol{v}_{k-1} + \bar{\boldsymbol{F}}_{k-1}\right), \quad \boldsymbol{G}_{k-1} = T\bar{\boldsymbol{M}}_{k-1}^{-1} \,. \tag{5}$$

The control input $\boldsymbol{\tau}$ is assumed to remain constant over a sampling interval of T seconds, which is chosen low enough for the Euler approximation error to be negligible.

3   Control Scheme

The complete neuro-adaptive control system detailed in [1], and briefly revisited in this section is depicted in Fig. 2. Some variables in this figure are defined later in the article. At this point one should particularly note the modular architecture which enables the kinematic and dynamic control modules to be treated separately [7]. The kinematic controller computes the desired wheel velocities in order to minimize the robot tracking or stabilization error, according to the task at hand[1]. The cascaded dynamic controller, which in the case of this paper is neuro-adaptive, ensures that the robot truly tracks these velocities, by determining the torques required at the wheels.

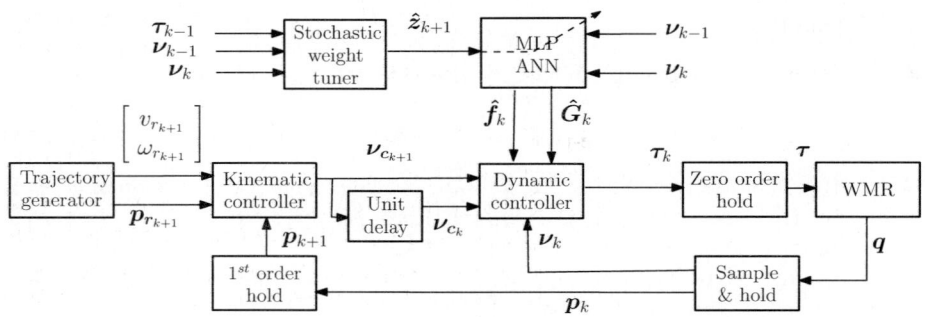

Fig. 2. MLP adaptive dynamic control scheme

3.1   Kinematic Control

To address the trajectory tracking problem we employ a discrete-time version of the trajectory tracking controller originally proposed in [9] and given by:

$$
v_{ck} = \begin{bmatrix} \frac{1}{r} & \frac{b}{r} \\ \frac{1}{r} & -\frac{b}{r} \end{bmatrix} \begin{bmatrix} v_{rk}\cos e_{3k} + k_1 e_{1k} \\ \omega_{rk} + k_2 v_{rk} e_{2k} + k_3 v_{rk} \sin e_{3k} \end{bmatrix} , \tag{6}
$$

where v_{ck} is the wheel velocity command vector computed by the kinematic controller, k_1, k_2, and k_3 are *positive* design parameters, v_{rk} and ω_{rk} are the translational and angular *reference* velocities respectively, and e_{1k}, e_{2k}, e_{3k} make up the tracking error vector defined as

$$
\boldsymbol{e}_k \triangleq \begin{bmatrix} e_{1k} \\ e_{2k} \\ e_{3k} \end{bmatrix} \triangleq \begin{bmatrix} \cos\phi_k & \sin\phi_k & 0 \\ -\sin\phi_k & \cos\phi_k & 0 \\ 0 & 0 & 1 \end{bmatrix} (p_{rk} - \boldsymbol{p}_k) , \tag{7}
$$

where $p_{rk} \triangleq [x_{rk}\ y_{rk}\ \phi_{rk}]^T$ denotes the reference pose vector.

---

[1] In this paper only the trajectory tracking problem is considerd.

3.2 Dynamic Functional-Adaptive Control

Robot Dynamics ANN Estimator

A two-layer MLP ANN with a sigmoidal hidden layer and a linear output layer is employed for the real-time approximation of the vector of nonlinear functions $\boldsymbol{f}_{k-1}$, the estimate of which is denoted by $\hat{\boldsymbol{f}}_{k-1}$ and given by

$$\hat{\boldsymbol{f}}_{k-1} = \begin{bmatrix} \boldsymbol{\phi}^T(\boldsymbol{x}_{k-1},\hat{\boldsymbol{a}}_k)\hat{\boldsymbol{w}}_{1k} \\ \boldsymbol{\phi}^T(\boldsymbol{x}_{k-1},\hat{\boldsymbol{a}}_k)\hat{\boldsymbol{w}}_{2k} \end{bmatrix}, \tag{8}$$

in the light of the following definitions and assumptions: [2]

1. $\boldsymbol{x}_{k-1}$ is the ANN input vector, and is set to $[\boldsymbol{v}_{k-1}\ \ 1]$.
2. $\boldsymbol{\phi}$ is the sigmoidal activation vector, whose ith element is given by $\phi_i = 1/\left(1+\exp(-\boldsymbol{s}_i^T\boldsymbol{x})\right)$, where $\boldsymbol{s}_i$ is the parameter vector of the ith neuron.
3. The sigmoidal parameter vectors are grouped in $\hat{\boldsymbol{a}}_k \triangleq \begin{bmatrix} \hat{\boldsymbol{s}}_{1_k}^T & \cdots & \hat{\boldsymbol{s}}_{L_k}^T \end{bmatrix}^T$.
4. L denotes the number of neurons in the network.
5. $\hat{\boldsymbol{w}}_{j_k}$ represents the synaptic weight vector of the connection between the neurons and the jth output element of the network, estimated in real-time.

Since $\boldsymbol{G}_{k-1}$ is a symmetric state-independent matrix, its estimate does not require an ANN, but is simply denoted by the parameters

$$\hat{\boldsymbol{G}}_{k-1} = \begin{bmatrix} \hat{g}_{1k-1} & \hat{g}_{2k-1} \\ \hat{g}_{2k-1} & \hat{g}_{1k-1} \end{bmatrix}, \tag{9}$$

where $\hat{g}_{1k-1}$ and $\hat{g}_{2k-1}$ are the unknown elements in $\hat{\boldsymbol{G}}_{k-1}$.

All the unknown parameters requiring estimation are grouped in a single vector $\hat{\boldsymbol{z}}_k \triangleq \begin{bmatrix} \hat{\boldsymbol{w}}_{1k}^T & \hat{\boldsymbol{w}}_{2k}^T & \hat{\boldsymbol{a}}_k^T & \hat{g}_{1k-1} & \hat{g}_{2k-1} \end{bmatrix}^T$. The true value of $\hat{\boldsymbol{z}}_k$ denoted by $\boldsymbol{z}_k^*$ is unknown, and in this work we opt to treat it as a random variable. Estimates $\hat{\boldsymbol{f}}_{k-1}$ and $\hat{\boldsymbol{G}}_{k-1}$ are employed in rewriting the WMR dynamic model (4) in the following state-space form:

$$\begin{aligned} \boldsymbol{z}_{k+1}^* &= \boldsymbol{z}_k^* \\ \boldsymbol{y}_k &= \boldsymbol{h}\left(\boldsymbol{x}_{k-1},\boldsymbol{\tau}_{k-1},\boldsymbol{z}_k^*\right) + \boldsymbol{\varepsilon}_k , \end{aligned} \tag{10}$$

where the vector field $\boldsymbol{h} = \hat{\boldsymbol{f}}_{k-1}(\boldsymbol{x}_{k-1},w_{1k}^*,w_{2k}^*,\boldsymbol{a}_k^*) + \hat{\boldsymbol{G}}_{k-1}(g_{1k-1}^*,g_{2k-1}^*)\boldsymbol{\tau}_{k-1}$, the measured output is denoted by $\boldsymbol{y}_k = \boldsymbol{v}_k - \boldsymbol{v}_{k-1}$, and $\boldsymbol{\varepsilon}_k$ is an independent zero-mean white Gaussian process with covariance matrix $\boldsymbol{R}_\varepsilon$, which accounts for measurement uncertainty. Since $\boldsymbol{h}$ is a nonlinear function of the unknown *random vector* $\boldsymbol{z}_k^*$, the extended Kalman filter (EKF) is used for its real-time stochastic estimation:

$$\begin{aligned} \hat{\boldsymbol{z}}_{k+1} &= \hat{\boldsymbol{z}}_k + \boldsymbol{K}_k \boldsymbol{i}_k \\ \boldsymbol{P}_{k+1} &= \boldsymbol{P}_k - \boldsymbol{K}_k \nabla_{hk} \boldsymbol{P}_k , \end{aligned} \tag{11}$$

where ∇_{hk} denotes the Jacobian matrix of $\boldsymbol{h}$ with respect to $\boldsymbol{z}_k^*$ evaluated at $\hat{\boldsymbol{z}}_k$, $\boldsymbol{P}_k$ is estimate's covariance matrix, $\boldsymbol{K}_k = \boldsymbol{P}_k \nabla_{hk}^T \left(\nabla_{hk}\boldsymbol{P}_k\nabla_{hk}^T + \boldsymbol{R}_\varepsilon\right)^{-1}$ is the Kalman gain matrix, and $\boldsymbol{i}_k = \boldsymbol{y}_k - \boldsymbol{h}\left(\boldsymbol{x}_{k-1},\boldsymbol{\tau}_{k-1},\hat{\boldsymbol{z}}_k\right)$ is the innovations vector.

---

[2] The ˆ and * notations denote *estimates* and *optimal* parameters respectively.

Using (8), (9), and the definition of h it can be shown that:

$$\nabla_{hk} = \begin{bmatrix} \boldsymbol{\phi}_{k-1}^T & \mathbf{0}^T & \cdots \hat{w_{1,i}}(\phi_i - \phi_i^2)\boldsymbol{x}^T \cdots & \tau_{rk-1} & \tau_{lk-1} \\ \mathbf{0}^T & \boldsymbol{\phi}_{k-1}^T & \cdots \hat{w_{2,i}}(\phi_i - \phi_i^2)\boldsymbol{x}^T \cdots & \tau_{lk-1} & \tau_{rk-1} \end{bmatrix},$$

where: $i = 1, \ldots, L$ and $\hat{w}_{j,i}$ denotes the ith element of the synaptic weight vector $\hat{\boldsymbol{w}}_{j_k}$ which connects the neurons to the jth network output; Notation-wise $\boldsymbol{\phi}_{k-1}$ implies that the activation function is evaluated for $\boldsymbol{x}_{k-1}$ and $\hat{\boldsymbol{a}}_k$; $\mathbf{0}$ denotes a zero vector having the same length as $\boldsymbol{\phi}_{k-1}$; ϕ_i and $\boldsymbol{x}$ correspond to time instant $(k-1)$; $\tau_{r_{k-1}}$, $\tau_{l_{k-1}}$ are the first and second elements of the torque vector $\boldsymbol{\tau}_{k-1}$ respectively.

Control Law

At each control iteration $\hat{\boldsymbol{f}}_k$ and $\hat{\boldsymbol{G}}_k$, generated by the MLP stochastic function estimator, are used in the following control law to ensure that the robot velocity vector $\boldsymbol{v}_k$ tracks the velocity command vector issued by the kinematic controller $\boldsymbol{v}_{ck}$:

$$\boldsymbol{\tau}_k = \hat{\boldsymbol{G}}_k^{-1}\left(\boldsymbol{v}_{ck+1} - \boldsymbol{v}_k - \hat{\boldsymbol{f}}_k + k_d\left(\boldsymbol{v}_{ck} - \boldsymbol{v}_k\right)\right), \tag{12}$$

where the design parameter $-1 < k_d < 1$. If we neglect the negligibly small inherent ANN approximation error (justified due to the *Universal Approximation Theorem* of ANNs) and $\boldsymbol{\varepsilon}_k$, (12) yields the following closed-loop dynamics: $\boldsymbol{v}_{k+1} = \boldsymbol{v}_{ck+1} + k_d\left(\boldsymbol{v}_{ck} - \boldsymbol{v}_k\right)$, clearly indicating that $|\boldsymbol{v}_{c_k} - \boldsymbol{v}_k| \to 0$ as $k \to \infty$.

4   Experimental Setup

The neuro-adaptive control scheme revisited in Sect. 3, was implemented successfully for the first time on a physical WMR, named NeuroBot and pictured in Fig. 3. This robot was recently designed and built by the authors to serve as a testbed for the development and validation of neuro-control algorithms.

NeuroBot is a differentially driven WMR. Each of the two 125mm diameter solid-rubber drive wheels, is independently driven by a 70W, 24V permanent magnet dc motor from *maxon motor*, which is equipped with a 113:1 reduction gearbox, and a 500 pulses per revolution incremental encoder. Each of the two motors is driven via the LMD18200 H-Bridge IC from *National Semiconductor*, which is controlled by a 20kHz pulse width modulation reference signal. The instantaneous current in each motor is measured using the HX 03-P/SP2 Hall effect current transducer from *LEM*, and filtered by a 4th-order continuous-time Bessel lowpass filter tuned for a corner frequency of 2kHz, and implemented using the MAX275 continuous-time filter IC from *Maxim Dallas Semiconductors*. NeuroBot is powered by four 12V, 9Ah sealed lead acid batteries (RM 12-9 HR) from *REMCO*.

The algorithms controlling NeuroBot are all implemented on a *MicroAutoBox* system from *dSPACE*. The *MicroAutoBox* is a compact stand-alone prototyping unit designed specifically for the rapid-prototyping of computationally demanding real-time control systems, typically requiring a number of analog/digital input and output channels to interface with ease with both sensors and actuators. A digital pole-placement torque

Fig. 3. NeuroBot: the experimental WMR

controller with integral action was designed and implemented in software to account for the motor dynamics. This inner torque control loop uses the current measurement as feedback and issues motor voltage commands to the motors. This ascertains that the actual torques at the wheels, which are proportional to currents in the motors, track precisely those issued by the outer loop control law (12). Naturally, this cascade approach imposes that the inner loop operates at a rate which is much faster than that of the outer loop. The sampling rates for the inner and outer loops where chosen to be 10kHz and 200Hz respectively.

A desktop computer is used to implement the control algorithms in *Simulink*® using the system blocks provided by the *dSpace Real-Time Interface*. *Real-Time Workshop*® is then used to automatically generate the required code which is then downloaded to the *MicroAutoBox* (which features also non-volatile memory) via the *dSpace Link Board* installed in the computer. From this point onwards the *MicroAutoBox* can be disconnected from the computer and the control code runs entirely on the *MicroAutoBox* which employs a multitasking approach to service each of the the two control loops in real-time. Vital information about the real-time execution of each task running on the *MicroAutoBox*, such as sampling times, priorities and execution times, can also be monitored via *ControlDesk*, developed for this purpose by *dSPACE*.

5   Experimental Results

NeuroBot was used to test and validate the MLP adaptive control scheme, reviewed in Sect. 3, for the first time on a real mobile robot. Moreover, the proposed adaptive controller was compared with its nonadaptive counterpart implemented via (12) but with $\hat{f}_{k-1}$ and $\hat{G}_{k-1}$ replaced by f_{k-1} and G_{k-1}.

This nonadaptive controller requires that the WMR parameters are known. For this reason NeuroBot was weighed and measured accordingly and the resulting parameter values were used to tune the nonadaptive controller. These parameters are: $b = 0.2295$m, $r = 0.0625$m, $d = 0$m, $m_c = 22$kg, $m_w = 1$kg, $I_c = 0.6320$kgm^2, $I_w = 0.002$kgm^2, and $I_m = 0.0029$kgm^2. Moreover, viscous friction was included by setting $F(\dot{q}) = F_c\dot{q}$, where F_c is a diagonal matrix of coefficients, with nominal diagonal values set to $[0.001, 0.001, 0.001, 0.18, 0.18]$. In contrast, the neuro-adaptive controller presented in this paper does not require any preliminary knowledge about the robot dynamics. Consequently the initial network parameter vector $\hat{z}_0$ was generated randomly. The MLP ANN contained 10 neurons (L=10).

A number of experimental results acquired from a typical experiment on NeuroBot as detailed above are presented in Fig. 4. Plot (a) depicts NeuroBot tracking a demanding reference trajectory (characterized by high linear and angular accelerations) for non-zero initial tracking error, when it is being controlled by the neuro-adaptive MLP scheme. It is clear that NeuroBot swiftly adapts to its own dynamics and simultaneously converges to the the reference trajectory. The WMR keeps tracking the trajectory with high precision for the rest of the experiment. This plot on its own validates experimentally the neuro-adaptive scheme presented in this paper for ultra-precise trajectory tracking. Plots (b) and (c) show the tracking errors $x_r - x$, $y_r - y$ and $\phi_r - \phi$ against time, corresponding to the same trajectory shown in Plot (a). From these plots it is also clear that the trajectory tracking errors are all reduced to zero in a few seconds and maintained there with unquestionable performance. Plot (d) shows the actual and reference angular wheel velocities v_r and v_l along the trajectory. The actual velocities are practically superimposed on the corresponding references. This implies that the adaptive dynamic controller achieves the wheel velocities requested by the kinematic controller with great precision, which ultimately leads to the good trajectory tracking performance depicted in the previous plots. Plots (e) and (f) compare the adaptive controller with its tuned and untuned nonadaptive counterparts, by depicting the position error norm $\sqrt{((x_r - x)^2 + (y_r - y)^2)}$ and the orientation error respectively for the three controllers following the same trajectory.

The tuned nonadaptive controller refers to the nonadaptive controller tuned with the true robot parameters. The untuned nonadaptive controller refers to the same controller but with $m_c = 10$kg, i.e. this controller believes that the robot weighs half its real weight. This scenario was included to examine the effects of uncertain and/or time-varying robot dynamics on the performance of nonadaptive dynamic controllers. From the two plots it is clear that the performance of the nonadaptive controller deteriorates from the tuned to the untuned case. This indicates the incapability of nonadaptive controllers in handling misinformation about the robot dynamics. More impressive is the fact that the adaptive controller outperforms even the tuned nonadaptive controller. We attribute this to the fact that perfect modelling is practically impossible, and since the adaptive controller uses no predefined dynamic model of the WMR, since it learns it in real-time, it does not suffer from the consequences of unmodelled dynamics and inexact model parameters.

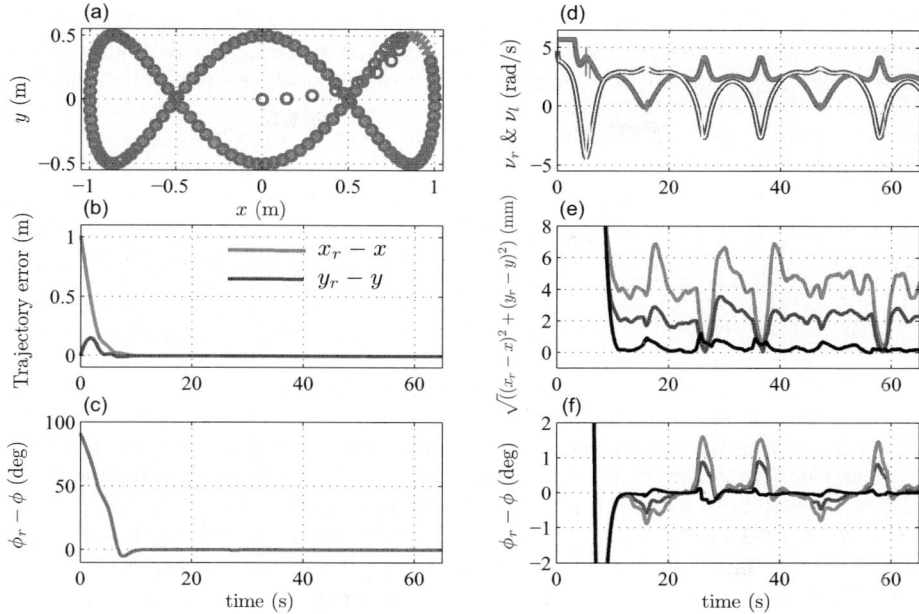

Fig. 4. (a): reference ($\times$) & actual trajectories ($\bigcirc$); (b): position errors; (c): orientation error; (d): angular velocities - right wheel (red), left wheel (yellow) & their references (green & blue respectively); (e): adaptive (black), tuned nonadaptive (blue), untuned nonadaptive (red); (f): as for (e) but orientation errors.

The inner torque control loop, operating at a sampling rate of 10kHz, takes approximately 14μs to execute. The outer speed loop, with a sampling rate of 200Hz, executes in 700μs and 350μs for the adaptive and nonadaptive cases respectively. This implies that the adaptive controller requires double the execution time of its nonadaptive counterpart. This was expected since the adaptive algorithm is more computationally demanding because it includes estimation as opposd to a nonadaptive controller. Nonetheless, the execution time of the neuro-adaptive controller is still as little as 14% of the total sampling period, implying that commercially available hardware is well endowed to handle the increased level of computational load brought about by real-time neuro-adaptive control.

6   Conclusion

The contribution of this paper comprises the experimental validation of a MLP adaptive dynamic control scheme, originally proposed in [1] for the trajectory tracking of mobile robots. In addition the recently designed robotic testbed NeuroBot is introduced, and the associated implementation issues are briefly discussed. The experimental results presented in this paper not only validate the employed neuro-adaptive control scheme in practice, but also demonstrate the great improvements in performance over non-adaptive schemes in the face of uncertain and/or time-varying robot dynamics. In addition, NeuroBot proved to be a very good research testbed and will continue to be used for the validation and development of innovative controllers for mobile robots.

References

1. Bugeja, M.K., Fabri, S.G.: Multilayer perceptron adaptive dynamic control for trajectory tracking of mobile robots. In: Proc. 32nd Annual Conf. of the IEEE Ind. Electronics Society (IECON 2006), Paris, France, November 2006, pp. 3798–3803 (2006)
2. Canudas de Wit, C., Khennoul, H., Samson, C., Sordalen, O.J.: Nonlinear control design for mobile robots. In: Zheng, Y.F. (ed.) Recent Trends in Mobile Robots, Robotics and Automated Systems, ch. 5, pp. 121–156, World Scientific (1993)
3. Corradini, M.L., Ippoliti, G., Longhi, S., Michelini, S.: Neural networks inverse model approach for the tracking problem of mobile robots. In: Proc. The 9th International Workshop on Robotics (RAAD 2000), Maribor, Slovenia (June 2000)
4. Corradini, M.L., Orlando, G.: Robust tracking control of mobile robots in the presence of uncertainties in the dynamic model. Journal of Robotic Systems 18(6), 317–323 (2001)
5. D'Amico, A., Ippoliti, G., Longhi, S.: A radial basis function networks apporach for the tracking problem of mobile robots. In: Proc. IEEE/ASME International Conference on Advanced Intelligent Mechatronics, Como, Italy, pp. 498–503 (2001)
6. Dixon, W.E., Dawson, D.M., Zergeroglu, E., Behal, A.: Adaptive tracking control of a wheeled mobile robot via an uncalibrated camera system. 31(3), 341–352 (2001)
7. Fierro, R., Lewis, F.L.: Control of a nonholonomic mobile robot: Backstepping kinematics into dynamics. In: Proc. IEEE 34th Conference on Decision and Control (CDC 1995), New Orleans, LA, December 1995, pp. 3805–3810 (1995)
8. Fierro, R., Lewis, F.L.: Control of a nonholonomic mobile robot using neural networks. IEEE Trans. Neural Netw. 9(4), 589–600 (1998)
9. Kanayama, Y., Kimura, Y., Miyazaki, F., Noguchi, T.: A stable tracking control method for an autonomous mobile robot. In: Proc. IEEE International Conference of Robotics and Automation, Cincinnati, OH, May 1990, pp. 384–389 (1990)
10. Oubbati, M., Schanz, M., Levi, P.: Kinematic and dynamic adaptive control of a nonholonomic mobile robot using RNN. In: Proc. IEEE Symposium on Computational Intelligence in Robotics and Automation (CIRA 2005), Helsinki, Finland (June 2005)
11. Wang, T.-Y., Tsai, C.-C.: Adaptive trajectory tracking control of a wheeled mobile robot via lyapunov techniques. In: Proc. 30th Annual Conference of the IEEE Industrial Electronics Society, Busan, Korea, November 2004, pp. 389–394 (2004)

Path Planning and Tracking Control for an Automatic Parking Assist System

Emese Szádeczky-Kardoss and Bálint Kiss

Dept. of Control Engineering and Information Technology Budapest University of Technology and Economics, Hungary
szadeczky@iit.bme.hu, bkiss@iit.bme.hu

Summary. This paper presents two major components of an automatic parking assist system (APAS). The APAS maps the environment of the vehicle and detects the existence of accessible parking place where the vehicle can park into. The two most important tasks the APAS must then realize are the design of a feasible path geometry and the tracking of this reference in closed loop such that the longitudinal velocity of the vehicle is generated by the driver and the controller determines the steering wheel angle which is realized by an Electronic Power Assist Steering (EPAS). We present in details the solution of the continuous curvature path planning problem and the time-scaling based tracking controller. These components are part of an implemented APAS on a commercial passenger car including an intelligent EPAS.

Keywords: automatic parking assist system, continuous curvature path planning, tracking control, time-scaling.

1 Introduction

An important goal of automatic vehicle control is to improve safety and driver's comfort. The APASs provide this for parking maneuvers which are usually performed at low velocities. The APAS collects first information about the environment of the vehicle (position of the obstacles) which are necessary to find a parking lot and to complete a safe parking maneuver [11].

One may distinguish fully or semi-automated APASs according to the driver's involvement during the maneuver. In the fully automatic case the APAS influences both the steering angle and the longitudinal velocity of the car. In the case of vehicles without automatic gear the semi-automatic APAS is the only available option. In this case the driver needs to generate the longitudinal velocity of the car with an appropriate management of the pedals, while the APAS controls the steering wheel.

There exists now several APASs on the market. The Aisin Seiki Co. Ltd. has developed a semi-automatic parking system for Toyota [1] where a camera observes the environment of the car. The Evolve project resulted a fully automatic APAS for a Volvo type vehicle based on ultrasonic sensor measurements [5]. The Volkswagen Touran may be ordered with an option that also assists parking maneuvers [17]. This solution also uses ultrasonic sonars for the semi-automatic parking maneuvers. We do not address in this paper the map making and parking lot detection procedures and the signal processing problems related to the map making from distance measurement, and the position

H. Bruyninckx et al. (Eds.): European Robotics Symposium 2008, STAR 44, pp. 175–184, 2008.
springerlink.com

and orientation estimation of the car. The first task is realized in our setup using ultrasonic sensors and the second problem is solved by the use of the angular velocity data of the ABS sensors of the car.

To design an APAS, a mathematical model of the car has to be given. Several vehicle models are presented in the literature including kinematic and dynamic models. Let us suppose that the kinematic models such as the ones reported in [4, 13, 2] describe in a satisfactory way the behavior of the vehicle at low velocities where parking maneuvers are executed.

If the model of the vehicle is known, then the path planning method and the tracking controller algorithm can be designed based on the motion equations. To plan the motion one can use deterministic [9] or probabilistic methods [15, 10] as well. To get a continuous curvature path special curves should be used [7]. Algorithms based on soft computing methods can also calculate the reference path [18, 3].

In the case of the semi-automatic systems the design of the tracking controller is more involved than in the fully automatic case. The tracking controller in the fully automatic case may influence the behavior of the car using two inputs whereas one losses one input, namely the longitudinal velocity in the semi-automated case where the car velocity is generated by the driver hence the velocity profile during the execution of the parking maneuver may be considerably different from the one used for the path planning. This problem is addressed by [8] using time-scaling.

Our goal was to develop a parking assist system which can operate both in fully and semi-automatic mode in three different parking situations (parking in a lane, parking in a row, and diagonal parking [12]).

In this paper we present first the components of the system (Sect. 2). The kinematic vehicle model is described in Sect. 3 and the path planning and tracking controller algorithms are detailed in Sects. 4–6. Section 7 presents some results based on tests on a real car and a short summary concludes the paper.

2   Components of the System

To ensure autonomous behavior, several tasks have to be solved: the system should be able to detect obstacles in its environment; it has to measure or estimate its position and orientation; the reference motion has to be planned, and finally, this reference should be tracked as accurately as possible. These tasks are performed by separate interconnected subsystems which are depicted in Fig. 1.

The ABS sensors of the vehicle can detect the displacement of the wheels of the car. Based on these data, an estimator calculates the actual position and orientation of the car in a fixed word coordinate frame. This estimated state is used by the mapping and controller modules. To draw a map, additional data are also required about the environment. Ultrasonic sensors are used to measure the distances to the surrounding obstacles. Based on these distance measurements a map can be created. One may then use simple algorithms to detect accessible parking places (if any) on the map.

During the motion planning a reference path is calculated which connects the initial and the desired final configurations in one step (i.e. without changing the driving direction). In this planning phase some constraints (e.g. the non-holonomic behavior

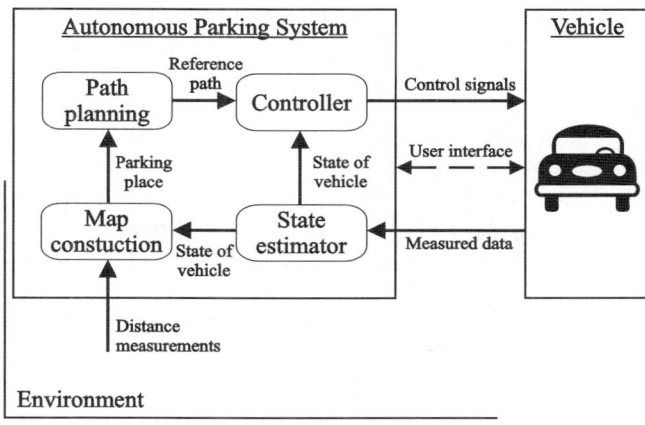

Fig. 1. Components of the Automatic Parking Assist System

described by the model, collision avoidance, maximal values of the actuator signals) have to be taken into consideration. Finally, the tracking control algorithm is used to track the reference path.

3 Kinematic Vehicle Model

Both the path planning and tracking control algorithms use the kinematic model of the vehicle. To calculate the geometry of the path a slightly extended model is used in order to ensure the continuous curvature property.

As it is usual in the literature, the reference point of the car is the midpoint of the rear axle denoted by R. The configuration of the vehicle (q) is described by four state variables: position of the reference point R (x, y), the orientation of the car ψ, and the curvature (κ), which is the inverse of the turning radius. Supposing that the Ackermann steering assumptions hold true, the motion of the vehicle can be described by the kinematic model of the bicycle fitted on the longitudinal symmetry axis of the car (see Fig. 2):

$$
\dot{q} = \begin{bmatrix} \dot{x} \\ \dot{y} \\ \dot{\psi} \\ \dot{\kappa} \end{bmatrix} = \begin{bmatrix} \cos\psi \\ \sin\psi \\ \kappa \\ 0 \end{bmatrix} v + \begin{bmatrix} 0 \\ 0 \\ 0 \\ 1 \end{bmatrix} \sigma .
\tag{1}
$$

The longitudinal velocity of the car is denoted by v and the velocity of the change in curvature is given by σ. Let us denote the axle space by b. Then we get the following relationship between the curvature and its derivative:

$$
\kappa = \frac{\tan\delta}{b}, \quad \sigma = \dot{\kappa} = \frac{\dot{\delta}}{b\cos^2\delta},
\tag{2}
$$

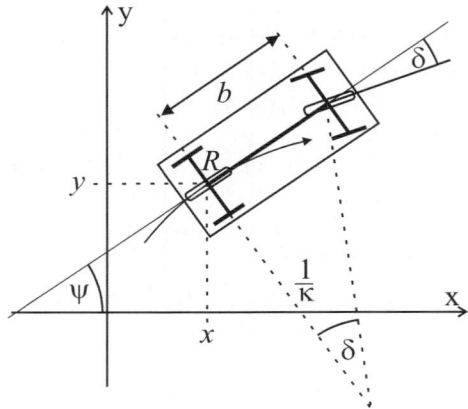

Fig. 2. Kinematic model of the vehicle in the $x - y$ plane

where δ is the angle between the front wheel of the bicycle and the longitudinal axis of the car. Both the curvature and the curvature derivative are limited, i.e.

$$|\kappa| \leq \kappa_{max} , \quad |\sigma| \leq \sigma_{max} . \tag{3}$$

To design the tracking controller we use a simpler version of (1) without the curvature.

$$\dot{q} = \begin{bmatrix} \dot{x} \\ \dot{y} \\ \dot{\psi} \end{bmatrix} = \begin{bmatrix} \cos \psi \\ \sin \psi \\ \frac{\tan \delta}{b} \end{bmatrix} v . \tag{4}$$

In case of the semi-autonomous system the controller cannot consider the longitudinal velocity of the vehicle as a system input since it is generated by the driver and cannot be influenced by the controller. In this case we denote this driver velocity by v_d. (We suppose, that this v_d velocity can be measured or well estimated.)

$$\dot{q} = \begin{bmatrix} \dot{x} \\ \dot{y} \\ \dot{\psi} \end{bmatrix} = \begin{bmatrix} \cos \psi \\ \sin \psi \\ \frac{\tan \delta}{b} \end{bmatrix} v_d . \tag{5}$$

4   Path Planning

The task of the path planning method is to determine the geometry of the reference path. Our goal is to have a reference path with continuous curvature which avoids stopping the car while steering the front wheels. Further constraints are the maximal limit on the curvature and on its time derivative, since the vehicle is not able to turn with arbitrary small turning radius and the change of the turning radius (or the curvature) is also limited by the applied EPAS system.

To plan such a path which fulfills the above mentioned constraints we use three different path primitives namely straight lines, circular segments, and continuous curvature turns (CC turns) [7]. The curvature is zero along a straight line, in a circular segment it has a nonzero constant value, which does not exceed a given maximum limit, and the curvature varies linearly with the arc length in the CC turns.

The geometry of the straight lines or circular segments can be described easily if their parameters are known (e.g. lengths, turning radius). The calculation is more complicated in the case of the CC turns. If the motion is started from the initial configuration $q_0 = [0,0,0,0]^T$, the velocity profile is constant (e.g. $v = 1$), and the curvature changes with the allowable maximum σ_{max} value, then the configurations in a CC turn can be described by the following equations:

$$x = \sqrt{\pi/\sigma_{max}} C_F \left(\sqrt{\kappa^2/(\pi\sigma_{max})} \right) , \tag{6}$$

$$y = \sqrt{\pi/\sigma_{max}} S_F \left(\sqrt{\kappa^2/(\pi\sigma_{max})} \right) , \tag{7}$$

$$\psi = \kappa^2/(2\sigma_{max}) , \tag{8}$$

$$\kappa = \sigma_{max}t , \tag{9}$$

where C_F and S_F denote the Fresnel integrals, which cannot be given in a closed form. Differentiating (6–9) we can see that these equations fulfill the kinematic model of the vehicle given in (1).

Using simple mathematical operations for (6–9) we can get the configurations for motions from different initial configurations with arbitrary constant velocity. If the values of the Fresnel integrals in (6–7) can be calculated beforehand, then the remaining computations can be performed in real time [16].

Now we describe how to use the three path primitives to get the reference trajectory for parking in a lane. In this case we use seven segments to put the path together (see Fig. 3). Without loss of generality we suppose that the vehicle starts the motion backwards from the $q_s = [0,0,0,\kappa_s]^T$ initial configuration. First it moves along a CC turn, until the maximal curvature (or the minimal turning radius) is reached. Then it turns along a circle with the minimal turning radius, the corresponding turning angle is denoted by φ_1. After the next CC turn the curvature becomes 0, and the car turns in the opposite direction with a CC turn, a circular motion with turning angle φ_2 and one more CC turn, such that the curvature becomes 0 again. The path ends with a straight line segment whose length is denoted by l.

Such a path has three parameters: the turning angles in the circular segments (φ_1, φ_2) and the length of the straight line (l). If these parameters are known, the geometry of the path can be calculated and the goal configuration can be determined:

$$q_g = \begin{bmatrix} x_g \\ y_g \\ \psi_g \\ \kappa_g \end{bmatrix} = \begin{bmatrix} f_1(\varphi_1, \varphi_2, l) \\ f_2(\varphi_1, \varphi_2, l) \\ f_3(\varphi_1, \varphi_2) \\ 0 \end{bmatrix} . \tag{10}$$

The values of the path parameters can be determined from the desired goal configuration (q_d) by solving the $q_d = q_g$ equation. So the full reference path can be calculated, and the reference values for the configuration ($q_{ref} = [x_{ref}, y_{ref}, \psi_{ref}, \kappa_{ref}]^T$) and its time derivatives can be determined.

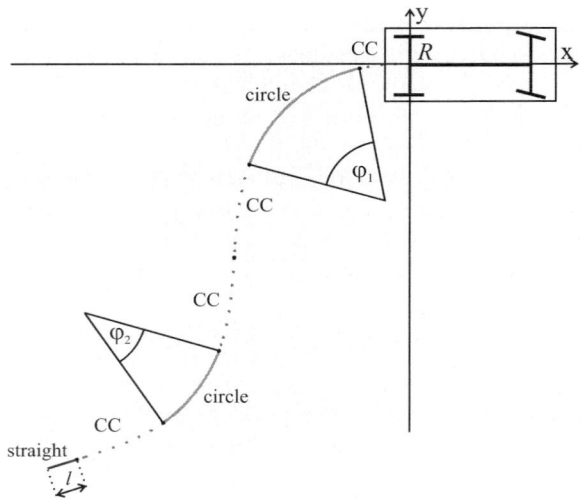

Fig. 3. Reference path for parking in a lane

5   Time-Scaling

During the path planning we considered a preliminary reference velocity profile and to avoid involved calculations we supposed that it is constant (e.g. $v = 1$). Since the driver will generate another velocity profile ($v_d \neq v$), it is enough if one is able to track the geometry of the reference of the designed path and the time distribution along the path will be adapted in real-time to the velocity profile generated by the driver.

In the above equations the states of the configuration q were functions of time t, where $\dot{t} = 1$. In a more general form we have a state equation

$$\dot{q}(t) = f(q(t), u(t), w(t)) , \tag{11}$$

where u is the vector of the inputs and w denotes the external signals. In our case, which is given by (5), we have $u = \delta$ and $w = v_d$.

We introduce a new scaled time denoted by τ such that τ is used to modify the time distribution along the path. We suggest that the relationship between t and τ should not only depend on the states of the car (as it is usually made in the literature [14]), but also on a new external input, denoted by u_s which is the so-called scaling input:

$$\frac{\mathrm{d}t}{\mathrm{d}\tau} = \frac{1}{\dot{\tau}} = g(q, u_s, w) . \tag{12}$$

Using this time-scaling (11) can be expressed with respect to the time τ:

$$q' = \frac{\mathrm{d}q}{\mathrm{d}\tau} = \frac{\mathrm{d}q}{\mathrm{d}t}\frac{\mathrm{d}t}{\mathrm{d}\tau} = g(q, u_s, w)f(q, u, w) . \tag{13}$$

The prime denotes differentiation according to τ, hence $\tau' = 1$.

The time-scaling defined in (12) has to satisfy some conditions:

- $\tau(0) = t(0) = 0$, since the original and the scaled trajectories should start from the same initial configuration;
- $\dot{t} > 0$, since time cannot stand or rewind.

During the time-scaling we modify the time distribution along the reference path, which was planned in time τ:

$$x_{ref}(t) = x_{ref}(\tau) , \tag{14}$$
$$\dot{x}_{ref}(t) = x'_{ref}(\tau)\dot{t} , \tag{15}$$
$$\ddot{x}_{ref}(t) = x''_{ref}(\tau)\dot{t}^2 + x'_{ref}(\tau)\ddot{t} . \tag{16}$$

The further derivatives and the other state variables can be calculated in t in a similar way. It can be seen from (14) that the time-scaling does not change the geometry of the reference path, only the velocity and the further derivatives are modified if $\dot{t} \neq 1$.

6   Tracking Control

In this section only some key features of the tracking controller are discussed, the entire algorithm can be found in details in [8]. The literature suggests several solutions [3, 6] to control the two-input kinematic car given in (4). These methods ensure exponential tracking of the reference path. In our one-input case these algorithms cannot be used without modification since our controller cannot influence the velocity of the vehicle. Our idea is to complement the lost velocity input by the time-scaling input as in (13).

For, the following time-scaling function can be used:

$$\frac{dt}{d\tau} = \frac{u_s}{v_d} . \tag{17}$$

In this case the model equation given in (5) which evolves according to t can be transformed using the time-scaling, and we get

$$q' = \begin{bmatrix} x' \\ y' \\ \psi' \end{bmatrix} = \begin{bmatrix} \cos\psi \\ \sin\psi \\ \frac{\tan\delta}{b} \end{bmatrix} u_s . \tag{18}$$

This scaled model has now two inputs (u_s and δ), hence one of the controllers described in the literature can be used for tracking. The selected method will compute δ and u_s according to the tracking error between the real and the scaled reference trajectories. This δ input is used to control the EPAS steering system and the scaling input u_s influences the time-scaling. The time-scaling function and its derivatives, which are required for (14–16), can be calculated using the following relationships, which are based on (17)

$$\tau(t) = \int_0^t \frac{v_d}{u_s} d\vartheta , \quad \tau(0) = t(0) = 0 , \tag{19}$$
$$v_d = \dot{t}u_s , \tag{20}$$
$$\dot{v}_d = \ddot{t}u_s + \dot{t}\dot{u}_s . \tag{21}$$

If the signs of u_s and v_d are the same than the time-scaling function satisfies the $\dot{t} > 0$ condition. If one of the two signals equals 0, the car is not controllable. This occurs at the very beginning and at the end of the motion.

The scheme of the closed loop control is depicted in Fig. 4. First the path planning module calculates the reference path in τ. In the next step this reference is scaled based on the longitudinal velocity of the car v_d, which is generated by the driver, and based on the scaling input u_s, which is calculated by the controller. After the time-scaling we have the scaled reference in t. The controller determines its outputs using the difference between the real and the scaled reference trajectories. So the inputs of the vehicle are the longitudinal velocity v_d generated by the driver, and δ, which is calculated by the controller. The output of the car is the position of the reference point and the orientation.

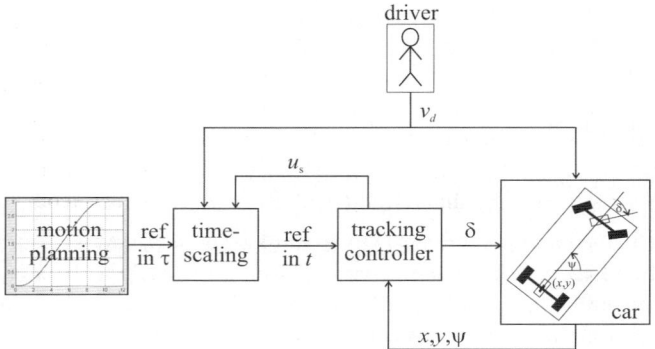

Fig. 4. Scheme of the tracking controller with time-scaling

7   Results

We implemented the presented methods on a Ford Focus type passenger car using an EPAS provided by ThyssenKrupp to realize the steering angle for the front wheels.

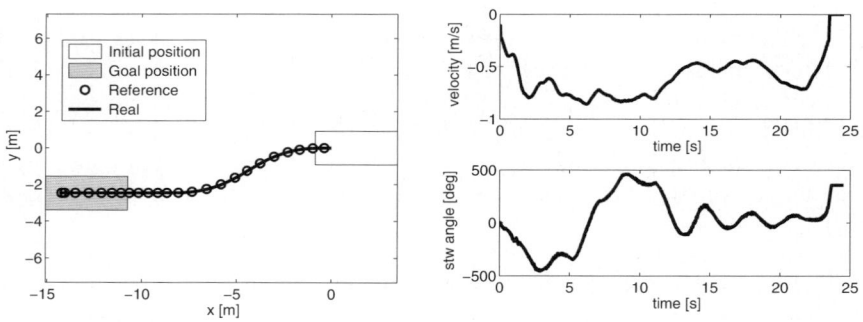

Fig. 5. Results: path in the $x - y$ plane, inputs of the car v_d, δ

Both the path planner and tracking controller were realized on a dSPACE hardware (dedicated Autobox) mounted in the car and connected to the EPAS actuator and to the CAN bus to read the ABS signals.

The presented motion is a continuous curvature backward parking maneuver with $q_s = [0,0,0,0]^T, q_g = [-2.5\,\mathrm{m}, -14.2\,\mathrm{m}, 0, 0]^T$. The results are depicted in Fig. 5. The system is able to track the reference path, such that the velocity of the car is not constant, as it was supposed during the path planning. The system works in different parking maneuvers similarly.

We also studied the performance of the tracking controller in the case of initial errors. We achieved exponential tracking.

8 Conclusions

This paper discussed two components of an APAS. The presented path planning method can calculate a continuous curvature path in real time. The time-scaling tracking controller is able to drive the car along the reference path such that the driver generates the car velocity. The time-scaling function can be calculated from the velocity of the car and from a scaling input, which is calculated by the controller based on the closed loop behavior. The presented algorithms were tested in a real car with encouraging results.

Acknowledgments. The research was partially funded by the Advanced Vehicles and Vehicle Control Knowledge Center under grant RET 04/2004 and by the Hungarian Science Research Fund under grant OTKA K71762.

References

1. AISIN SEIKI Co. Ltd. Intelligent Parking Assist, http://www.aisin.com
2. Boissonnat, J.-D., Cérézo, A., Lebond, J.: Shortest paths of bounded curvature in the plane. Journal of Intelligent and Robotics Systems 11, 5–20 (1994)
3. Cuesta, F., Ollero, A.: Intelligent Mobile Robot Navigation. Springer Tracts in Advanced Robotics, vol. 16. Springer, Heidelberg (2005)
4. Dubins, L.E.: On curves of minimal length with a constraint on average curvature, and with prescribed initial and terminal positions and tangents. American Journal of Mathematics 79(3), 497–516 (1957)
5. Evolve, http://www.ikp.liu.se/evolve
6. Fliess, M., Lévine, J., Martin, P., Rouchon, P.: Flatness and Defect of Nonlinear Systems: Introductory Theory and Examples. International Journal of Control 61(6), 1327–1361 (1995)
7. Fraichard, T., Scheuer, A.: From Reeds and Shepp's to Continuous-Curvature paths. IEEE Transaction on Robotics and Automation 20(6), 1025–1035 (2004)
8. Kiss, B., Szádeczky-Kardoss, E.: Tracking control of the orbitally flat kinematic car with a new time-scaling input. In: Proceedings of the IEEE Conference on Decision and Control, pp. 1969–1974 (2007)
9. Laugier, C., Fraichard, T., Garnier, P.: Sensor-based control architecture for a car-like vehicle. Autonomous Robot 6(2), 165–185 (1999)
10. LaValle, S.M., Kuffner, J.J.: Rapidly-exploring random trees: Progress and prospects. In: Proceedings of the 2000 International Workshop on the Algorithmic Foundations of Robotics, pp. 293–308 (2000)

11. Nissan Around View Monitor, http://www.nissan-global.com
12. Paromtchik, I.E., Laugier, C.: Motion generation and control for parking an autonomous vehicle. In: Proceedings of the IEEE International Conference on Robotics and Automation, April 1996, pp. 3117–3122 (1996)
13. Reeds, J.A., Shepp, L.A.: Optimal paths for a car that goes both forwards and backwards. Pacific Journal of Mathematics 145(2), 367–393 (1990)
14. Sampei, M., Furuta, K.: On time scaling for nonlinear systems: Application to linearization. IEEE Transactions on Automatic Control 31, 459–462 (1986)
15. Svestka, P., Overmars, M.H.: Motion planning for carlike robots using a probabilistic learning approach. International Journal of Robotics Research 16(2), 119–143 (1997)
16. Szádeczky-Kardoss, E., Kiss, B.: Continuous-curvature turns in motion planning for car like mobile robots. In: Proceedings of the International Carpathian Control Conference, June 2007, pp. 684–687 (2007)
17. Volkswagen Park Assist, http://www.volkswagen.com
18. Zhao, Y., Collins, E.G., Dunlap, D.: Design of genetic fuzzy parallel parking control systems. Proceedings of the American Control Conference 5, 4107–4112 (2003)

Performance Evaluation of Ultrasonic Arc Map Processing Techniques by Active Snake Contours

Kerem Altun and Billur Barshan

Department of Electrical and Electronics Engineering, Bilkent University, Bilkent,
TR-06800 Ankara, Turkey
{kaltun,billur}@ee.bilkent.edu.tr

Summary. Active snake contours are considered for representing the maps of an environment obtained by different ultrasonic arc map (UAM) processing techniques efficiently. The mapping results are compared with the actual map of the room obtained with a very accurate laser system. This technique is a convenient way to represent and compare the map points obtained with different techniques among themselves, as well as with an absolute reference. It is also applicable to map points obtained with other mapping techniques.

1 Introduction

Ultrasonic sensors have been widely used in robotic applications due to their accurate range measurements, robustness, low cost, and simple hardware interface. When coupled with intelligent processing, they provide a useful alternative to more complex laser and camera systems, especially when it is not possible to use these systems in some environments due to surface characteristics or insufficient ambient light. Despite their advantages, the frequency range at which air-borne ultrasonic transducers operate is associated with a large beamwidth that results in low angular resolution and uncertainty in the location of the echo-producing object. Thus, having an intrinsic uncertainty of the actual angular direction of the range measurement and being prone to various phenomena such as multiple and higher-order reflections and cross-talk between transducers, a considerable amount of modeling, processing, and interpretation of ultrasonic data is necessary.

Most commonly, the large beamwidth of the transducer is accepted as a device limitation that determines the angular resolving power of the system, and the reflection point is assumed to be along the line-of-sight (LOS) of the transducer. According to this naive approach, a simple mark is placed along the LOS at the measured range, resulting in inaccurate maps with large angular errors and artifacts. In earlier work, basically, there have been two approaches to the representation of ultrasonic data: feature-based and grid-based. Grid-based approaches do not attempt to make difficult geometric decisions early in the interpretation process unlike feature-based approaches that extract the geometry of the sensor data as the first step. As a first attempt, several researchers have fitted line segments to ultrasonic data as features that crudely approximate the room geometry [10, 18, 13]. This approach proved to be difficult and brittle because straight

H. Bruyninckx et al. (Eds.): European Robotics Symposium 2008, STAR 44, pp. 185–194, 2008.
springerlink.com

lines fitted to time-of-flight (TOF) data do not necessarily match or align with the world model, and may yield many erroneous line segments. Improving the algorithms for detecting line segments and including heuristics does not really solve the problem. A more physically meaningful representation is the use of *regions of constant depth* (RCDs) as features. RCDs are circular arcs which are natural features of the raw ultrasonic TOF data from specularly reflecting surfaces, first reported in [16], and further elaborated on in [9]. Alternatively, the angular uncertainty in the range measurements has been represented by ultrasonic arc maps (UAMs) [6] that preserve more information (see Fig.1(c) for a sample UAM). This is done by drawing arcs spanning the beamwidth of the sensor at the measured range, representing the angular uncertainty of the object location and indicating that the echo-producing object can lie anywhere on the arc. Thus, when the same transducer transmits and receives, all that is known is that the reflection point lies on a circular arc of radius r. More generally, when one transducer transmits and another receives, it is known that the reflection point lies on the arc of an ellipse whose focal points are the transmitting and receiving elements. The arcs are tangent to the reflecting surface at the actual point(s) of reflection.

Several techniques have been proposed to process these UAMs (Table 1), that result in more accurate maps of the environment. These techniques are summarized in Section 2. Each processed UAM results in a collection of (usually a large number of) data points, represented as a black-on-white image. In [4], the DM technique is newly proposed, and a comparison of these techniques is provided based on three different error criteria. In this paper, we propose a method to compactly and efficiently represent the resulting map points that will also make it convenient to assess the accuracy of the different UAM processing techniques. Basically, active snake curves will be fitted to the results of processing the UAM by each technique and a comparison with a very accurate laser map (considered as absolute reference) will be provided.

Table 1. Different UAM processing techniques

1	Point marking (PM) [16]	5	Bayesian update (BU) [1]
2	Voting and thresholding (VT) [3]	6	Triangle based fusion (TBF) [15]
3	Directional maximum (DM) [4]	7	Arc transversal median (ATM-org) [8]
4	Morphological processing (MP) [6]	8	Modified ATM (ATM-mod) [4]

2   UAM Processing Techniques

This section summarizes various techniques for processing the UAM constructed from raw ultrasonic TOF measurements. Detailed descriptions of the methods can be found in [4], or respective references indicated in the subsections or in Table 1.

2.1   Point Marking (PM)

This is the simplest approach, mentioned above, where a mark is placed along the LOS at the measured range [16]. This method produces reasonable estimates for the locations of objects if the arc of the cone is small. This can be the case at higher frequencies of operation where the corresponding sensor beamwidth is small or at nearby ranges. Since every arc is reduced to a single point, this technique cannot eliminate any of the outlying TOF readings. The resulting map is usually inaccurate with large angular errors and artifacts.

2.2   Voting and Thresholding (VT)

In this technique, each pixel stores the number of arcs crossing that pixel, resulting in a 2-D array of occupancy counts for the pixels [3]. By simply thresholding this array and zeroing the pixels lower than the threshold, artifacts can be eliminated and the map is extracted.

2.3   Directional Maximum (DM)

This technique is based on the idea that in processing the acquired range data, there is a direction-of-interest (DOI) associated with each detected echo. Ideally, the DOI corresponds to the direction of a perpendicular line drawn from the sensor to the nearest surface from which an echo is detected. However, in practice, due to the angular uncertainty of the object position, the DOI can be approximated as the LOS of the sensor when an echo is detected. Since prior information on the environment is usually unavailable, the DOI needs to be updated while sensory data are being collected and processed on-line [4].

In the implementation, the number of arcs crossing each pixel of the UAM is counted and stored, and a suitable threshold value is chosen, exactly the same way as in the VT method. The novelty of the DM method is the processing done along the DOI. Once the DOI for a measurement is determined using a suitable procedure, the UAM is processed along this DOI as follows: The array of pixels along the DOI is inspected and the pixel(s) exceeding the threshold with the maximum count is kept, while the remaining pixels along the DOI are zeroed out. If there exist more than one maxima, the algorithm takes their median (If the number of maxima is odd, the maxima in the middle is taken; if the number is even, one of the two middle maxima is randomly selected.) This way, most of the artifacts of the UAM can be removed.

2.4   Morphological Processing (MP)

The processing of UAMs using morphological operators was first proposed in [6]. This approach exploits neighboring relationships and provides an easy to implement yet effective solution to ultrasonic map building. By applying binary morphological operators, one can eliminate the artifacts of the UAM and extract the surface profile.

2.5   Bayesian Update Scheme for Occupancy Grids (BU)

Occupancy grids were first introduced by Elfes, and a Bayesian scheme for updating their probabilities of occupancy and emptiness was proposed in [1] and verified by ultrasonic data. Starting with a blank or completely uncertain occupancy grid, each range measurement updates the probabilities of emptiness and occupancy in a Bayesian manner. The reader is referred to [1] for a detailed description of the method and [4] for its implementation in this work.

2.6   Triangulation-Based Fusion (TBF)

The TBF method is primarily developed for accurately detecting the edge-like features in the environment based on triangulation [15]. The triangulation equations involved are not suitable for accurately localizing planar walls.

Unlike the previously introduced grid-based techniques, the TBF method extracts the features of the environment by using a geometric model suitable for edge-like features. In addition, TBF considers a sliding window of ultrasonic scans where the number of rows of the sliding window corresponds to the number of ultrasonic sensors fired, and the number of columns corresponds to the number of most recent ultrasonic scans to be processed by the algorithm. TBF is focused on detection of edge-like features located at ≤ 5 m. The other methods consider all of the arcs in the UAM corresponding to all ranges, and are suitable for detecting all types of features.

2.7   Arc-Transversal Median (ATM)

The ATM algorithm requires both extensive bookkeeping and considerable amount of processing [8]. For each arc in the UAM, the positions of the intersection(s) with other arcs, if they exist, are recorded. For arcs without any intersections, the mid-point of the arc is taken to represent the actual point of reflection (as in PM). If the arc has a single intersection, the algorithm uses the intersection point as the location of the reflecting object. For arcs with more intersections, the median of the positions of the intersection points with other arcs is chosen to represent the actual point of reflection. In [8], the median operation is applied when an arc has *three or more* intersection points. If there is an even number of intersections, the algorithm uses the mean of the two middle values (except that arcs with two intersections are ignored). It can be considered as a much improved version of the PM approach.

We have also implemented a modified version of the algorithm (ATM-mod) where we ignored arcs with no intersections. Furthermore, since we could not see any reason why arcs with two intersections should not be considered, we took the mean of the two intersection points.

3   Fitting Active Snake Contours to UAMs

A snake, or an active contour [14] can be described as a continuous deformable closed curve. It is commonly used in image processing for edge detection or image segmentation [14, 17]. The deformation is controlled by external and internal forces. External forces depend on the image and they try to stretch or shrink the curve to fit to the data, whereas internal forces impose elasticity and rigidity constraints on the curve. We define a snake as a parametrized closed curve $\mathbf{v}(s) = (x(s), y(s)), s \in [0,1]$, whose energy is given by the functional

$$E_{snake} = \int_0^1 \left(E_{int}(\mathbf{v}(s)) + E_{ext}(\mathbf{v}(s)) \right) ds \qquad (1)$$

where $x(s)$ and $y(s)$ are periodic functions representing the x and y coordinates of the snake and s is the normalized arc length parameter of the snake curve. The internal energy is given by

$$E_{int}(\mathbf{v}(s)) = \frac{1}{2} \left(\alpha \left\| \frac{d(\mathbf{v}(s))}{ds} \right\|^2 + \beta \left\| \frac{d^2(\mathbf{v}(s))}{ds^2} \right\|^2 \right) \tag{2}$$

where α is the elasticity parameter and β is the rigidity parameter, taken as constants. The external energy will be denoted by $E_{ext}(\mathbf{v}(s)) = P(\mathbf{v}(s))$, where P is a potential function that depends on the image data.

In general, the selection of the potential function varies depending on the application. However, it must be minimum on the image edges if the snake is used for edge detection. Kass et al. suggest using the negative of the image gradient magnitude as a potential function [14]. However this is only feasible if the snake is initialized close to the image boundaries. Filtering the image with a Gaussian low-pass filter is also suggested in the same paper to increase the capture range of the snake, but this causes the edges to become blurry, thus reducing the accuracy. If the image is a black-on-white one (as is the case in this study), then the image intensity can be used as the potential function, either as itself or convolved with a Gaussian blur [11]. Obviously this method also suffers from the same drawbacks stated above. Another solution proposed in [12] is using a distance map to increase the capture range of the contour, which is the approach used in this study.

Approaches that do not use a potential function as the external energy term also exist in literature [5]. This relaxes the constraint that the external forces pulling the snake towards the edges should be conservative, i.e., derived from a potential field. Xu et al. define a non-conservative force field representing the external forces and use force balance equations rather than energy-based approach to solve the problem [5]. However, this idea is not used in our work.

Having chosen a potential function, the goal is to find the curve that minimizes the energy functional in Eqn. (1). This problem can be solved by using calculus of variations. The minimizing curve must satisfy the following Euler-Lagrange equation [14]:

$$\alpha \frac{d^2(\mathbf{v}(s))}{ds^2} - \beta \frac{d^4(\mathbf{v}(s))}{ds^4} - \nabla P(\mathbf{v}(s)) = 0 \tag{3}$$

For some cases it may be possible to solve this equation analytically, but a general analytical solution does not exist. The general practice is to initialize an arbitrary time-dependent snake curve $\mathbf{v}(s,t)$. Eqn. (3) is then set equal to the time derivative of the snake, where a solution will be found when the time derivative vanishes. That is,

$$\alpha \frac{\partial^2 \mathbf{v}}{\partial s^2} - \beta \frac{\partial^4 \mathbf{v}}{\partial s^4} - \nabla P(\mathbf{v}) = \frac{\partial \mathbf{v}}{\partial t} \tag{4}$$

These equations are then discretized for a numerical solution. Furthermore, the snake is treated as a collection of discrete points joined by straight lines, and a snake curve is initialized on the image. Approximating the derivatives by finite differences, the evolution equations of the snake reduce to [17]:

$$x_{t+1} = (A + \gamma I)^{-1} \left(\gamma x_t - \kappa \left. \frac{\partial P}{\partial x} \right|_{(x_t,y_t)} \right) \tag{5}$$

$$y_{t+1} = (A + \gamma I)^{-1} \left(\gamma y_t - \kappa \left. \frac{\partial P}{\partial y} \right|_{(x_t,y_t)} \right) \tag{6}$$

for all points (x,y) on the snake. Here t is the current time (or iteration) step, γ is the Euler step size, κ is the external force weight, I is the identity matrix of appropriate size and A is a pentadiagonal banded matrix that depends on α and β. The sizes of the matrices A and I are determined by the number of points on the snake, which may change as the algorithm is executed. The variables x_t and y_t represent the coordinates of the discrete points on the snake at time t.

4   Experiments

The different techniques, listed in Table 1, are considered for processing the UAMs. Each of these techniques results in a different set of points to which a snake curve is fitted.

The potential function used in this study is based on the Euclidean distance transform, as suggested in [12]. As stated before, processed UAMs in our case are represented as black pixels (i.e., $I(x,y) = 0$) on white background (i.e., $I(x,y) = 1$), where $I(x,y)$ is the intensity of the image. Euclidean distance transform is defined for all points on the image as the Euclidean distance to the nearest black pixel. That is, the potential function is selected as

$$P(x,y) = \min_{\{(x',y') | I(x',y')=0\}} \sqrt{(x-x')^2 + (y-y')^2} \tag{7}$$

for all points (x,y) on the image. Note that the value of the potential function is zero for the points on the extracted map and increases gradually when (x,y) moves away from the map points. The Euclidean distance transform is computationally costly, and a number of algorithms and other distance transforms have been proposed in the literature to approximate it [7, 2]. However, in this study the Euclidean distance transform is implemented in its original form.

An example image of a room acquired with a laser system is shown in Fig. 1(a). This is the original laser data which is quite accurate, and is used as the absolute reference to compare the methods given in Table 1. The corresponding distance map is shown in Fig. 1(b), which is drawn by scaling the values of the potential function to be between 0 and 255. Fig. 1(c) shows the raw UAM for the room.

The values for the parameters in Eqns. (5) and (6) are selected as $\alpha = \gamma = 1$, $\beta = 0.1$ and $\kappa = 2.5$. Selecting $\beta = 0.1$ enforces the second derivative in the energy term to have less weight, thus allowing sharp corners in the snake. The snake curves fit to the laser and the processed UAMs are given in Fig. 2. The blue curves are the snakes fitted

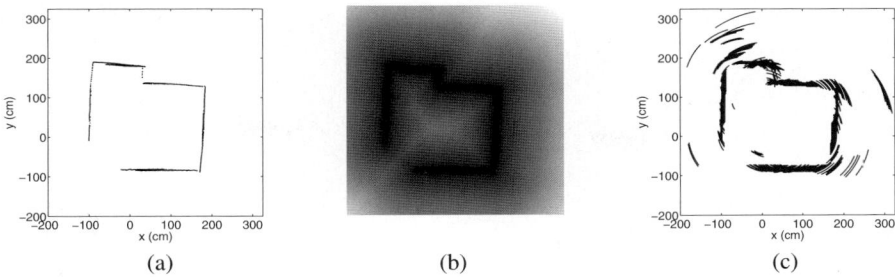

(a)   (b)   (c)

Fig. 1. (a) Laser map of the room, (b) its distance map, (c) and the UAM

to these UAM data. The red curves are the snakes fitted to the laser data, and it is superimposed on the processed UAMs for better visualization. There was an opening on the lower-left corner of the room from which no ultrasonic data were collected. Therefore, the part of the snake curve in that region is not drawn in the figure and not included in the error calculations.

The snake is initialized as a circle whose center is at (30, 55) having a radius of 185 units so that it encompasses the room boundary. We allow the snake to converge to outlier points caused by crosstalk and/or multiple reflections to provide a better evaluation of the methods. Then, the snake is evolved for a fixed amount of iterations (currently 100), determined experimentally to ensure that each UAM snake converges to the map. After each iteration, the points on the snake are checked for uniformity. The distance between any two neighboring points is maintained between 2–4 units, also determined experimentally. That is, the points are destroyed or created as required by this constraint, after each iteration.

The snake fit to the laser data is referred to as C_{laser} from now on in this text. The snakes fitted to the processed UAM data will be referred to as C_i, where i denotes the index of the method given in Table 1. Thus the ith snake is represented as a collection of points $(x_{ij}, y_{ij}), j = 1, \ldots, N_i$ where N_i is the total number of points on snake C_i.

An error measure is defined to determine the closeness between the laser snake C_{laser} and processed UAM snake C_i. It is calculated by finding the distance of the nearest point on snake C_{laser} for all points on snake C_i and averaging these distances. First, a distance function is defined for points on a given snake C_i as:

$$d_{i/laser}(x_{ij}, y_{ij}) = \min_{k=1,\ldots,N_{laser}} \sqrt{(x_{ij} - x_k)^2 + (y_{ij} - y_k)^2}, \quad j = 1, \ldots, N_i \qquad (8)$$

where k is an index for points on snake C_{laser} and N_{laser} is the total number of points on the snake C_{laser}. Then, the error is given as:

$$e_i = \frac{1}{N_i} \sum_{j=1}^{N_i} d_{i/laser}(x_{ij}, y_{ij}) \qquad (9)$$

The errors for the different methods are tabulated in Table 2. According to the results, ATM-mod and DM techniques have the smallest errors, and MP and BU perform the worst. The remaining techniques are comparable to the PM method.

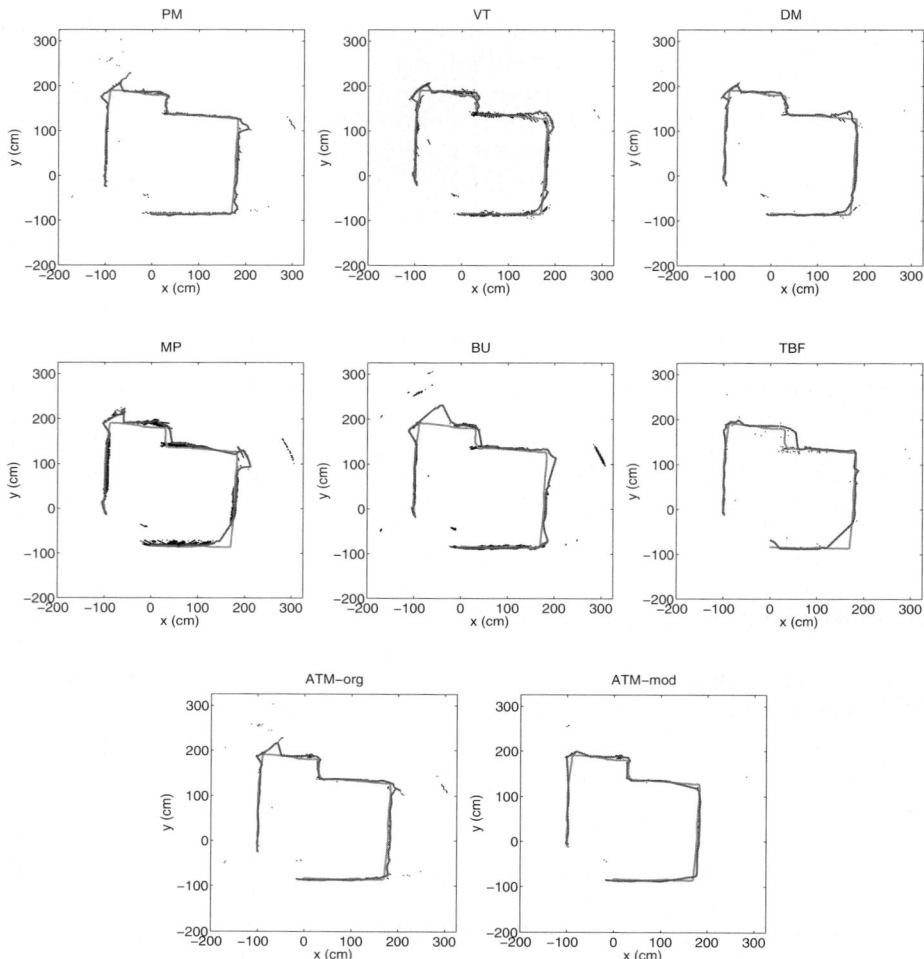

Fig. 2. Results of snake curve fitting for all the UAM processing techniques. Red curves correspond to the snake fitted to laser data and the blue curves are the snakes fitted to the map points.

Table 2. Error values of the different techniques

Method	PM	VT	DM	MP	BU	TBF	ATM-org	ATM-mod
Error	4.321	4.218	3.478	6.806	6.871	4.730	4.427	3.403

DM and ATM-mod methods eliminate most of the artifacts resulting from crosstalk and multiple and higher-order reflections (Fig. 2) so that the corresponding snake curves follow the laser data very closely. MP and BU methods cannot eliminate those artifacts as much, resulting in larger errors. Same applies to the PM method; it can not eliminate the outlier points and could have resulted in large error values. We should note here that the snake was initialized outside the boundaries of the room. Initializing the snake inside the room would also be possible. In this case, the spurious points outside the boundaries would not affect the snake curve as much, allowing it to follow the boundaries of the room more closely. However, this would not result in a fair comparison between the techniques. In a practical application when comparison and evaluation of the techniques is not an issue, this might be a good choice to eliminate the erroneous points outside the boundary.

5 Conclusion

We have presented a technique to compactly and efficiently represent the maps obtained by processing the UAMs using different techniques. The representation of the map points with snake curves makes it easy and convenient to compare maps obtained with different techniques among themselves, as well as with an absolute reference. This approach can be applied to other mapping techniques. Our current work involves using Kohonen's self-organizing maps for the same purpose that takes into account all the outlier points.

Acknowledgment

Kerem Altun is supported by The Scientific and Technological Research Council of Turkey (TÜBİTAK) with a doctoral scholarship. This work is supported by TÜBİTAK under grant number EEEAG-105E065.

References

1. Elfes, A.: Sonar based real-world mapping and navigation. IEEE Transactions on Robotics and Automation RA-3(3), 249–265 (1987)
2. Rosenfeld, A., Pfaltz, J.L.: Distance functions on digital pictures. Pattern Recognition 1(1), 33–61 (1968)
3. Barshan, B.: Ultrasonic surface profile determination by spatial voting. Electronics Letters 35(25), 2232–2234 (1999)
4. Barshan, B.: Directional processing of ultrasonic arc maps and its comparison with existing techniques. International Journal of Robotics Research 26(8), 797–820 (2007)
5. Xu, C., Prince, J.L.: Snakes, shapes and gradient vector flow. IEEE Transactions on Image Processing 7(3), 359–369 (1998)
6. Başkent, D., Barshan, B.: Surface profile determination from multiple sonar data using morphological processing. International Journal of Robotics Research 18(8), 788–808 (1999)
7. Borgefors, G.: Distance transformations in digital images. CVGIP 34(3), 344–371 (1986)
8. Choset, H., Nagatani, K., Lazar, N.: The arc-transversal median algorithm: a geometric approach to increasing ultrasonic sensor azimuth accuracy. IEEE Transactions on Robotics and Automation 19(3), 513–522 (2003)
9. Leonard, J.J., Durrant-Whyte, H.F.: Directed Sonar Sensing for Mobile Robot Navigation. Kluwer Academic Publishers, Boston (1992)
10. Crowley, J.L.: Navigation for an intelligent mobile robot. IEEE Transactions on Robotics and Automation RA-1(1), 31–41 (1985)

11. Cohen, L.D.: On active contour models and balloons. CVGIP: Image Understanding 53(2), 211–218 (1991)
12. Cohen, L.D., Cohen, I.: Finite element methods for active contour models and balloons for 2-D and 3-D images. IEEE Transactions on Pattern Analysis and Machine Intelligence 15(11), 1131–1147 (1993)
13. Drumheller, M.: Mobile robot localization using sonar. IEEE Transactions on Pattern Analysis and Machine Intelligence PAMI-9(2), 325–332 (1987)
14. Kass, M., Witkin, A., Tersopoulos, D.: Snakes: Active contour models. International Journal of Computer Vision 1(4), 321–331 (1987)
15. Wijk, O., Christensen, H.I.: Triangulation-based fusion of sonar data with application in robot pose tracking. IEEE Transactions on Robotics and Automation 16(6), 740–752 (2000)
16. Kuc, R., Siegel, M.W.: Physically-based simulation model for acoustic sensor robot navigation. IEEE Transactions on Pattern Analysis and Machine Intelligence PAMI-9(6), 766–778 (1987)
17. Menet, S., Saint-Marc, P., Medioni, G.: Active contour models: Overview, implementation and applications. In: Proceedings of the IEEE International Conference on Systems, Man and Cybernetics, November 1990, pp. 194–199 (1990)
18. Gex, W., Campbell, N.: Local free space mapping and path guidance. In: Proceedings of IEEE International Conference on Robotics and Automation, March 1987, pp. 424–431 (1987)

Planning Robust Landmarks for Sensor Based Motion

Michel Taïx[1,2], Abed C. Malti[1,2], and Florent Lamiraux[1]

[1] LAAS-CNRS, University of Toulouse, Toulouse, France
{taix,amalti,florent}@laas.fr
[2] University Paul Sabatier, Toulouse, France

Summary. Our work is focused on defining a generic approach for planning landmark based motion. In previous works we have introduced the formal basis for this and showed simulation results. In this paper we first demonstrate the relevance of our work with experiments on a real robot. Then, on the base of these results we introduce new strategy for planning selecting landmarks in order to improve the robustness of the navigation task in a cluttered environment.

1 Introduction

Path planning in a reference map for a robot produces a non-collision continuous path in the robot configuration space [7, 8, 2, 9]. However, the execution of this path in a real environment remains problematic for two main reasons. The first difficulty lies in the inaccuracy of the environment map used to plan the path, and the second is that the navigation task in cluttered environments requires precise localization. Many approaches have been proposed to solve these two problems :localizing the robot along the path with respect to local landmarks [15], reactive methods to avoid unexpected obstacles [12, 13, 1, 6], path planning with uncertain approaches [5, 4]. In our work, we aim to introduce sensor-landmark constraints along a geometric path to solve these problems. In a previous work [10, 11] we have introduced a generic approach to correct a planned geometric path. The idea was to plan sensor-landmark primitives to perform sensor based motion along a path. Instead of planning a path in a first stage and following it in a second stage, we will produce a sequence of sensor-landmark based motions, each sensor-landmark pairs are weighted. These weights distinguish the sensor-landmark pairs from the most to the least relevant. For example when passing through a door the most relevant landmarks are the two sides of the door. The goal is not to localize the robot but to give input for sensor-based motion controller.

In this paper we present the first validation conducted upon a real robot platform. These experiments lead us to propose several criteria in order to improve the selection process of the landmarks. These improvements involve the robustness of the localization process and the success of the matching process. In section 2 and 3 we give the definition of a landmark based motion and how we plan this motion. In section 4 we present a parking manoeuvre conducted on a mobile robot. In section 5 we describe the landmark selection that will allow to improve the robustness of the localization. In

H. Bruyninckx et al. (Eds.): European Robotics Symposium 2008, STAR 44, pp. 195–204, 2008.
springerlink.com

section 6 we introduce the landmark selection that takes into account the success of the matching process. Finally, in section 7, we give the navigation task experiment to show the relevance of our work.

2   Definition of a Landmark-Based Motion

Landmark: a landmark can be any geometric feature in the workspace. Let us denote by $\mathbb{L}$ the configuration space of a landmark L. We denote by l the configuration of L.

Sensor: a sensor S is a mobile device that maps one or several landmarks to a feature in the image space. Let us denote by $\mathbb{S} = SE(2)$ or $SE(3)$ the configuration space of sensor S. We will denote by s the configuration of S.

Sensing a landmark: the perception of landmark L by sensor S is a mapping between a pairs of configuration (sensor, landmark) and a feature in the image space $I_{S,L}$.

$$P_{S,L} : \mathbb{S} \times \mathbb{L} \rightarrow \quad I_{S,L}$$
$$(s,l) \quad \rightarrow P_{S,L}(s,l)$$

Localization: each pair (S_i, L_j) of sensors and landmarks gives rise to a *localization equation* where $im \in I_{S_i,L_j}$ is the image of L_j in S_i. l_j is supposed to be known and im is measured. The unknown of this equation is the configuration $\mathbf{q}$ of the robot:

$$P_{S_i,L_j}(s_i(\mathbf{q}),l_j) = im \tag{1}$$

The linearization of equation (1) around $\mathbf{q}_0$ leads to:

$$\frac{\partial P_{S_i,L_j}}{\partial s}(s_i(\mathbf{q}_0),l_j).\frac{\partial s_i}{\partial \mathbf{q}}(\mathbf{q}_0).(\mathbf{q}-\mathbf{q}_0) = im - im_0 \tag{2}$$

im_0 is the image of L_j in S_i at $\mathbf{q}_0$. This equation expresses the approximation of order 1 of the relation between a variation of configuration about $\mathbf{q}_0$ and the variation of the image of each landmark in the corresponding sensor.

Weighting localization: this equation can be written for all m sensor-landmark pairs. A weight ω can be associated to each sensor-landmark pair k:

$$w_k\frac{\partial P_{S_i,L_j}}{\partial s}(s_i(\mathbf{q}_0),l_j).\frac{\partial s_i}{\partial \mathbf{q}}(\mathbf{q}_0).(\mathbf{q}-\mathbf{q}_0) = w_k.im - w_k.im_0 \tag{3}$$

This weight expresses the importance of the sensor-landmark pair (collision and/or localization) along the path and is part of the motion control task specifications. Thus,from linear equations (3) we can build a linear system of equations by weighting each equation by a positive real number ω_j. We thus get the following linear system:

$$W(\mathbf{q} - \mathbf{q}_0) = IM - IM_0 \tag{4}$$

The least square solution of this system given by $(W^+$ is the pseudo-inverse of $W)$:

$$\hat{\mathbf{q}} = \mathbf{q}_0 + W^+(IM - IM_0) \tag{5}$$

Landmark-Based Motion: given a mobile robot with n sensors S_i and an environment with p landmarks L_j, a *Landmark-Based Motion, LBM,* is defined by:

1. a reference collision-free path:

$$\gamma : [0, U] \rightarrow \mathscr{C}$$
$$u \rightarrow \gamma(u)$$

   where $[0, U]$ is an interval (U length of the free path),

2. m continuous positive real valued functions $w_1, ..., w_m$:

$$w_k : [0, U] \rightarrow \mathbb{R}^+$$
$$u \rightarrow w_k(u)$$

such that w_k is associated to a pair (S_i, L_j).

   The developments conducted in this section can be summarized as follows. Localizing a mobile robot using landmarks involves solving a system of equations that relate the configuration of the robot with the images of the landmarks in the sensors of the robot. If the system is over-constrained, localization involves finding a configuration that minimizes a weighted sum of residues. If the landmarks are at exactly the same position in the model map as in the real map, the choice of weights will have no effect on the result. However, if the map of landmarks is not exact, the choice of weights will have a big influence. That is why in our approach, we suggest using these weights as a tool for planning landmark-based motions.

3   Landmark Based Motion Planning

At a first step, a geometric non-collision path $\gamma(u)$ is planned in the configuration space of the model map from an initial configuration to a final one. This path is computed by the probabilistic path planner Move3D which is developed in our laboratory [14]. Now it is necessary to calculate the weight of the sensor-landmark pairs during the path in the model map.

3.1   Weight Computing

In this work, the weight is an intrinsic specification of the robotic task associated to the execution of the geometric planned path. Thus for a given configuration along this path, the weight of any sensor-landmark pair has to represent its importance in relation to the environment and the path in order to avoid collisions and to satisfy the result of the geometric path planning stage. We define a weight of a sensor-landmark pair as a positive continuous function in the configuration space as:

$$w : \mathbb{S} \times \mathbb{L} \rightarrow \mathbb{R}^+$$
$$(s, l) \mapsto w(s, l)$$

This function vanishes of the sensor range view. It represents:

- the visibility of the landmark (distance and orientation for example)
- the danger of collision with this landmark (collision distance)

3.2   The Construction of a Landmark Based Motion

For a static configuration we can draw comparisons between landmarks and depict the most relevant, so as to execute localization. Now it is necessary to plan the best N_L sensor-landmark pairs along the path. The landmarks which have good properties of localization or presenting a risk of collision with the path will have an important weight and will be selected thus automatically. We know that the minimum number of landmarks required to localize the robot is 2 in dimension two. In practice this number is too low because the equations of landmarks can be dependent and it is then necessary to consider a bigger number of landmarks. A number from 4 to 5 landmarks is sufficient in practice. It is thus enough to select in every $\mathbf{q}$, N_L sensor-landmark pairs having the best weight.

In general, the inputs of *LBM* algorithms are as follows:

1. a model map of the environment,
2. the set of sensors $\mathscr{S}$ and landmarks $\mathscr{L}_{env}$,
3. the non-collision geometric path $\gamma(u)$, $u \in [0, U]$,
4. the number of maximum best landmarks N_L.

The output is a landmark based motion *LBM* composed of $\gamma(u)$ and a set of weighed sensor-landmark pairs. In basic terms, for a given sensor, this algorithm associates to every part of the path the best N_L landmarks that have the highest values of weights.

4   Parking Manoeuvre

We integrate the software *LBM* as a module in the generic architecture control of the Hilare 2 platform [3].

At a first stage, a geometric non-collision path is planned in this map from an initial to a final configuration so that the robot will be able to enter the car park, as show in figure (1). The second stage involves planning the landmark based motion with the generic platform we developed. Along the geometric planned path, the four best

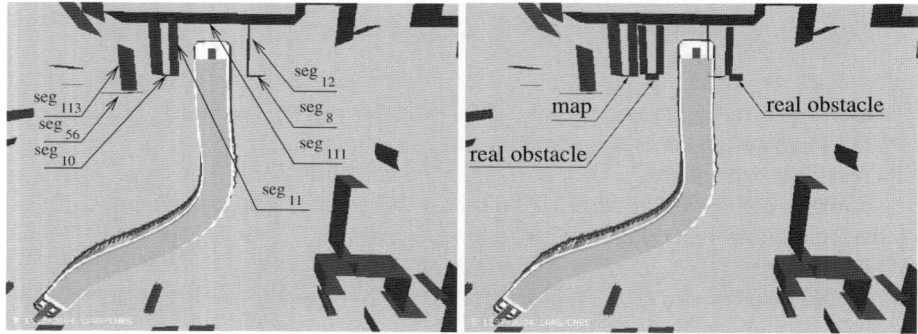

Fig. 1. The left figure shows the geometric free path planned to park the robot and the segments use to build the *LBM*. The right one shows the path executed to enter the shifted car park using the planned *LBM*.

sensor-landmark pairs are selected according to their weights. Before executing the landmark based motion, the car park is shifted right to modify the real environment in relation to the reference map. During the run of the movement, the robot corrects its path with regard to the new placement of the parking and then the task is led with success.

The scenario of this experiment shows the relevance of the formalism in a local area when the reference map is inaccurate. However, in a wider area, where the robot executes a navigation task, a landmark selection based on highest values of weights is restrictive for two main reasons:

1. It does not take into account the conditioning of the localization system (4).
2. It does not take into account the success of the matching process.

5   Landmark Selection with Regular Matrix Condition

In the previous experiment, when the robot starts the last stage of the parking it uses the three segments seg_{11}, seg_{12} and seg_{111}. If we remove the segment seg_{111}, the weighted localization matrix in (4) becomes ill-conditioned. Typically, this case happens when the robot navigates in a long corridor, segments seg_{11}, seg_{12} became the side of the corridor. So by taking into account solely both sides the weighted localization system is ill-conditioned. In this case, the localization process will produce a big jump in the value of the configuration of localization $\hat{\mathbf{q}}$ in relation to the current position. The pseudo-code algorithm 1 avoids such undesirable situations.

Algorithm 1. *LBM* with matrix condition

Data: $\mathscr{L}_{env}$, $K_{d_{max}}$, *LBM*

Result: *LBM*

begin

    for *each piece of LBM along* γ **do**

        $\mathscr{L}_{select} \leftarrow$ get_landmark();

        $K_d \leftarrow$ condition_number($\mathscr{L}_{select}$);

        if $K_d \geq K_{d_{max}}$ **then**

            $\mathscr{L}_{\mathscr{H}} \leftarrow$ improve_condition($\mathscr{L}_{env}$,$\mathscr{L}_{select}$,$K_{d_{max}}$);

        LBM $\leftarrow$ add_landmark($\mathscr{L}_{\mathscr{H}}$);

end

The main input of this algorithm is the maximum condition number $K_{d_{max}}$ that is defined as the highest value we accept for the ratio between the highest and the smallest singular value of the localization matrix (this parameter is a good information for localization process). Thus, for each piece along a geometric path of a pre-constructed landmark based motion *LBM*, the condition number K_d is computed for the corresponding landmarks $\mathscr{L}_{select}$. If its value is higher than the maximum condition number, then the algorithm looks for visible landmarks that can improve the matrix conditioning and add them to the landmark based motion. The function *improve_condition* uses new landmark of $\mathscr{L}_{env}$ (not included in $\mathscr{L}_{select}$), $\mathscr{L}_{\mathscr{H}}$, to decrease K_d and include it in the *LBM*.

We have no guarantee of convergence of the algorithm towards a solution but generally, the number of landmark-sensor pairs N_L used in the pre-calculation of *LBM* is less than the maximum number of useful sensor. In the case where there is no other landmarks, it is not possible to localize the robot with respect to landmarks, and therefore to realize the sensor based motion.

Finally, we obtain a set of landmarks that is less sensitive with respect to small errors between the reference map and the real environment.

6   Landmark Selection with Matching Condition

The second important issue for successfully a landmark based motion is the ability to retrieve continuously the selected landmarks in the real environment. Recognizing those landmarks is absolutely required for mainly two reasons:

- to avoid big gaps in the computation of successive localization errors,
- those landmarks have to be taken into account during the achievement of the robotic task since they are considered as relevant.

Using the most relevant landmarks in the localization process is a good idea to give them more importance than other visible landmarks during the motion execution. However, this reasoning reduces the probability of retrieving them especially in the case where the environment is rich of landmarks and where the selected primitives do not constitute a recognizable shape in relation to the environment.

We give a generic pseudo-code algorithm 2 that allows to pick up those landmarks used to help the success of the matching process. This algorithm takes as inputs a landmark based motion, *LBM*, constructed as described in previous sections and the landmarks of the reference map $\mathscr{L}_{env}$. In a first stage, the algorithm introduces some perturbations on the configurations of the selected landmarks $\mathscr{L}_{select}$ so that it simulates errors in the reference map. In the second stage, it tries to match the perturbed landmarks $\mathscr{L}_{real}$ with those of the reference map by function $Matching(\mathscr{L}_{real}, \mathscr{L}_{env})$. In the case where the matching is successfully then it concludes that the selected landmarks for the current piece of path constitutes a recognizable shape (same matching algorithm has been used to plan *LBM* than in real execution, only the data are different). In contrast to this situation, if one of the selected landmarks is not matched then the algorithm adds others visible landmarks. The function *Pick_visi_landmark* takes a new landmark-sensor pair include in $\mathscr{L}_{env}$ but not in $\mathscr{L}_{select}$ to try to construct a recognizable shape of selected landmarks. This operation is repeated while $Matching(\mathscr{L}_{real}, \mathscr{L}_{env})$ fails or no other one exists. It is important to insist on the fact that the landmarks added by this algorithm are used to help the success of the matching process, but they are not used in the localization process.

The improvement presented in the two last sections was integrated on our generic framework. Actually, the landmark based motion planner we are developing selects the landmarks that are the most relevant in relation to:

1. the danger of collision and their visibility,
2. the conditioning of the localization matrix,
3. the success of the matching process.

Algorithm 2. *LBM* with matching condition

Data: $\mathcal{L}_{env}$, *LBM*
Result: *LBM*
$L, \mathcal{L}_{real} \leftarrow Null$;
begin

    for *each piece of LBM along* γ **do**

        $\mathcal{L}_{match} \leftarrow Null$;

        $\mathcal{L}_{select} \leftarrow$ Get_landmark();

        $\mathcal{L}_{real} \leftarrow$ Perturbation($\mathcal{L}_{select}$);

        while *Matching($\mathcal{L}_{real}$, $\mathcal{L}_{env}$) fails* **do**

            $L \leftarrow$ Pick_visi_landmark($\mathcal{L}_{env}$, $\mathcal{L}_{select}$);

            $\mathcal{L}_{match} \leftarrow \mathcal{L}_{match} \oplus \{L\}$;

            $\mathcal{L}_{real} \leftarrow$ Perturbation($\mathcal{L}_{select} \oplus \mathcal{L}_{match}$);

            $LBM \leftarrow$ Add_matching_landmark(LBM, $\mathcal{L}_{match}$);

end

The selected landmarks according to the two first one criteria are used both in the matching process and in the localization process. The landmarks selected according to the last criterion are used solely in the matching process. This last version of our software has been tested and validated on the mobile robot Hilare 2 towing a trailer by realizing a navigation task. This experiment is detailed in the next section.

7 A Navigation Task in a Cluttered Environment

In the corridors of our laboratory we plan a geometric free path with Move3D (see figure 2). Thereafter, the produced path and the reference map of the environment are used by the landmark based motion planner to select the most relevant landmarks according to the three criteria presented above.

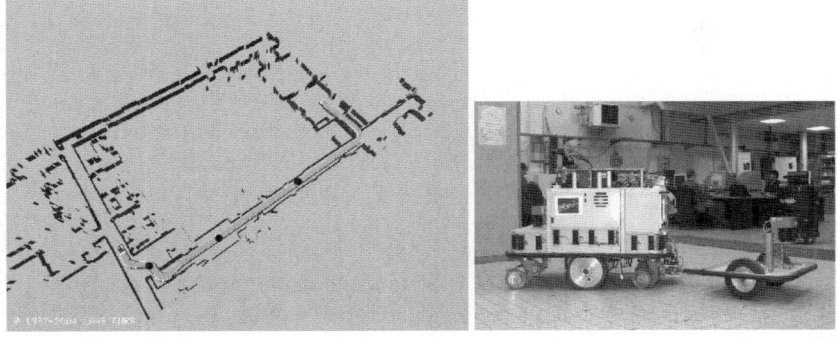

Fig. 2. The planned free path. Points represent three configurations for Hilare 2 robot.

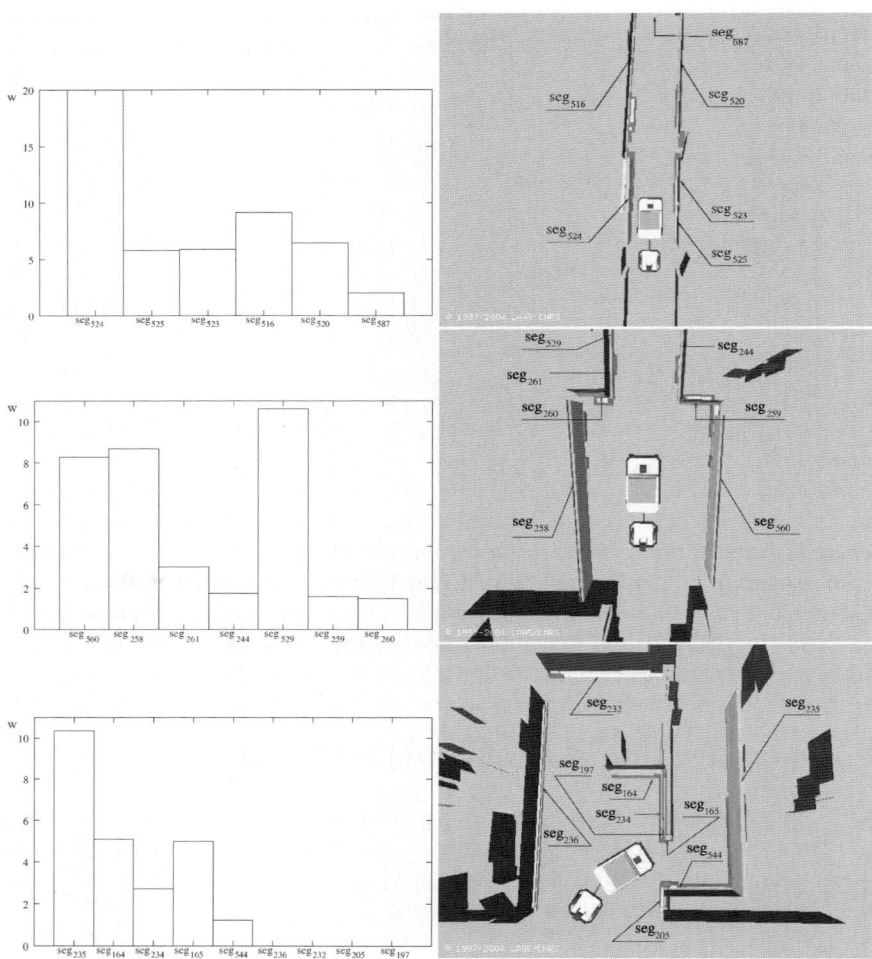

Fig. 3. The left figures shows the instantaneous weights associated to selected landmarks viewed by the front sensor for three configurations of the right figure. The green segments are selected by the initial *LBM* algorithm. The purple segments are selected by the criterion of algorithm 1 to improve the conditioning of the localization matrix. The white segments are selected by the criterion of algorithm 2 to help the matching process. The yellow color illustrate the segments matched with success.

The navigation task we describe involves some difficulties that have to be raised:

- The reference map we have is not exact. Indeed, we pick up some errors in terms of distances between walls in the reference map and in the real environment (the difference is about an average value of some ten centimeters).

- The size of the robot in relation to the free space of the environment is a critic issue for the achievement of the navigation task. Indeed, unlike the parking manoeuvre (c.f. section 4) where the robot has a large free space, here the passages are narrowed and the manoeuvres are geometrically very constrained.

Although this difficulty constrained hardly the achievement of the navigation task, Hilare 2 drives with success the circuit using the landmark based motion to correct its path. Here after, we show and comment some of the critical passages. The figure (3) shows the details for the three configurations oto the left of the figure 2.

The bottom figure shows the first critical crossing. Even if the odometric error is not important at this step of the navigation, the errors of the map can generate failure in the experiment. The segments seg_{164}, seg_{165}, seg_{234}, seg_{235} and seg_{544}, are used for localization. The segments seg_{197}, seg_{205}, seg_{232} and seg_{236} are selected only to help the matching process (this is why their weights in left figure are zero). The passage being narrower in reality than in the map, the segments seg_{165} and seg_{544} have big values of weights to ensure a safe crossing.

The middle figure illustrates a passage where it is necessary to take into account landmarks that improve the conditioning of the localization matrix. The algorithm 1 allows to select segments seg_{259} and seg_{260} for this purpose.

The top figure represents the classical situation of a long corridor. Because of the limitation in the perception of the robot sensor's, the sole available landmarks are those of both sides of the corridor. To avoid jumps in the result of localization in such a situation, we correct the position only following the crosswise. Following the lengthwise, the robot continues its path without any correction until it senses the end of the corridor, the weight of seg_{587} (at the end of the corridor) is taken into account.

After the analysis of this experiment, the main issue that attracts our attention for future works concerns the incoherences between the reference map and the real environment. This study involves the formulation of such a problem in relation to a planned geometric free path in order to take a decision whether one has to correct this path in order to correct the map errors or to plan another one. Further works could be led about jumps in the localization caused by the unexpected appearance or disappearance too early or too late of some selected landmarks.

8   Conclusion

In this paper, we presented the first experiment we have conducted upon the mobile robot Hilare 2 towing a trailer. This experiment raises two main issues : the improvement of the conditioning of the localization matrix and the success of the matching process. Thereby, we have developed two algorithms that allowed to select further landmarks to overcome the lacks raised by these two issues. This improvements were integrated to our software and validated across a navigation task. The success of these experiments is very encouraging for future works on the link between path planning and real motion which requires procedures of localization and control.

References

1. Arras, K.O., Persson, J., Tomatis, N., Siegwart, R.: Real-time obstacle avoidance for polygonal robots with a reduced dynamic window. In: IEEE Int. Conf. on Robotics and Automation (2002)
2. Choset, H., Lynch, K.M., Hutchinson, S., Kantor, G.A., Burgard, W., Kavraki, L.E., Thrun, S. (eds.): Principles of Robot Motion: Theory, Algorithms, and Implementations. MIT Press, Cambridge (2005)
3. Fleury, S., Herrb, M., Chatila, R.: Genom: a tool for the specification and the implementation of operating modules in a distributed robot architecture. In: IEEE/RSJ Int. Conf. on Intelligent Robots and Systems (1997)
4. Le Fort-Piat, N., Collin, I., Meizel, D.: Planning robust displacement missions by means of robot-tasks and local maps. Robotics and Autonomous Systems 20, 99–114 (1997)
5. Khatib, M., Bouilly, B., Siméon, T., Chatila, R.: Indoor navigation with uncertainty using sensor based motions. In: IEEE Int. Conf. on Robotics and Automation (1995)
6. Lamiraux, F., Bonnafous, D.: Reactive trajectory deformation for nonholonomic systems: Application to mobile robots. In: IEEE Int. Conf. on Robotics and Automation (2002)
7. Latombe, J.C.: Robot Motion Planning. Kluwer Academic Publishers, Boston (1991)
8. Laumond, J.-P. (ed.): Robot Motion Planning and Control. Lectures Notes in Control and Information Sciences, vol. 229. Springer, Heidelberg (1998)
9. LaValle, S.M.: Planning Algorithms. Cambridge University Press, Cambridge (2006)
10. Malti, A.C., Lamiraux, F., Taïx, M.: Sensor landmark motion planning in mobile robots. In: IEEE/RSJ Int. Conf. on Intelligent Robots and Systems, Sendai, Japan (2004)
11. Malti, A.C., Lamiraux, F., Taïx, M.: Sensor landmark succession for motion planning along a planned trajectory. In: IEEE Conf. on Mechatronics and Robotics, Aachen, Germany (2004)
12. Minguez, J., Montano, L., Siméon, T., Alami, R.: Global nearness diagram navigation (gnd). In: IEEE Int. Conf. on Robotics and Automation (2001)
13. Quinlan, S., Khatib, O.: Elastic bands: Connecting path planning and control. In: IEEE Int. Conf. on Robotics and Automation (1993)
14. Siméon, T., Laumond, J.-P., Lamiraux, F.: Move3d : a generic platform for path planning. In: IEEE Int. Symp. on Assembly and Task Planning (2001)
15. Victorino, A.C., Rives, P.: A relative motion estimation by combining laser measurement and sensor based control. In: IEEE/RSJ Int. Conf. on Intelligent Robots and Systems (2002)

Postural Control on a Quadruped Robot Using Lateral Tilt: A Dynamical System Approach

Luiz Castro, Cristina P Santos[1], Miguel Oliveira[1], and Auke Ijspeert[2]

[1] Industrial Electronics' Department, University of Minho
   cristina@dei.uminho.pt
[2] School of Computer and Communication Sciences, Department of Computer Science, EPFL,
   Swiss Federal Institute of Technology
   auke.ijspeert@epfl.ch

Summary. Autonomous adaptive locomotion over irregular terrain is one important topic in robotics research. Postural control, meaning movement generation for robot legs in order to attain balance, is a first step in this direction. In this article, we focus on the essential issue of modeling the interaction between the central nervous system and the peripheral information in the locomotion context. This issue is crucial for autonomous and adaptive control, and has received little attention so far. This modeling is based on the concept of dynamical systems whose intrinsic robustness against perturbations allows for an easy integration of sensory-motor feedback and thus for closed-loop control. Herein, we focus on achieving balance without locomotion.

The developed controller is modeled as discrete, sensory driven corrections of the robot joint values in order to achieve balance. The robot lateral tilt information modulates the generated trajectories thus achieving balance. The system is demonstrated on a quadruped robot which adjusts its posture until reducing the lateral tilt to a minimum.

1 Introduction

Autonomous adaptive locomotion over irregular terrain is one important topic in the robotics research. Generating trajectories in autonomous robotics, including legged robots, is still a complex, unsatisfactory solved problem. Despite an intensive research in the field ([3],[7],[1]), adaptation to unpredicted changes is still an open problem. In order to develop autonomous robot systems able to deal with less knowledge of terrain irregularity, it is required a tight coupling of planning, sensing and execution.

The work presented in this article is part of a larger project which aims at developing a closed loop control architecture based on dynamical systems for the autonomous generation, modulation and planning of complex motor behaviors for legged robots with many DOFs. Our approach is partly inspired from the biological concept of CPGs ([4]) and by the concepts of force fields ([5]) found in biology, as we believe this bio-inspiration enables to develop new flexible and robust technical solutions to the locomotion problem. We apply autonomous differential equations to model how behaviors related to locomotion are programmed in the oscillatory feedback systems of CPGs in the nervous systems.These systems are solved using numerical integration.

This dynamical systems approach model for CPGs presents multiple interesting properties comparatively to other methods based on finite-state machines, sine-based

H. Bruyninckx et al. (Eds.): European Robotics Symposium 2008, STAR 44, pp. 205–214, 2008.
springerlink.com   © Springer-Verlag Berlin Heidelberg 2008

trajectories, ZMP-based ([11]) or heuristic control laws such as the Virtual Model Control ([1]). These include: low computation cost which is well-suited for real time; the stability properties of the limit cycle behavior (i.e. perturbations are quickly forgotten); intrinsic robustness against small perturbations; the smooth online modulation of trajectories through changes in the dynamical systems parameters and phase-locking between the different oscillators for different DOFs. Further, these systems, once coupled, produce coordinated multidimensional rhythms of motor activity, under the control of simple input signals.

In order to tackle the complexity inherent to the design of dynamical systems, we choose a modular approach to build our model on the hypothesis that complex movements can be generated through the sequencing and/or superposition of simpler movement primitives implemented as simple, stable discrete and rhythmic dynamical systems. This modular approach is supported by current neurological and human motor control findings. Further, this approach enables to generate rhythmic and discrete movements, as well as their superposition.

As a main application and a first step, we address postural control without locomotion, considering robot lateral tilt information online acquired by accelerometers. Trajectories are modulated online according to these feedback pathways thus achieving balance, i.e, movements for the robot legs are generated in order to reduce this lateral tilt to a minimum. This task involves closed-loop control and we will thus particularly focus on the integration of sensory-motor information in the architecture. The controller is composed of two embedded dynamical discrete and rhythmic modules. The discrete module specifies the offset of the rhythmic movement. In a next step, we will extend this work to the achievement of adaptive quadruped locomotion in unknown, rough terrain that we model as discrete, sensory driven corrections of a basic rhythmic motor pattern for locomotion.

We present results that show how the developed controller successfully generates the required joint movements in order to reduce the lateral tilt.

Control approaches based on CPGs and nonlinear dynamical systems are widely used in robotics to achieve tasks which involve rhythmic motions including autonomous adaptive dynamic walking over irregular terrain ([6],[2]), juggling ([8]), drumming ([9]), and basis field approaches for limb movements ([5]). Quadruped walking control using CPGs exploring sensory feedback integration into the locomotion control has been extensively explored by Hiroshi Kimura and his colleagues. Herein, we address postural control in the framework of dynamical systems with superposition of discrete and rhythmic movements. We build on previous work, where controllers were developed for combining discrete and rhythmic motor primitives in drumming and dancing tasks ([9],[10]). In this article, we focus the issue of modeling the interaction between the central nervous system and the peripheral information. This issue is crucial for autonomous and adaptive control, and has received little attention so far. The intrinsic robustness of the dynamical systems approach against perturbations allows for an easy integration of sensory-motor feedback and thus for closed-loop control. The proposed work tries to serve these purposes and focus on the integration of sensory-motor information in the developed dynamical architecture.

In this article, we will first present the dynamical systems to model the rhythmic and discrete movements. In Section 3, we present how we achieve postural control in an AIBO robot using lateral tilt information by applying the developed dynamical systems. In Section 4, we present the results obtained. We conclude by discussing the main results we obtained and the work we are currently working on.

2   Dynamical System

In this section we present our model of the MPG (Motor Pattern Generator) used to generate the trajectories for one DOF. Two motor primitives generate the rhythmic and the discrete movements. The rhythmic movement is turned off such that only postural control is taken care, without locomotion generation. It exists because the system is conceived to generate locomotion in a posteriori phase. The rhythmic movement is generated by an Hopf oscillator. The discrete primitive is generated by a stable differential equation such that it integrates sensory information in the controller that generates the trajectories.

2.1   Architecture of the MPG

The control mechanism of a generic MPG is built on the hypothesis that complex movements can be generated through the superimposition and sequencing of simpler motor primitives implemented as a discrete and a rhythmic movement. Discrete movement is incorporated to the final trajectory as an offset of the rhythmic movement. Trajectory is modulated by particular choices of the dynamical control parameters (see [9] and [10] for details).

The MPG for a DOF i is divided in two dynamical subsystems, one generating the discrete part of the movement (y_i) and another generating the rhythmic part (x_i, z_i). The generated trajectories, x_i, are sent online for each DOF. The lower level control is done by PID controllers.

2.2   MPG Rhythmic Movement

To generate the rhythmic movements we apply the following dynamical system

$$\dot{x}_i = \beta \left(\mu_i - r_i^2 \right) (x_i - y_i) - \omega z_i \tag{1}$$

$$\dot{z}_i = \beta \left(\mu_i - r_i^2 \right) z_i + \omega (x_i - y_i) \tag{2}$$

where $r_i = \sqrt{(x_i - y_i)^2 + z_i^2}$.

These eqs. describe an Hopf oscillator, where μ_i controls the amplitude of the oscillations, ω and β controls the speed of convergence to the limit cycle. This Hopf oscillator contains a bifurcation from a fixed point (when $\mu_i < 0$) to a structurally stable, harmonic limit cycle with radius $R = \sqrt{\mu_i}$ and relaxation to the limit cycle given by $\frac{1}{2\beta\mu_i}$, for $\mu_i > 0$. The fixed point x_i has an offset given by y_i, which is the state variable of the discrete system. Thus, the resulting position x_i, modifies according to the y_i variable as specified by the discrete movement. For $\mu_i < 0$ the system exhibits a stable fixed point at $x_i = y_i$. This Hopf oscillator describes a rhythmic motion which can be switched on or off by simply setting μ_i to positive or negative values, respectively. Moreover, the amplitude of the movement is specified by μ_i and its frequency by ω_i.

Currently, rhythmic motion is switched off by setting μ_i to a negative value, so that the attractor of the Hopf oscillator is no longer a limit cycle but a fixed point specified by the offset, i.e. by y variable. Relaxation time for the rhythmic system is given by $\frac{1}{2\beta\mu_i}$ parameter. The y variable evolution will be specified and explained in the next subsection. This easy control of the different patterns generated is an advantage of the proposed architecture.

In Fig. 1, y variable (dashed line) is considered to have constant values during some time intervals. Initially $y = 9$, it is decreased to 2 at $t = 5.6$s and at instant $t = 11.3$s y is set to -9. The resulting x trajectory (solid line) converges asymptotically to the current value of y (dashed line). By modifying on the fly the offset values (y variable), one can easily modulate the generated trajectories. Whatever the change is, the system converges almost immediately to the new solution of the system. Further, notice the smoothness of the trajectory when the parameters are changed.

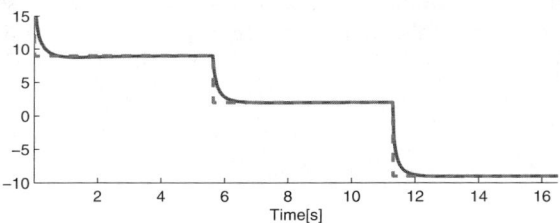

Fig. 1. Trajectory modulation through changes in the y values (offset) for dynamical system defined by eqs. 1 and 2, when rhythmic motion is turned off ($\mu_i < 0$).

2.3   MPG Discrete Movement

To generate the discrete movements, we define a nonlinear dynamical system whose solution, given by y_i, is the offset of the output x_i (eq. 1). This discrete system is designed to keep balance, such that by measuring the robot lateral tilt, the controller is able to maintain lateral stability reducing this tilt to a minimum.

It is important that this discrete movement generator applies to the control of a real robot. Thus, the generated movement must be able to: 1) smoothly adapt to the control parameters and 2) allow trajectory modulation through changes in these control parameters. In our case, the roll is not fixed but changes according to the robot movement during postural control. Therefore, we apply differential equations to model the discrete movement. The discrete movement is generated by the following dynamical system

$$\dot{y}_i = k_{j,i} \, f(\phi) + \alpha \, (y_i - M_i) \, e^{-\frac{(y_i - M_i)^2}{2\sigma^2}} + \alpha (y_i - D_i) \, e^{-\frac{(y_i - D_i)^2}{2\sigma^2}}, \tag{3}$$

where ϕ is the robot lateral tilt, $k_{j,i}$ (j = left, right) is a static gain and f is defined as a linear function of the body's lateral tilt and is given by:

$$f(\phi) = \begin{cases} 0, & -0.2 < \phi < 0.2 \text{ (degrees)} \\ 0.8\,\phi, & \text{elsewhere} \end{cases} \qquad (4)$$

A dead-zone was defined in order to deal with sensor noise. In this dynamical system, two repellors specify joint limits (M_i and D_i). These repellors are multiplied by a Gaussian function which delimits their range of action around the joint limits.

3   AIBO Postural Control Using Robot Lateral Tilt

In this section we show how we apply the presented MPG to achieve quadruped postural control using lateral tilt of the body on a real ers-7 AIBO robot.

3.1   Robotic Setup

We use an AIBO dog robot, which is a 18 DOFs quadruped robot made by Sony. The robot stands up on two platforms, one fixed and another moveable (see Fig. 2(a)). We control the swing and knee of the four AIBO legs, that is 8 DOFs of the robot, 2 DOFs in each leg: elevator and knee joints. For $i = 1,3$ (Swing, Knee) we control FLL[i], FRL[i], for fore legs and HLL[i], HRL[i] for hind legs. The other DOFs are not used for the moment, and remain fixed to an appropriately chosen value during the experiments. Fig. 2(b) shows a schematic view of the AIBO and the controlled DOFs.

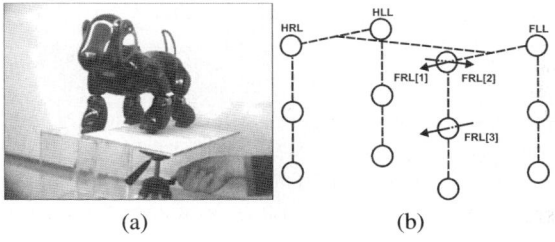

(a)   (b)

Fig. 2. (a) Real AIBO robot mounted over two platforms, one fixed an the other moveable. (b) Scheme of the AIBO controlled DOFs.

The AIBO has a set of 3-axis accelerometers, built into its body. They enable us to calculate the lateral tilt of the robot body.

3.2   The Overall Architecture

We use one generic MPG for each controlled DOF. In order to ensure phase-locked synchronization between the different DOFs of the robot, we couple the different MPGs together. We bilaterally couple the Hopf oscillators of the MPGs, those couplings being illustrated by right-left arrows on fig. 3 and unilaterally couple each swing MPG to the corresponding Knee MPG. For the swing joints, we modify Eqs. 1 and 2 of all the DOFs as follows:

$$\begin{bmatrix} \dot{x}_{i[1]} \\ \dot{z}_{i[1]} \end{bmatrix} = \begin{bmatrix} \beta\mu_i & \omega \\ -\omega & \beta\mu i \end{bmatrix} \begin{bmatrix} x_{i[1]} - y_{i[1]} \\ z_{i[1]} \end{bmatrix} - \beta r_{i[1]}^2 \begin{bmatrix} x_{i[1]} - y_{i[1]} \\ z_{i[1]} \end{bmatrix} + \sum_{j \neq i} \mathbf{R}(\theta_{i[1]}^{j[1]}) \begin{bmatrix} x_{j[1]} - y_{j[1]} \\ z_{j[1]} \end{bmatrix}$$

For the knee joints, we modify Eqs. 1 and 2 of all the knee DOFs as follows:

$$\begin{bmatrix} \dot{x}_{i[3]} \\ \dot{z}_{i[3]} \end{bmatrix} = \begin{bmatrix} \beta\mu i & \omega \\ -\omega & \beta\mu i \end{bmatrix} \begin{bmatrix} x_{i[3]} - y_{i[3]} \\ z_{i[3]} \end{bmatrix} - \beta r_{i[3]}^2 \begin{bmatrix} x_{i[3]} - y_{i[3]} \\ z_{i[3]} \end{bmatrix} + \frac{1}{2} \mathbf{R}(\psi_{i[3]}^{j[1]}) \begin{bmatrix} x_{j[1]} - y_{j[1]} \\ z_{j[1]} \end{bmatrix}$$

where $r_i[k]$ is the norm of vector $(x_i[k], z_i[k])^T$ ($k = 1,3$). The linear terms are rotated onto each other by the rotation matrices $\mathbf{R}(\theta_{i[1]}^{j[1]})$ and $\mathbf{R}(\psi_{i[3]}^{j[1]})$, where $\theta_{i[1]}^{j[1]}$ is the desired relative phase among the i[1]'s and j[1]'s MPGs and $\psi_{i[3]}^{j[1]}$ is the desired relative phase among the i[3]'s and j[1]'s MPGs (i, j = FLF, FRL, HLL, HRL). In our case, we set these values according to table 1, which defines the phases required for performing a walking gait (we exploit the fact that $\mathbf{R}(\theta) = \mathbf{R}^{-1}(-\theta)$). The $\psi_{i[3]}^{j[1]}$ were all set to $-90°$. Due to the properties of this type of coupling among oscillators, the generated trajectories are always smooth and thus potentially useful for real-world implementations such as trajectory generation in a robot.

Table 1. Phase differences between swing oscillators ($i[1]$) used to perform a walking gait

	$\theta_{FLL-FRL}$	$\theta_{FLL-HLL}$	$\theta_{FLL-HRL}$	$\theta_{FRL-HLL}$	$\theta_{FRL-HRL}$	$\theta_{HLL-HRL}$
(°)	-180	-270	-90	-90	90	180

For offset y_i specification, we measure the lateral tilt of the body, ϕ. We want to stretch the legs towards which the robot is tilted, and fold the other legs, thus reducing the robot lateral tilt and keeping the body parallel to the ground. This is achieved by reducing the swing and knee joint values of the former and increasing these joint values for the later.

The effect of this offset joint change should be of opposite effect on the left and right legs of the robot, but should also influence similarly both legs. The $k_{j,i}$ static gain ($j =$ left, right) (eq. 3) is set symmetrically for the robot left and right legs, and is given by

$$k_{left,i} = -C_i \tag{5}$$
$$k_{right,i} = C_i, \tag{6}$$

where C_i is set according to the joints range of each leg such that joints change with the same velocity.

The controller architecture is depicted in fig. 3. The MPG generates discrete and rhythmic trajectories, as well as their superposition. Within the MPG, the discrete

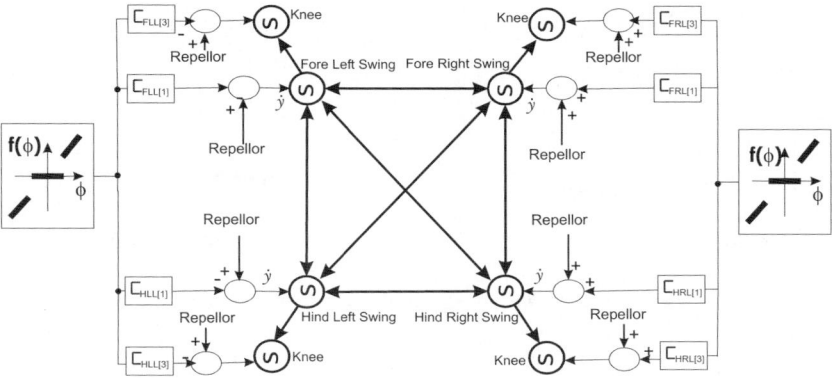

Fig. 3. Controller architecture. The lateral tilt value and the joint limits modulate the y and x trajectories.

system specifies an offset for the rhythmic movement. In this particular situation, the rhythmic motion is turned off, so that the attractor of the Hopf oscillator is a fixed point specified by the offset, i.e, by the discrete movement.

Trajectories generated by this architecture are modulated by sensory feedback, according to the lateral tilt of the body. This tilt is linearly transformed by a f function such that it specifies a rate of change for the robot joints. A larger ϕ results in stronger rates of change, $\dot{y}_i$. y_i define the fixed points towards which the MPG Hopf oscillators will converge. The final trajectories x_i specify the planned joint values needed to reduce the lateral tilt to a minimum. These are sent online for each DOF and the lower level control is done by PID controllers. Because motion is sufficiently slow there is no need to apply inverse dynamics.

4 Results

In this section, we describe two experiments done in a real AIBO robot. The robot stands in a moveable platform and we forced some changes on the tilt of the robot's body. At each sensorial cycle, sensory information is acquired, dynamic equations are calculated and integrated thus specifying servo positions. The robot control loop is measured and has 8 milliseconds. The dynamics of the CPGs are numerically integrated using the Euler method with a fixed time step of 1 ms. Parameters were chosen in order to respect feasibility of the experiment and are given in table 2. We recorded the actual trajectories from the joints incremental encoders $\tilde{\mathbf{x}}$ and the planned trajectories $\mathbf{x}$.

4.1 First Experiment

To show the behavior of the system, we start with a very simple experiment, in which the robot is first inclined of $\sim 6°$ to its left side and, after sometime, it is again inclined of $\sim 5°$ to the same side. We expect the system to react to each of these tilt changes

Table 2. Parameter values used in the experiments

β	ω (rad s^{-1})	μ_i	$\frac{1}{2\beta\mu_i}$ (s)	α (s^{-1}))	σ
1.38	1.0472	−36	0.1	5000	0.05

by reducing it to values belonging to a small region around zero, as defined by the dead-zone. Right and left legs are expected to exhibit a symmetric behavior.

In fig. 4 we can see the result of the experiment in the real dog. We depict actual trajectories $\tilde{\mathbf{x}}_{FLL[3]}$ (solid line) and the planned trajectories $\mathbf{x}_{FLL[3]}$ (dashed line) of the fore leg knees. The controller reduces considerably the tilt in the first 2 seconds, from $\sim 6°$ to $\sim 2°$. When the robot is inclined to the left (positive lateral tilt), $\dot{y}_{FLL[3]}$ is negative (middle panel in the left side of fig. 4), meaning that the offset $y_{FLL[3]}$ is reduced and the leg extends reducing the tilt. Because at $t = 9.4$s the inclination is slightly smaller than at $t = 1$s, $\dot{y}_{FLL[3]}$ has a smaller value. Comparing the right and the left knee $\mathbf{x}_{i[3]}$ trajectories, we see that the system behave as expected, having symmetric trajectories.

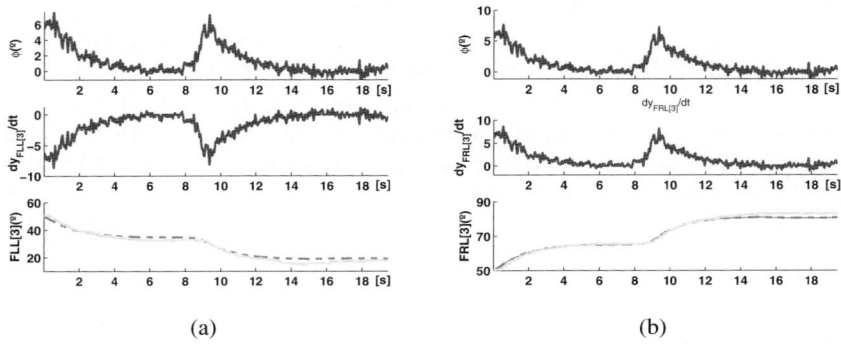

(a)   (b)

Fig. 4. Real experiment. (a) Fore Left Knee (b) Fore Right Knee. Up panel: lateral tilt (ϕ); Middle panel: $\dot{y}_{i[3]}$; Bottom panel: $\mathbf{x}_{i[3]}$ (dashed line) and $\tilde{\mathbf{x}}_{i[3]}$ (solid line) ($i = $ FLL, FRL).

Note that despite the noisy sensorial information, the resultant trajectories are smooth. Further, the sensors are able to follow the planned trajectories as expected.

4.2   Second Experiment

In this experiment, the robot is subjected to more abrupt tilt changes and these happen during the controller recover. At $t \sim 4$s we inclined the robot of $\sim 10°$ to the left direction and we expect the system to react to this perturbation by stretching the left legs and folding the right ones, and as such reducing the tilt. The robot is again inclined at $t \sim 5.8$s to its left side by $\sim 12°$, but this change happens before the system had reached a balanced position. Finally, at $t \sim 8s$, the robot is again inclined but towards its right side by $\sim 8°$. We expect the system to react to this change in order to reach the equilibrium. Further, right and left legs are expected to exhibit a symmetric behavior.

Fig. 5, depicts the obtained results. The robot successfully reacts to lateral tilt changes by reducing this to a minimum. The trajectories are symmetric for right and left legs. When the robot is inclined to the left (positive lateral tilt), $\dot{y}_{FLL[3]}$ is negative, thus $y_{FLL[3]}$ is reduced and the robot extends this leg. At $t \sim 8s$, the robot is inclined to its right and lateral tilt becomes negative. Thus, $\dot{y}_{FLL[3]}$ is positive and $y_{FLL[3]}$ is increased meaning the fore left leg folds. Balanced position is achieved at $t \sim 12s$.

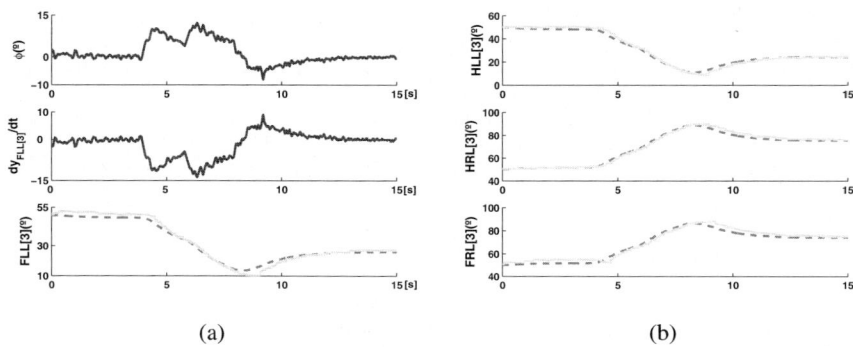

(a)   (b)

Fig. 5. a) Fore Left knee. Up panel: lateral tilt (ϕ); Middle panel: $\dot{y}_{FLL[3]}$; Bottom panel: planned $x_{FLL[3]}$ (dashed line) and $\tilde{x}_{FLL[3]}$ (solid line) trajectories. (b) Planned $x_{i[3]}$ (dashed line) and $\tilde{x}_{i[3]}$ (solid line) trajectories (i =HLL, HRL, FRL).

5 Conclusions

In this article, we have presented a controller for correcting posture in an irregular terrain, where a MPG implemented as two embedded dynamical systems was able to generate discrete mode of movement. Online trajectory modulation is achieved through the inclusion of feedback loops that enable to take external perturbations into account (lateral tilt), such that when the environment changes, the system online adjusts the dynamics of trajectory generation. Moreover, due to the properties of dynamical systems the trajectory will always remain smooth.

As an application we apply a network of MPGs coupled together to the control of a quadruped robot (an ers7 AIBO). The robot was subjected to external perturbations that change its lateral tilts, measured by the built-in accelerometers. The MPGs network uses this information to compensate for the tilt changes and reduce them to near zero. The obtained results have been quite satisfactory. This controller showed to be fast enough since the robot can quickly recover from the induced physical inclinations. The controller also proved to be efficient according to the obtained results.

Presently, we are extending this work to compensate not only the lateral tilt but also the pitch inclination and merging both information in a single controller. We are also working on the integration of the touch sensors' information in order to assure contact of the robots' end effectors with the platform before performing a posture correction. Further, we are extending this work to combine this approach in order to obtain locomotion.

References

1. Pratt, J., Chew, C., Torres, A., Dilworth, P., Pratt, G.: Virtual Model Control: An intuitive approach for bipedal locomotion. The Int. J. of Robotics Research 20(2), 129–143 (2001)
2. Fukuoka, Y., Kimura, H., Cohen, A.: Adaptive dynamic walking of a quadruped robot on irregular terrain based on biological concepts. Int. J. of Robotics Research 3–4, 187–202 (2003)
3. Ijspeert, A., Nakanishi, J., Schaal, S.: Learning attractor landscapes for learning motor primitives. Advances in Neural Information Processing Systems 15, 1547–1554 (2003)
4. Delcomyn, F.: Neural basis for rhythmic behaviour in animals. Science 210, 492–498 (1980)
5. Giszter, S., Mussa-Ivaldi, F., Bizzi, E.: Convergent force fields organized in the frog's spinal cord. J. of Neuroscience 13, 467–491 (1993)
6. Taga, G.: Emergence of bipedal locomotion through entrainment among the neuro-musculo-skeletal system and the environment. Physica D 75(1-3), 190–208 (1994)
7. Blickhan, R.: The spring-mass model for running and hopping. J. Biomechanics 22(11–12), 1217–1227 (1989)
8. Bühler, M., Koditscheck, S.: Planning and control of a juggling robot. Int. J. of Robotics Research 13(2), 101–118 (1994)
9. Degallier, S., Santos, C.P., Righetti, L., Ijspeert, A.: Movement Generation using Dynamical Systems: a Drumming Humanoid Robot. In: Humanoids 2006: IEEE-RAS International Conference on Humanoids Robots, Genova, Italy, December 4-6 (2006)
10. Santos, C.P., Ferreira, M., Oliveira, M., Pires, A., Dégallier, S., Ijspeert, A.: Choreography generation for a quadruped robot using dynamical systems. Autonomous Robots (submitted)
11. Vukobratovic, M., Borovac, B.: Zero-moment point - thirty five years of life. International Journal of Humanoid Robotics 1(1), 157–173 (2004)

Propose of a Benchmark for Pole Climbing Robots

Mahmoud Tavakoli, Lino Marques, and Anibal T. de Almeida

Institute for Systems and Robotics - University of Coimbra,3030-290 Coimbra, Portugal
mahmoodtavakoli@gmail.com, {lino,anibal}@isr.uc.pt

Summary. Development of climbing robots was a challenging area during last decade and re-ceived an increased attention in recent years. On the other hand benchmarking is considered an important factor for robotic researches as it can reduce unnecessary efforts and orient re-searches to the proper direction. In this paper a set of benchmarks and testing methodologies for pole climbing robots are proposed.

Keywords: Pole climbing robots, Benchmarking,Testing Methodology.

1 Introduction

The current practice of publishing research results in robotics makes it extremely diffi-cult not only to compare results of different approaches, but also to asses the quality of the research presented by the authors. Though for pure theoretical articles this may not be the case, typically when researchers claim that their particular algorithm or system is capable of achieving some performance, those claims are intrinsically unverifiable, either because it is their unique system or just because a lack of experimental details, including working hypothesis [1].

Benchmarking can principally reduce the research efforts by preventing ineffective researches and as a result of benchmarking competitions can be designed to have com-parative results from different robots. The importance of Robotic Benchmarking was mainly discusses by Angel P. del Pobil in [2].

According to Wikipedia, benchmarking is a process used in management, in which organizations evaluate various aspects of their processes in relation to best practice, usually within their own sector. A short definition of benchmarking proposed in [3]. It includes three essential aspects of benchmarks:

1. Task: the robot has to perform a given mission, e.g., it actually has to do some-thing.
2. Standard:the benchmark is accepted by a significant set of experts in the field.
3. Precise Definition: the task is described exactly, especially the execution Environ-ment, the mission goal, and limiting constraints.

According to Dillmann [5], this definition lacks one important feature of bench-marks, which is a numerical evaluation of the performance. Without that, it is only possible to decide whether or not a given system is able to perform a mission. What we need in fact is to "develop performance metrics" [4] for a given application [5].

There are also disadvantages connected to the introduction of benchmarks. As soon as benchmarks enter the field and are widely respected, researchers and manufacturers

H. Bruyninckx et al. (Eds.): European Robotics Symposium 2008, STAR 44, pp. 215–222, 2008.
springerlink.com

are likely to compare and optimize their products to the benchmarks rather than to the real application areas. Whenever there exists a gap between the benchmark and the real world, optimization towards the benchmark test will not necessarily improve the system's performance in the real application [5]. So it is important to consider real applications when designing a benchmark. To address this problem, benchmarks proposed in this paper are focused on achieving of practical and useful tasks in Industrial applications.

Definition of benchmarks for robotic applications consists of definition of 3 interrelated elements (Fig. 1), which are the robot, the environment and the mission. A working environment should be fully defined, and then the robot mission should be defined based on the environment. The robot mission itself consists of some tasks. Then the functionality of different robots can be measured according to the level of success in achievement of each task on the defined environment.

In this paper basic benchmarks for pole climbing robots are proposed, the mission is defined and a fully defined structure for tests is also presented. But standardization of this mainly depends of acceptance by more experts in this area.

2   Climbing Robots

Climbing robots have received an increased attention in recent years due to their potential application in several areas, namely: in the construction and maintenance of tall buildings and bridges; in the shipyard production and for general operation in dangerous areas with difficult access to humans, like nuclear facilities.

Development of climbing robots was a challenging area during last decade. Different types of climbing robots were developed for climbing over flat or curved surfaces. For holding the robot attached to the surface they used suction cups [6][7][8][9][10] (or magnets [11][12]). Also robots whose end-effectors match engineered features of the environment like fences or porous materials or bars were developed[13][14][15][16][17]. Different kind of robots were also developed for climbing inside pipes or ducts [18][19]

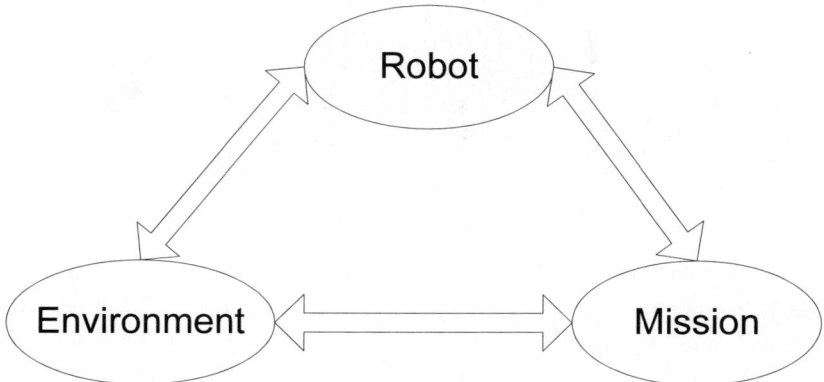

Fig. 1. Interrelation of the elements in benchmarking process

or climbing over poles [20][21][22][23]. Figure 2 shows a climbing robot designed to climb over poles and pipes [22].

Climbing robots can be divided to two main groups: Wall Climbing Robots (WCRs) and Pole Climbing Robots (PCRs).Even though both WCRs and PCRs have common similarities in problems, they also have several differences and could not be discussed together.

During last decade, most of the research in this area was focused on WCRs and only a few PCRs were developed. A factor contributing to this fact is certainly the higher difficulty in designing PCR, since while WCR can use standard grippers like vacuum cups, to stick to the wall, PCRs need special grippers with dedicated design.

3 Benchmarking

Benchmarking for Pole Climbing Robots is of great importance because of increasing interests on this research area. In the current research we tried to propose definition of a benchmark for a PCR system, for a given task. To achieve this two other parameters should be well defined, the working area of the PCR and desired task that the robot should achieve. The working area of the PCR is the structure that the robot should climb.

Fig. 2. A Pole Climbing Robot [22]

3.1 Working Area of the Robot (Environment)

According to authors' experiences on design of climbing robots [22] [24], and [25], geometry and size of the working can have a huge impact in complexity of the robot. Structures can vary from a simple straight pole, to poles with bends and branches, and even with changes on the cross section size.

As stated earlier, benchmarks which proposed in this paper are focused on achievement of practical and useful tasks in Industrial applications. For this the structure should be similar to real human made 3D structures e.g. pipe structures which are used in petrochemical plants. These kinds of structures usually are not a straight pole, and include bent section and branches. On the other hand robots which can just climb over straight poles are very different with those which can overcome bents and branches. They are less complicated in several technical aspects and can not be categorized with those which can pass bent and branches. So the structure for benchmarking should include bent section and branches. Angle of bent sections is also an important parameter. Usually in human made plants, poles and tubes have bent with 90°, while in few cases it can be between 45° to 90°. So structure is limited to have angles between 45 to 90 degrees.

The size of the structure cross section can also affect the level of complexity of the robot. The size of the structure can be defined in a certain range, but if the outer diameter of the cross section profile is more than 400 mm, the robot can be designed like wall climbing robots and take advantageous of vacuum or magnet gripper, which is not the case.

We should not consider structures designed specifically to facilitate the robot operation, instead the robot should be designed to operate in common real structures.

According to these considerations a structure was designed and built (as shown in Fig. 3). As it can be seen in the picture, the structure includes bent sections of 45 and 90 degrees. The outer diameter of the pipes is 219 mm.

3.2   Definition of the Mission

A mission is defined for the robot to perform it on the defined environment. The mission defined as:

The robot manipulator should climb over structure and continue climbing in bent section and branches. The robot manipulator should also be able to reach to any position in the structure and scan all surface of the structure, since for practical application e.g. NDT test of welding on the pipe it is necessary that robot be able to reach every position on the structure.

Then the mission is divided to some submissions which are called tasks. The following set of standard tasks is defined:

Start climbing from one side of the structure
Scanning some areas of the structure
Descending from the other side of the structure.
Scanning the entire surface of the structure and finding the defected areas and publishing a report of that.

3.3   Metrics and Quantitative Evaluations

To evaluate the performance of robots, quantitative metrics for different parameters should be considered. Some parameters considered to be used for the performance evaluation of robots are:

Fulfilment of the mission and quality of achievement
Speed of the robot

Fig. 3. Developed structure for Pole Climbing Robot Benchmarking

Fault tolerance of the robot
Self attachment to the pole
Ability to move in floor
Level of automation.

Fulfillment of the Mission, Quality of Performance

Fulfilment of the mission and quality of performance are the most important parameter of evaluation. They can include 50 percent of the total score. If the robot can just climb from one side of the structure, scan some parts and descend from the other side of the pole, it means that it fulfills the mission. The quality of performance can be measured in several methods. A suggestion can be distributing some defected weldings in the structure, and the robot should find and register position of the defected weldings. It can be done by an ultrasonic NDT probe, which can be installed on the robot manipulator. The quantity of non detected items or falsely detected items will determine the score. Table 1 shows metrics for evaluation of all parameters.

Speed of the Robot

Speed of climbing is an important factor, because it is a fundamental parameter in industrial applications. Scoring method should be normalized so that the fastest robot gains 10 scores and followers gain fewer scores.

Fault Tolerance

As the robot should work in high altitudes, it is important to consider the performance of the robot in case of power fails. Scoring method depends on the level of safety considered for such case. If developers claim that robot can stay attached to the structure and maintain the position in the structure in case of power failure, then it should be tested and scored.

Power failure can occur for controller power supply or main power supply. The robot gains 10 scores if the robot is tolerant to both cases. If it is tolerant to one case and not the other it gains 5 scores.

Ability to Move on Floor and Self Attachment to the Pole

It will be quite interesting if the robot can move on the floor, with wheels or legs. In this case the robot can perform bigger missions. For instance it can start to test a set of structures, because it can move on floor and locate each structure independently.

On the other hand to this another parameter should be considered: The robot should be able to attach itself to the pole without the help of human. It mostly depends on the design of the gripper. Posses of each of these items can gain 5 scores.

Level of the Automation

Level of the automation is also of great importance. The most autonomous robot is the one which can perform a mission without having the geometry and dimensions of the structure. It means that the robot also make the "World Modeling". This robot gains 10 scores. A robot can be considered Semi-autonomous if it can perform the mission autonomously when it has the geometry and dimensions of the structure in advance. A robot which is controlled manually e.g. by a joystick is considered non autonomous and don't gain any score.

Other Parameters

10 score is left for the referees to decide according to other parameters. Parameters like Modular design, Simplicity, creativity and etc. are some examples.

4   Summary

In this paper benchmarking for pole climbing robots discussed and a special benchmark for Pole climbing robots suggested. A structure for PCR benchmarking was also designed and developed. The designed benchmark mainly considers industrial application for pole climbing robots to overcome the gap between industrial applications

Table 1. Metrics for evaluation of PCRs performance

Item	Evaluation Method	Total Scores
Fulfillment of the mission	The robot can climb over the structure from one side and descend from the other part.	30
Quality of Performance	10 defected weldings should be recognized and their position should be registered. Each successful register gains 2 scores.	20
Speed of mission achievement	The fastest robot gain 10 points and the slowest one gain 0 points.	10
Fault tolerance	The robot gains 10 scores if the robot is tolerant to failure on main power and controller power. If it is tolerant just to one of them it gains 5 scores.	10
Ability to move on floor	Yes=5 points No =0 Points	5
Self attachment to the pole	Yes=5 points No=0 Points	5
Level of Automation	Fully Autonomous:10 Semi Autonomous:5 Non autonomous:0	10
Other Parameters Modular design Simplicity, creativity	Without metrics	10

and laboratory prototypes. Scoring system proposed in this paper is a suggestion and can be discussed and changed as some parameter might be considered of more or less importance.

References

1. del Pobil, A.P.: Benchmarks in Robotics Research. In: IROS 2006 Workshop on benchmarks in robotics research, Beijing, China (2006)
2. del Pobil, A.P.: Why do We Need Benchmarks in Robotics Research? In: IROS 2006 Workshop on benchmarks in robotics research, Beijing, China (2006)
3. Hanks, S., Pollack, M.E., Cohen, E.: Benchmarks, Test Beds, Controlled Experimentation, and the Design of Agent Architectures. AI Magazine 14(4) (1993)
4. Jacoff, A., Messina, E., Evans, J.: Reference Test Arenas for Autonomous Mobile Robots. In: 14th International FLAIRS Conference (2001)
5. Dillmann, R.: Benchmarks for Robotics Research. EURON (April 2004),
   http://www.cas.kth.se/euron/eurondeliverables/
   ka1-10-benchmarking.pdf
6. Dulimarta, H., Tummala, R.L.: Design and control of miniature climbing robots with non-holonomic constraints. In: World Congress on Intelligent Control and Automation, Shanghai, P.R.China (2002)
7. Nagakubo, A., Hirose, S.: Walking and running of the quadruped wall climbing robot. In: IEEE Int. Conf. on Rob. And Aut., vol. 2, pp. 1005–1012 (1994)

8. Rachkov, M.: Control of climbing robot for rough surfaces. In: Int. Workshop on Robot Motion and Control, pp. 101–105 (2002)
9. Ryu, S.W., Park, J.J., Ryew, S.M., Choi, H.R.: Self-contained wall-climbing robot with closed link mechanism. In: IEEE/RSJ Int. Conf. on Int. Rob. And Sys, Maui, HI (2001)
10. Yan, W., Shuliang, L., Dianguo, X., Yanzheng, Z., Hao, S., Xuesban, G.: Development and application of wall-climbing robots. In: IEEE Int. Conf. on Rob. And Aut. Detroit, MI (1999)
11. Grieco, J.C., Prieto, M., Armada, M., de Santos, P.G.: A six-legged climbing robot for high payloads. In: IEEE Int. Conf. on Cont. App., Trieste, Italy,
12. Hirose, S., Nagabuko, A., Toyama, R.: Machine that can walk and climb on floors, walls, and ceilings. In: ICAR 1991, Pisa, Italy, pp. 753–758 (1991)
13. Bevly, D., Dubowsky, S., Mavroidis, C.: A simplified Cartesian-computed torque controller for highly geared systems and its application to an experimental climbing robot. ASME J. of Dynamic Systems, Measurement, and Control 122(1), 27–32 (2000)
14. Xu, Y., Brown, H., Friendman, M., Kanade, T.: Control system of the selfmobile space manipulator. IEEE Tr. on Cont. Sys. Tech. 2(3), 207–219 (1994)
15. Yim, M., Homans, S., Roufas, K.: Climbing with snake-robots. In: IFAC Workshop on Mobile Robot Technology, Jejudo, Korea (2001)
16. Amano, H., Osuka, K., Tarn, T.J.: Development of vertically moving robot with gripping handrails for fire fighting. In: IEEE/RSJ Int. Conf. on Int. Rob. And Sys. Maui, HI (2001)
17. Balaguer, C., Giménez, A., Pastor, J., Padrón, V., Abderrahim, M.: A climbing autonomous robot for inspection applications in 3D complex environments. Robotica 18, 287–297 (2000)
18. Neubauer, W.: A spider-like robot that climbs vertically in ducts or pipes. In: Int. Conf. on Int. Rob. And Sys., Munich, Germany, pp. 1178–1185 (1994)
19. Roßmann, T., Pfeiffer, F.: Control of an eight legged pipe crawling robot. In: Int. Symp. on Experimental Robotics, pp. 353–346 (1997)
20. Almonacid, M., Saltarén, R., Aracil, R., Reinoso, O.: Motion planning of a climbing parallel robot. IEEE Tr. on Rob. And Aut. 19(3), 485–489 (2003)
21. Ripin, Z., Soon, T.B., Abdullah, A., Samad, Z.: Development of a low-cost modular pole climbing robot. In: TENCON, Kuala Lumpur, Malaysia, vol. 1, pp. 196–200 (2000)
22. Tavakoli, M., Zakerzadeh, M.R., Vossoughi, G.R., Bagheri, S.: A hybrid Pole Climbing and Manipulating Robot with Minimum DOFs for Construction and Service Applications. Journal of Industrial Robot (March 2005)
23. Baghani, A., Ahmadabadi, M., Harati, A.: Kinematics Modelling of a Wheel-Based Pole Climbing Robot (UT-PCR). In: IEEE International Conference on Robotics and Automation, Barcelona (2005)
24. Tavakoli, M., Marques, L., de Almeida, A.: Pole climbing and manipulating robots: Assessment of different design categories. In: Proc. 37th Intl. Symp. on Robotics, Munich, Germany (2006)
25. Tavakoli, M., Zakerzadeh, M.R., Vossoughi, G.R., Bagheri, S., Salarieh, H.: A Novel Serial/Parallel Pole Climbing/Manipulating Robot: Design, Kinematic Analysis and Workspace Optimization with Genetic Algorithm. In: 21st International Symposium on Automation and Robotics in Construction, Jeju island, Korea

Rat's Life: A Cognitive Robotics Benchmark

Olivier Michel[1], Fabien Rohrer[2], and Yvan Bourquin[1]

[1] Olivier Michel, Cyberbotics Ltd., PSE C - EPFL, 1015 Lausanne, Switzerland
{Olivier.Michel,Yvan.Bourquin}@cyberbotics.com
[2] Fabien Rohrer, Swiss Federal Institute of Technology in Lausanne (EPFL), 1015 Lausanne, Switzerland
Fabien.Rohrer@epfl.ch

Summary. This paper describes Rat's Life: a complete cognitive robotics benchmark that was carefully designed to be easily reproducible in a research lab with limited resources. It relies on two e-puck robots, some LEGO bricks and the Webots robot simulation software. This benchmark is a survival game where two robots compete against each other for resources in an unknown maze. Like the rats in cognitive animal experimentation, the e-puck robots look for feeders which allow them to live longer than their opponent. Once a feeder is reached by a robot, the robot draws energy from it and the feeder becomes unavailable for a while. Hence, the robot has to further explore the maze, searching for other feeders while remembering the way back to the first ones. This allows them to be able to refuel easily again and again and hopefully live longer than their opponent. ...

1 Why We Need Cognitive Robotics Benchmarks

1.1 Introduction

Most scientific publications in the area of robotics research face tremendous challenges: comparing the achieved result with other similar research results and hence convincing the reader of the quality of the research work. These challenges are very difficult because roboticists lack common tools allowing them to evaluate the absolute performance of their systems or compare their results with others. As a result, such publications often fail at providing verifiable results, either because the studied system is unique and difficult to replicate or they don't provide enough experimental details so that the reader could replicate the system accurately.

Nevertheless, some of these publications become the de facto state of the art and this makes it extremely difficult to further explore these research areas, and hence to demonstrate advances in robotics research.

This matter of fact is unfortunately impairing the credibility of robotics research. A number of robotics researchers proposed to develop series of benchmarks to provide a means of evaluation and comparison of robotics research results [1, 2, 3, 7, 17].

Although a few robotics benchmarks already exist, the only robotics benchmarks that are widely known and practiced are implemented as robot competitions.

1.2 Not All Robot Competitions Are Suitable Benchmarks

Popular robot competitions include Robocup [5] and FIRA [14], where various teams of robots play soccer, MicroMouse [9], where wheeled robots have to solve a maze,

H. Bruninckx et al. (Eds.): European Robotics Symposium 2008, STAR 44, pp. 223–232, 2008.
springerlink.com

FIRST [10], Eurobot [11] and Robolympics [12], where robots compete in many disciplines, the AAAI Robot Competition [13], where robots have to solve different tasks in a conference environment, and the DARPA Grand Challenge [15] and the European Land-Robot Trial [16], where unmanned ground and aerial vehicles race against each other.

Although some of these competitions clearly focus on education and are more intended to students and children rather than researchers (FIRST, Eurobot), others competitions (Robocup, FIRA, AAAI) are more intended to researchers. Such competitions are useful as they can provide elements of comparison between different research results. However one of the major problem is that the rules often change across the different editions of the same competition. Hence it is difficult to compare the progress achieved over time. Also these competitions are very specific to particular problems, like Robocup is focused mostly on robot soccer and has arguably a limited interest for cognitive robotics [4]. Moreover, in most cases, and especially in the Robocup case, installing a contest setup is expensive and takes a lot of resources (many robots, robot environment setup, room, maintenance, controlled lighting conditions, etc.).

1.3   Going Further with Cognitive Robotics Benchmarks

Among all the benchmarks we reviewed which are mostly robot competitions, none of them provides both stable rules with advanced cognitive robotics challenges and an easy setup. This paper proposes a new robotics benchmark called "Rat's Life" that addresses a number of cognitive robotics challenges while being cheap and very easy to setup for any research lab. The aim of this benchmark is to foster advanced robotics and AI research.

Comparing to soccer playing contests (RoboCup, FIRA), the Rat's Life benchmark is more bio-inspired as it focuses on foraging and survival. Also, it is more likely to contribute to scientific advances in Learning and Self Localization And Mapping (SLAM) as mazes are initially unknown to the robots. Moreover, it allows the researchers to focus on a single agent (competiting against another) rather than a whole team of agents, making the problem somehow simpler to handle. Finally, it is cheaper.

2   Benchmark Requirements

In order to be useful a benchmark has to be practiced by a large number of the best researchers trying to push further the current state of the art. This can be achieved by proposing a scientifically and practically appealing series of benchmarks that will convince researchers to invest their time with these tools. Hence the Rat's Life benchmark is trying to achieve a number of objectives:

2.1   Scientifically Appealing

To be scientifically interesting, a benchmark has to address a number of difficult challenges in robotics. The Rat's Life benchmark focuses on cognitive robotics and addresses advanced research topics such as image processing, learning, navigation in an unknown environment, landmark recognition, SLAM, autonomy management, game strategies, etc.

2.2   Cheap and Easy to Setup

The benchmark should be easily practicable by any researcher. Hence it has to be cheap and easy to setup. All the components should be easily available. The Rat's Life benchmark costs no more than EUR 2000, for two e-puck robots and many LEGO components (including a LEGO NXT unit and four LEGO distance sensors). It requires only a table to setup a LEGO maze of 114x114 cm.

2.3   Accurate

Accuracy is a very important aspect of a benchmark. The environment, robots and evaluation rules should be defined very carefully in an exhaustive manner. This way, the benchmark is accurately replicable and hence different results obtained with different instances of the setup in different research lab can be compared to each others.

2.4   Comparable

Finally, a benchmark is useful if users can compare their own results to others and thus try to improve the state of the art. Hence a benchmark should keep a data base of the solutions contributed by different researchers, including binary and source code of the robot controller programs. These different solutions should be ranked using a common performance metrics, so that we can compare them to each other.

3   Standard Components

The Rat's Life benchmark is based on three standard affordable components: the e-puck mobile robot, LEGO bricks and the Webots robot simulation software (free version).

3.1   The E-Puck Mobile Robot

The e-puck mini mobile robot was originally developed at the EPFL for teaching purposes by the designers of the successful Khepera robot. The e-puck hardware and software is fully open source, providing low level access to every electronic device and offering unlimited extension possibilities. The robot is already equipped with a large number of sensors and actuators (figure 1). It is well supported by the Webots simulation software with simulation models, remote control and cross-compilation facilities. The official e-puck web site [20] gathers a large quantity of information about the robot, extension modules, software libraries, users mailing lists, etc. The robot is commercially available from Cyberbotics [19] for about EUR 570.

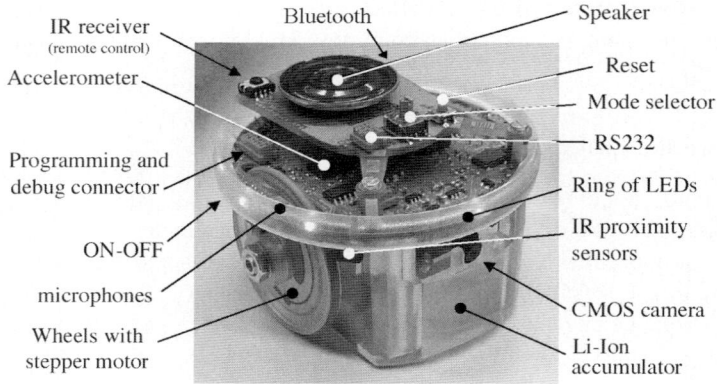

IR receiver (remote control)
Accelerometer
Bluetooth
Speaker
Reset
Mode selector
RS232
Programming and debug connector
ON-OFF
microphones
Wheels with stepper motor
Ring of LEDs
IR proximity sensors
CMOS camera
Li-Ion accumulator

Fig. 1. The e-puck robot

Fig. 2. The Rat's Life maze: LEGO bricks, e-puck robots and a feeder device (left) and its simulated counterpart (right)

3.2   LEGO Bricks

The LEGO bricks are used to create an environment for the e-puck robot. This environment is actually a maze which contains "feeder" devices (see next sections) as well as visual landmarks made up of patterns of colored LEGO brick in the walls of the maze (see figure 2). These landmarks are useful hints helping the robot to navigate in the maze. Since LEGO models are easily demountable, the maze is easily reconfigurable so that the users can create different instances of the maze according to the specifications of the benchmark.

All the maze, landmarks and the feeder devices are properly defined in a LEGO CAD file in LXF format using the LEGO digital designer software freely available from the LEGO factory web site [18]. The corresponding LXF files are freely available on the Rat's Life web site [8].

Thanks to the LEGO factory system, users can very easily order a box containing all the LEGO bricks necessary to build the environment of the robots.

3.3   The Webots Robot Simulation Software

Webots [6] is a commercial software for fast prototyping and simulation of mobile robots. It was originally developed at the Swiss Federal Institute of Technology in Lausanne (EPFL) from 1996 and has been continuously developed, documented and supported since 1998 by Cyberbotics Ltd. Over 500 universities and industrial research centers worldwide are using this software for research and educational purposes. Webots has already been used to organize robot programming contests (ALife contest and Roboka contest).

Although Webots is a commercial software, a demo version is freely available from Cyberbotics's web site [19]. This demo version includes the complete Rat's Life simulation. So, anyone can download, install and practice the simulation of the Rat's Life benchmark at no cost.

4   Rat's Life Benchmark Description

This paper doesn't claim to be a technical reference for the Rat's Life benchmark. Such a technical reference is available on the Rat's Life web site [8].

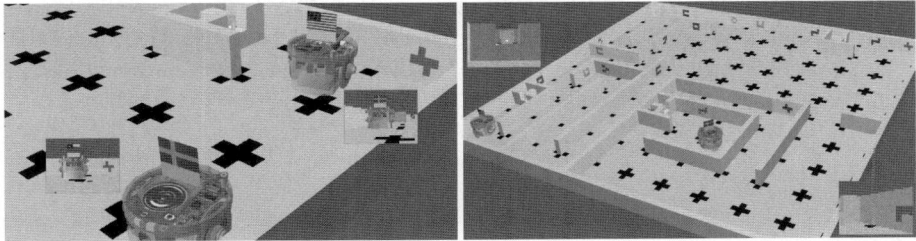

Fig. 3. Closeup of the Rat's Life simulated robots in Webots (left) and general overview (right)

4.1   Software-Only Benchmark

The Rat's Life benchmark defines precisely all the hardware necessary to run the benchmark (including the robots and their environment). Hence the users of the benchmarks don't have to develop any hardware. Instead, they can focus on robot control software development only. This is similar to the Robocup standard league where the robot platforms (Aibo robots) and the environment is fully defined and the competitors are limited to develop control software only. This has the disadvantage of preventing hardware research and is constraining the contest to the defined hardware only. However, it has the great advantage of letting the users focus on the most challenging part of cognitive robotics, i.e., the control software.

4.2   Configuration of the Maze

For each evaluation, the maze is randomly chosen among a series of 10 different configurations of the maze. In each configuration, the walls, landmarks and feeder are placed at different locations to form a different maze. Each configuration also has 10 different possible initial positions and orientations for the two robots. One of them is chosen randomly as well. This makes 100 possible initial configurations. This random configuration of the maze prevents the robots from having a prior knowledge of the maze, and forces them to discover their environment by exploring it. This yields to much more interesting robot behaviors. A possible configuration is depicted on figure 3 (right).

4.3   Virtual Ecosystem

The Rat's Life benchmark is a competition where two e-puck robots compete against each other for resources in a LEGO maze. Resources are actually a simulation of energy sources implemented as four feeder devices. These feeder devices are depicted on figure 4. They are made up of LEGO NXT distance sensors which are controlled by a LEGO NXT control brick. They display a red light when they are full of virtual energy. The e-puck robots can see this colored light through their camera and have to move forward to enter the detection area of the distance sensor. Once the sensor detects the robot, it turns its light off to simulate the fact that the feeder is now empty. Then, the robot is credited an amount of virtual energy corresponding to the virtual energy that was stored in the feeder. This virtual energy will be consumed as the robot is functioning and could be interpreted as the metabolism of the rat robot. The feeder will remain empty (i.e., off) for a while. Hence the robot has to find another feeder with a red light on to get more energy before its energy level reaches 0. When a robot runs out of virtual energy (i.e., its energy level reaches 0), the other robot wins.

Fig. 4. A full feeder facing an e-puck robot (left) and an empty one (right)

Biological Comparison

This scenario is comparable to an ecosystem where the energy is produced by feeders and consumed by robots. The feeders could be seen as plants, slowy growing from the energy of the sun, water and ground and producing fruits whereas the robots could be seen as rats, foraging fruits. Since the fruits produced by a single plant are not sufficient to feed a rat, the rat has to move around to find more plants.

Electronics Comparison

Although the e-puck robots and feeder devices used in the Rat's contest are electronic devices, the energy is actually simulated for convenience reasons. However, it could be possible to deal with real electrical energy: the feeder would correspond to photovoltaic solar docking stations accumulating electrical energy over time and the robots could recharge their actual battery from these stations. However, such a system would be more complex to setup from a practical and technological point of view. This is why we decided to use virtual energy instead.

4.4 Robotics and AI challenges

Solving this benchmark in an efficient way requires the following cognitive capabilities:

- Recognize a feeder (especially a full one) from a camera image.
- Navigate to the feeder and dock to it to grab energy.
- Navigate randomly in the maze while avoiding to get stuck.
- Remember the path to a previously found feeder and get back to it.
- Optimize energy management.
- Try to prevent the other robot from getting energy.

This translates into a number of control software techniques, namely image processing, motor control, odometry, landmark based navigation, SLAM, autonomy management, game theory. Most of these techniques are still open research areas where new progress will benefit directly to robotics and AI applications. Both bio-inspired (neural networks, generic algorithms, learning) and traditional approaches (control theory, environment mapping) are concerned as no assumption is made on the technologies used to implement the controllers. Moreover, because of its similarities with experiments with rodents, the Rat's Life contest may be a very interesting benchmark for testing different bio-inspired models, such as place cells, grid cells, spatial learning, conditioning, etc.

   The best robots are expected to be able to somehow fully memorize the maze they explore with the help of the landmarks, to rapidly find their way to the feeders, to maintain an estimation of the status of every feeder and to develop a strategy to prevent the opponent from recharging.

4.5 Online Contest

Real World and Simulation

The Rat's Life contest is defined both as a real environment and a simulation. However, the same control programs, written in C or Java programming language, can run on both the simulation and the real robots. To run the control program on the real robots, there are actually two options. The user can either execute the controller program on a computer remote controlling the robot or cross-compile it and execute it on the real robot. In the first case, the program running on the computer remote controls the real robot by reading the sensor values from and sending the motor commands to the Bluetooth connection with the robot. In the second case, the control program is executed directly on the real robot. All the necessary software tools for remote control and cross-compilation are integrated within the Webots software, making the transfer from the simulation to the real robot a very easy process. This way, the same controller program can control both the real and the simulated robot.

Participation to the Contest

In order to participate in the online contest, the competitors can download the free version of Webots from Cyberbotics' web site [19]. They can program the simulated e-puck robots to perform in the simulated maze. Then, they have to register a contestant account on the contest's web site [8]. Once open, this account allows the competitors to upload the controller programs they developed with the free version of Webots. Participation to the contest is totally free of charge.

Ranking System

Every business day (i.e., Monday to Friday) at 12 PM (GMT) a competition round is started in simulation and can be watched online from the Rat's Life web site [8]. A hall of fame displays a table of all the competitors registered in the data base and who submitted a robot controller program. If there are N competitors in the hall of fame, then $N - 1$ matches are played. The first match of a round opposes the last entry, i.e., number N at the bottom of the hall of fame, to the last but one entry, i.e., number $N - 1$. If the robot number N wins, then the position of these two robots in the hall of fame are switched. Otherwise no change occurs in the hall of fame. This procedure is repeated with the new robot number $N - 1$ (which may have recently changed due to the result of the match) and robot number $N - 2$. If robot number $N - 1$ wins, then it switches its position with robot number $N - 2$, otherwise nothing occurs. This is repeated with robots number $N - 3$, $N - 4$, etc. until robots number 2 and 1, thus totaling a number of $N - 1$ matches.

This ranking algorithm is similar to the bubble sort. It makes it possible for a newcomer appearing initially at the bottom of the ranking, to progress until the top of the ranking in one round. However, any existing entry cannot loose more than one position in the ranking during one round. This prevents a rapid elimination of a good competitor (which could have been caused by a buggy update of the controller program for example).

And the Winner Is...

The contest is open for a fixed period of time. During this period of time, new contestants can register and enter the contest. The contestants can submit new versions of their controller program any time until the closing date. Once the closing date is reached. The top entry of the hall of fame is declared to be the "winner of the simulated Rat's Life benchmark" and its authors will receive a prize for this. Moreover the top 5 competitors will be selected for a real world series of 4 rounds (i.e., 16 matches). The winner of these real world rounds will be declared to be the "winner of the real world Rat's Life benchmark" and will receive a prize.

The real world rounds should however occur during an international conferences or robotics competition to ensure that a large number of people, including a scientific committee, attends the event and can check that nobody is cheating the benchmark.

The contest will run continuously over years so that we can measure the progress and performances of the robot controllers over a fairly long period of robotics and AI research.

5   Expected Outcomes and Conclusions

Thanks to the Rat's Life benchmark, it will become possible to evaluate the performance of various approaches to robot control for navigation in an unknown environment, including various SLAM and bio-inspired models. The performance evaluation will allow us to make a ranking between the different control programs submitted, but also to compare the progresses achieved over several years of research on this problem. For example, we could compare the top 5 controller programs developed in 2008 to the top 5 controller programs developed in 2012 and evaluate how much the state of the art progressed.

The control program resulting from the best robot controllers could be adapted to real world robotics applications in the areas of surveillance, mobile manipulators, UAV, cleaning, toys, etc. Also, interesting scientific comparisons with biological intelligence could be drawn by opposing the best robot controllers to a real rat (or a rat-controlled robot) in a similar problem. Similarly, we could also oppose the best robot controllers to a human (possibly a child) remote controlling the robot with a joystick and with limited sensory information coming only from the robot sensors (mainly the camera).

We hope that this initiative is a step towards a more general usage of benchmarks in robotics research. By its modest requirements, simplicity, but nevertheless interesting challenges it proposes, the Rat's Life benchmark has the potential to become a successful reference benchmark in cognitive robotics and hence open the doors to more complex and advanced series of cognitive robotics benchmarks.

References

1. Baltes, J.: A benchmark suite for mobile robots. In: IROS 2000 conference proceeding, vol. 2, pp. 1101–1106 (2000)
2. Serri, A.: A Lego robot for experimental benchmarking of robust exploration algorithms. Circuits and Systems 3, 163–166 (2004)
3. Eaton, M., Collins, J.J., Sheehan, L.: Toward a benchmarking framework for research into bio-inspired hardware-software artefacts. In: Artificial Life and Robotics, Japan, March 2001, vol. 5(1), pp. 40–45. Springer, Heidelberg (2001)
4. Collins, J.J., Eaton, M., Mansfield, M., Haskett, D., O'Sullivan, S.: Developing a benchmarking framework for map building paradigms. In: Proceedings of the 9th International Symposium on Artificial Life and Robotics, Oita, Japan, January 2004, pp. 614–617 (2004)
5. Kitano, H., Asada, M., Kuniyoshi, Y., et al.: RoboCup: the robot world cup initiative. In: IJCAI-95 workshop on entertainment and AI/ALife (1995)

6. Michel, O.: Webots: Professional Mobile Robot Simulation. Journal of Advanced Robotics Systems 1(1), 39–42 (2004), http://www.ars-journal.com/International-Journal-of-Advanced-Robotic-Systems/Volume-1/39-42.pdf
7. Dillmann, R.: Benchmarks for Robotics Research. EURON (April 2004), http://www.cas.kth.se/euron/eurondeliverables/ka1-10-benchmarking.pdf
8. Rat's Life contest, http://www.ratslife.org
9. Micromouse contest, http://www.micromouseinfo.com
10. FIRST Robotics competition, http://www.usfirst.org
11. Eurobot Robotics competition, http://www.eurobot.org
12. ROBOlympics, http://www.robolympics.net
13. AAAI Robot Competition and exhibition, http://palantir.cs.colby.edu/aaai07
14. The Federation of International Robot-soccer Association (FIRA), http://www.fira.net
15. DARPA Grand Challenge, http://www.darpa.mil/grandchallenge
16. European Land-Robot Trial ELROB, http://www.elrob.org/
17. EURON Benchmarking Activities, http://www.euron.org/activities/benchmarks/index.html
18. LEGO factory, http://factory.lego.com
19. Cyberbotics Ltd, http://www.cyberbotics.com
20. e-puck web site, http://www.e-puck.org

Reactive Trajectory Deformation to Navigate Dynamic Environments

Vivien Delsart and Thierry Fraichard

Inria Rhône-Alpes, LIG-CNRS, Grenoble Universities (FR)
vivien.delsart@inrialpes.fr

Summary. Path deformation is a technique that was introduced to generate robot motion wherein a path, that has been computed beforehand, is continuously deformed on-line in response to unforeseen obstacles. In an effort to improve path deformation, this paper presents a trajectory deformation scheme. The main idea is that by incorporating the time dimension and hence information on the obstacles' future behaviour, quite a number of situations where path deformation would fail can be handled. The trajectory represented as a space-time curve is subject to deformation forces both external (to avoid collision with the obstacles) and internal (to maintain trajectory feasibility and connectivity). The trajectory deformation scheme has been tested successfully on a planar robot with double integrator dynamics moving in dynamic environments.

1   Introduction

Where to move next? is a key question for an autonomous robotic system. This fundamental issue has been largely addressed in the past forty years. Many motion determination strategies have been proposed (see [8] for a review). They can broadly be classified into *deliberative* versus *reactive* strategies: deliberative strategies aim at computing a complete motion all the way to the goal, whereas reactive strategies determine the motion to execute during the next few time-steps only. Deliberative strategies have to solve a motion planning problem. They require a model of the environment as complete as possible and their intrinsic complexity is such that it may preclude their application in dynamic environments. Reactive strategies on the other hand can operate on-line using local sensor information: they can be used in any kind of environment whether unknown, changing or dynamic, but convergence towards the goal is difficult to guarantee.

To bridge the gap between deliberative and reactive approaches, a complementary approach has been proposed based upon *motion deformation*. The principle is simple: a complete motion to the goal is computed first using a priori information. It is then passed on to the robotic system for execution. During the course of the execution, the still-to-be-executed part of the motion is continuously deformed in response to sensor information acquired on-line, thus accounting for the incompleteness and inaccuracies of the a priori world model. Deformation usually results from the application of constraints both external (imposed by the obstacles) and internal (to maintain motion feasibility and connectivity). Provided that the motion connectivity can be maintained, convergence towards the goal is achieved.

H. Bruyninckx et al. (Eds.): European Robotics Symposium 2008, STAR 44, pp. 233–241, 2008.
springerlink.com

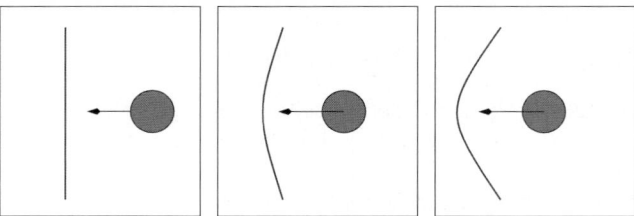

Fig. 1. Path deformation problem: in response to the approach of the moving disk, the path is increasingly deformed until it snaps (like an elastic band)

The different motion deformation techniques that have been proposed [9, 5, 1, 7, 10] all performs *path deformation*. In other words, what is deformed is a geometric curve, *ie* the sequence of positions that the robotic system is to take in order to reach its goal. The problem with path deformation techniques is that, by design, they cannot take into account the time dimension of a dynamic environment. For instance in a scenario such as the one depicted in Fig. 1, it would be more appropriate to leave the path as it is and adjust the velocity of the robotic system along the path so as to avoid collision with the moving obstacle (by slowing down or accelerating). To achieve this, it is necessary to depart from the path deformation paradigm and resort to *trajectory deformation* instead. A trajectory is essentially a geometric path parameterized by time. It tells us where the robotic system should be but also when and with what velocity. Unlike path deformation wherein spatial deformation only takes place, trajectory deformation features both *spatial and temporal* deformation meaning that the planned velocity of the robotic system can be altered thus permitting to handle gracefully situations such as the one depicted in Fig. 1.

The first trajectory deformation scheme has been proposed by one of the authors in [6]. It operates in two stages (collision avoidance and connectivity maintenance stages) and was limited to holonomic robotic systems. The contribution of this paper is a new trajectory deformation scheme, henceforth called Teddy (for Trajectory Deformer). It operates in one stage only and is explicitly designed to handle arbitrary nonholonomic and dynamic constraints.

The paper is organised as follows: Teddy is overviewed in §2. Its application to the case of a planar robot with double integrator dynamics (subject to velocity and acceleration bounds) is detailed in §3. Experimental results are then presented in §4.

2   Overview of the Approach

2.1   Notations and Definitions

Let $\mathscr{A}$ denote a robotic system operating in a workspace $\mathbf{W}$ ($\mathbb{R}^2$ or $\mathbb{R}^3$). $q \in \mathbf{C}$ denote a configuration of $\mathscr{A}$. The dynamics of $\mathscr{A}$ is described by a differential equation of the form:

$$\dot{s} = f(s, u)$$

where $s \in \mathbf{S}$ is the state of $\mathscr{A}$, $\dot{s}$ its time derivative and $u \in \mathbf{U}$ a control. $\mathbf{C}$, $\mathbf{S}$ and $\mathbf{U}$ respectively denote the configuration space, the state space and the control space of $\mathscr{A}$. Let $\xi : [0, t_f[\longrightarrow \mathbf{U}$ denote a control input, *ie* a time-sequence of controls. Starting from an initial state s_0 (at time 0) and under the action of a control input ξ, the state of $\mathscr{A}$ at time t is denoted by $\xi(s_0, t)$. A couple (s_0, ξ) defines a trajectory for $\mathscr{A}$, *ie* a curve in $\mathbf{W} \times \mathbf{T}$ where $\mathbf{T}$ is the time space.

For the sake of trajectory deformation, a trajectory will be discretized in a sequence of nodes. A node is a state-time, it is denoted by $n_i = (s_i, t_i)$. The discrete trajectory of $\mathscr{A}$ is $\Gamma_0 = \{n_0, n_1 \cdots n_N\}$ with n_0 (resp. n_N) the initial (resp. final) node of the trajectory.

2.2   Trajectory Deformation Principle

Teddy operates periodically: at time t_k, it takes as input the still-to-be-executed part of the trajectory $\Gamma_k = \{n_k, n_{k+1} \cdots n_N\}$ and an updated model of the workspace. This model includes the position of the obstacles of $\mathbf{W}$ at time t_k along with information about their future behaviour. Teddy then deforms Γ_k in response to the updated position and future behaviour of the obstacles and outputs a deformed trajectory $\Gamma_k' = \{n_k, n_{k+1}' \cdots n_N'\}$ with n_i' the updated node corresponding to n_{k+1}.

Like a particle placed in a force field, a node is displaced in response to the application of a force which is the combination of two kind of forces: external and internal. External forces are repulsive forces exerted by the obstacles of the environment, their purpose is to deform the trajectory in order to keep it collision-free. They are detailed in §2.3. Internal forces on the other hand are aimed at maintaining the feasibility and the connectivity of the trajectory, *ie* to ensure that the deformed trajectory still satisfies the kinematic and dynamic constraints of $\mathscr{A}$. They are detailed in §2.4.

In certain cases, the constraints imposed by the environment are such that the deformation process fails to produce a trajectory which remains collision-free and connected (for instance when the topology of $\mathbf{S} \times \mathbf{T}$ changes). Should this situation arise, a motion planner should be invoked to compute a new trajectory.

2.3   External Forces

External forces are repulsive forces exerted by the obstacles of the environment for collision avoidance purposes. They are derived from a potential function V_{ext}. To explicitly take into account the future behaviour of the moving obstacles, V_{ext} is defined in the space-time $\mathbf{W} \times \mathbf{T}$ (instead of $\mathbf{S} \times \mathbf{T}$ for efficiency reason). In a manner similar to [1], a set of points p_j are selected on the body of $\mathscr{A}$. Each node n_i of the trajectory Γ_k yield a set of control points $c_i^j = (p_j, t_i)$ in $\mathbf{W} \times \mathbf{T}$. For a control point c corresponding to the configuration q and the state s along the trajectory, V_{ext} is defined as:

$$V_{ext}(c) = \begin{cases} k_{ext}(d_0 - d_{wt}(c))^2 & \text{if } d_{wt}(c) < d_0 \\ 0 & \text{otherwise} \end{cases} \tag{1}$$

where $d_{wt}(c)$ is the distance from c to the closest obstacle in $\mathbf{W} \times \mathbf{T}$. d_0 is the region of influence around the obstacles and k_{ext} is the repulsion gain. d_{wt} is a distance function in $\mathbf{W} \times \mathbf{T}$. It is derived from the Euclidean distance by scaling the space versus the time dimension. In $\mathbb{R}^2$ for instance, the distance d_{wt} between (x_0, y_0, t_0) and (x_1, y_1, t_1) is given by:

$$d_{wt}^2 = w_s^2(x_1 - x_0)^2 + w_s^2(y_1 - y_0)^2 + w_t^2(t_1 - t_0)^2 \qquad (2)$$

with w_s (resp. w_t) the spatial (resp. temporal) weight. The force resulting from this potential function acting on c is then defined as:

$$\mathbf{F}_{ext}^{wt}(c) = -\nabla V_{ext}(c) = k_{ext}(d_0 - d_{wt}(c))\frac{\mathbf{d}}{||\mathbf{d}||} \qquad (3)$$

where $\mathbf{d}$ is the vector between c and the closest obstacle point. Now, $\mathbf{F}_{ext}^{wt}$ has to be mapped into $\mathbf{S} \times \mathbf{T}$. $\mathbf{F}_{ext}^{wt}$ is first mapped into $\mathbf{C} \times \mathbf{T}$ as follows:

$$\mathbf{F}_{ext}^{ct} = J_c^T(q, t)\mathbf{F}_{ext}^{wt}(c) \qquad (4)$$

where $J_c^T(q, t)$ represents the Jacobian at point c. The mapping into $\mathbf{S} \times \mathbf{T}$ that yields $\mathbf{F}_{ext}$ is carried out by leaving the additional state parameters unchanged.

2.4   Internal Forces

The external forces defined above push each node of the trajectory away from the obstacles if they are inside their influence region. Internal forces are introduced to ensure that the trajectory remains connected, *ie* that there exists a trajectory verifying the dynamics of $\mathcal{A}$ between two consecutive nodes of the trajectory. Trajectory connectivity is related to the concepts of forward and backward reachability. The set of states that are reachable from a given state s_0 are defined as:

$$\mathcal{R}(s_0) = \{s \in \mathbf{S}|\exists \xi, \exists t, \xi(s_0, t) = s\} \qquad (5)$$

Likewise, the set of states from which it is possible to reach a given state s_0 are defined as:

$$\mathcal{R}^{-1}(s_0) = \{s \in \mathbf{S}|\exists \xi, \exists t, \xi(s, t) = s_0\} \qquad (6)$$

Let n_-, n and n_+ denote three consecutive nodes of the trajectory Γ_k. Γ_k is connected at n iff $n \in \mathcal{R}(n_-)$ and $n_+ \in \mathcal{R}(n)$. In other words, n must belong to $\mathcal{R}(n_-) \cap \mathcal{R}^{-1}(n_+)$.

Now, two cases arise: in the first case, n_- and n_+ are connected. $\mathcal{R}(n_-) \cap \mathcal{R}^{-1}(n_+)$ is therefore nonempty and the purpose of the internal forces is to ensure that n remains within $\mathcal{R}(n_-) \cap \mathcal{R}^{-1}(n_+)$. In the second case, n_- and n_+ are no longer connected (it happens when the external forces pushes n_- and n_+ too far away from one another, or when their time components are no longer coherent). $\mathcal{R}(n_-) \cap \mathcal{R}^{-1}(n_+)$ is therefore empty and the purpose of the internal forces in this case is only to ensure that n remains within $\mathcal{R}(n_-)$ (priority being given to earliest connection). The internal forces for both cases are defined in the next two sections.

Case 1: n_- and n_+ Connected

To ensure that n remains within $\mathcal{R}(n_-) \cap \mathcal{R}^{-1}(n_+)$, a virtual spring is defined between n and H, the centroid of $\mathcal{R}(n_-) \cap \mathcal{R}^{-1}(n_+)$. It yields a potential function V_{int} defined in the space-time $\mathbf{S} \times \mathbf{T}$ as:

$$V_{int}(n) = k_{int}d_{st}^1(n)^2 \qquad (7)$$

where $d^1_{st}(n)$ is the distance between n and H. It is defined in a manner similar to d_{wt}. k_{int} is the attraction gain.

$$\mathbf{F}_{int}(n) = -\nabla V_{int}(n) = k_{int}d^1_{st}(n)\frac{\mathbf{d}_1}{||\mathbf{d}_1||} \tag{8}$$

where $\mathbf{d}_1$ is the vector between n and H.

Case 2: n_- and n_+ Disconnected

To ensure that n remains within $\mathscr{R}(n_-)$, a virtual spring is defined between n and $\mathscr{R}(n_-)$. It yields a potential function V_{int} defined in the space-time $\mathbf{S} \times \mathbf{T}$ as:

$$V_{int}(n) = k_{int}d^2_{st}(n)^2 \tag{9}$$

where $d^2_{st}(n)$ is the distance between n and the closest point of $\mathscr{R}(n_-)$. It is defined in a manner similar to d_{wt}. k_{int} is the attraction gain.

$$\mathbf{F}_{int}(n) = -\nabla V_{int}(n) = k_{int}d^2_{st}(n)\frac{\mathbf{d}_2}{||\mathbf{d}_2||} \tag{10}$$

where $\mathbf{d}_2$ is the vector between n and the closest point of $\mathscr{R}(n_-)$.

2.5   Total Force

Once both internal and external forces have been computed for a node n, the net force applied to it is:

$$\mathbf{F}(n) = \mathbf{F}_{ext}(n) + \mathbf{F}_{int}(n) \tag{11}$$

3   Case Study: Double Integrator System

To begin with, `Teddy` has been applied to the case of a 2D planar robot $\mathscr{A}$ with double integrator dynamics. A state of $\mathscr{A}$ is characterised by (p, v) that respectively denote the 2D position and velocity of $\mathscr{A}$ ($|v| \leq v_{max}$). The dynamics of $\mathscr{A}$ is given by:

$$\begin{pmatrix} \dot{p} \\ \dot{v} \end{pmatrix} = \begin{pmatrix} v \\ a \end{pmatrix} \tag{12}$$

with a the acceleration control applied to $\mathscr{A}$ ($|a| \leq a_{max}$).

The key point in the adaptation of `Teddy` to a particular robotic systems lies in the computation of the internal forces, *ie* on the computation of the sets $\mathscr{R}(n_-)$, $\mathscr{R}^{-1}(n_+)$, $\mathscr{R}(n_-) \cap \mathscr{R}^{-1}(n_+)$ and the centroid H of the latter.

In this case, the key to efficiency lies in *not* computing $\mathscr{R}(n_-)$ or $\mathscr{R}^{-1}(n_+)$. What is computed instead is fixed time-slices of these sets. For instance, determining whether n_- and n_+ are connected is carried out by checking if $\mathscr{R}(n_-, t_{int})$ and $\mathscr{R}^{-1}(n_+, t_{int})$ intersects, where

$$t_{int} = 1/2(t_+ - t_-)$$

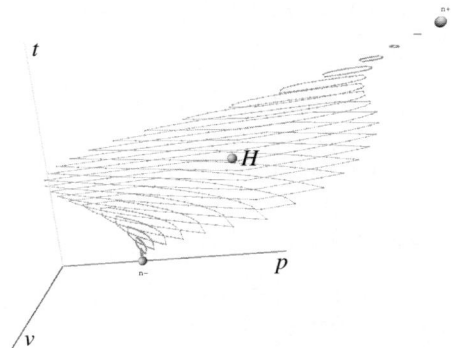

Fig. 2. $\mathscr{R}(n_-) \cap \mathscr{R}^{-1}(n_+)$ and its centroid H in the 1D case

and $\mathscr{R}(n_-, t_{int})$ (resp. $\mathscr{R}^{-1}(n_+, t_{int})$) is the set of states reachable from n_- at time t_{int} (resp. the set of states from which it is possible to reach n_+ from time t_{int}). Computing $\mathscr{R}(n_-, t_{int})$ is carried out in two steps: first, the range of reachable positions $[p_{\min}(t_{int}), p_{\max}(t_{int})]$ is determined by integrating (12). Then for a given position p in this range, the range of reachable velocities $[v_{\min}(p, t_{int}), v_{\max}(p, t_{int})]$ is computed. This way, we efficiently obtain an approximation of $\mathscr{R}(n_-, t_{int})$ (the same reasoning applies to $\mathscr{R}^{-1}(n_+, t_{int})$) that can be used to check the connectivity between n_- and n_+ and to compute H (the reader interested in more details on this part is referred to [3] or [4] for the English version). Fig. 2 depicts an example of a region $\mathscr{R}(n_-) \cap \mathscr{R}^{-1}(n_+)$ and its centroid H obtained by numerical computation in the 1D case.

4   Experimental Results

Teddy has been implemented in C++ and tested on an Intel Pentium 4 desktop PC (3GHz, 1GB RAM, Linux OS). Teddy has been evaluated in different scenarios featuring up to 10 circular obstacles moving randomly. At each time step, Teddy is provided with a new model of the environment and its future evolution. To better illustrate, the interest of Teddy, we have focused in this section on a simple "cutting" scenario similar to the one depicted in Fig. 1. This scenario has been selected because it is problematic for classical path deformation schemes.

Teddy relies upon a number of parameters to operate properly: the repulsion gain k_{ext}, the attraction gain k_{int} and the distance functions d_{wt}, d_{st}^1 and d_{st}^2. The two examples presented below have been selected to illustrate the importance of the distance function d_{wt} on the performance of Teddy. Recall that d_{wt} is used to determine the distance between a trajectory node and the closest obstacle in $\mathbf{W} \times \mathbf{T}$ (*cf* §2.3). In both examples, the initial trajectory had a duration of 20s and the discrete trajectory contained 320 nodes. Teddy would run at approximately 28Hz.

In the same situation, two very different deformation patterns can be obtained by properly selecting the weights w_s and w_t in (2). The first example is obtained by giving more weight to w_s thereby allowing more important spatial deformations to take place

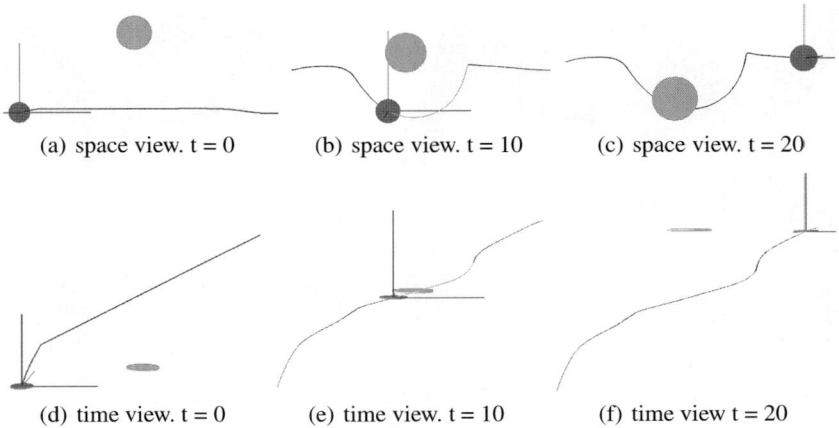

Fig. 3. Example 1 (spatial deformation): $\mathscr{A}$ is moving from the left to the right, the obstacle is moving downwards. The top snapshots depict the path at different time instant ($x \times y$ view). The bottom snapshots depict the velocity profile at the same instants ($x \times t$ view).

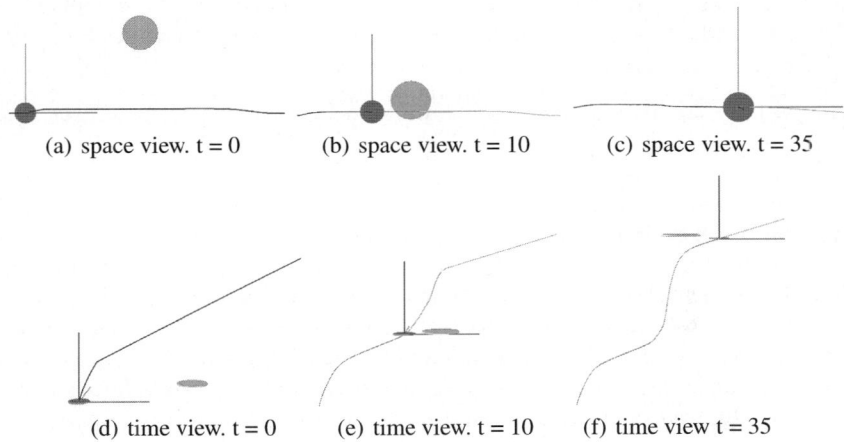

Fig. 4. Example 2 (temporal deformation): $\mathscr{A}$ is moving from the left to the right, the obstacle is moving downwards. The top snapshots depict the path at different time instant ($x \times y$ view). The bottom snapshots depict the velocity profile at the same instants ($x \times t$ view).

(Fig. 3). In this case, $\mathscr{A}$ has time to pass before the obstacle crosses its path. The path component of the trajectory is deformed downwards for safety reasons whereas the velocity component is only slightly modified.

The second example on the other hand is obtained by giving more weight to w_t thereby allowing more important temporal deformations to take place (Fig. 4). In this case, $\mathscr{A}$ let the obstacle cross its path before proceeding. The path component of the

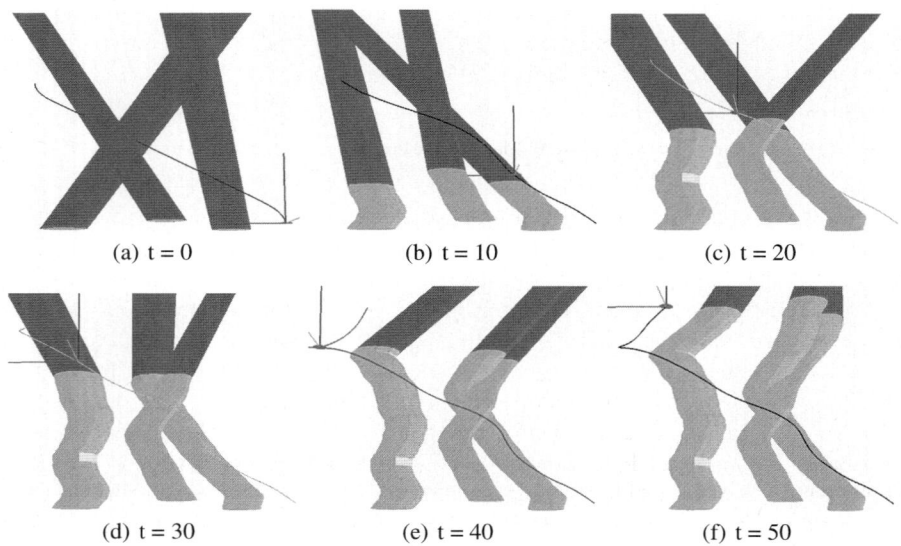

(a) t = 0   (b) t = 10   (c) t = 20

(d) t = 30   (e) t = 40   (f) t = 50

Fig. 5. Example 3 (general scenario): $\mathscr{A}$ is moving from the right to the left amidst three moving obstacles

trajectory is only slightly modified whereas the velocity component is largely deformed so as to allow $\mathscr{A}$ to slow down and stop in order to give way to the obstacle.

These two examples have shown the influence of the choice of the parameters in the final performance of Teddy. They have also illustrated the advantage of trajectory deformation versus path deformation. Fig 5 presents a more general scenario featuring three randomly moving obstacles. Each snapshot depicts the $\mathbf{W} \times \mathbf{T}$ space at different time instants. Each cylinder represents the motion of a given moving obstacle in $\mathbf{W} \times \mathbf{T}$. The lower part is the past (how the moving obstacle has moved), the upper past is the future (estimated future behaviour of the moving obstacles: assumed constant linear velocity)). In this experiment, the velocity of the obstacles was picked up randomly at each time step. One can see how the trajectory deforms itself both spatially and temporally as time passes by.

Table 1. Running time (in ms) of one deformation cycle as a function of the number of nodes and obstacles

number of	number of nodes				
obstacles	50	100	180	250	320
1	6	11	20	27	35
3	44	48	68	70	73
10	49	88	135	199	229

From a complexity point of view, Teddy's running time grows linearly with the number of nodes and the number of obstacles. Table 1 gives the running time of one deformation cycle for different numbers of nodes and obstacles.

5   Conclusion and Future Works

The paper has presented Teddy, a trajectory deformation scheme. Given a nominal trajectory reaching a given goal, Teddy deforms it reactively in response to updated information about the environment's obstacles. Teddy can handle robotic systems with arbitrary dynamics. It has been applied to the case of a 2D double integrator system and tested in various situations. Because, Teddy explicitly takes into account information on the future behaviour of the obstacles, it is able to handle situations that are problematic for classical path deformation schemes. In the future, it is planned to consider other robotic systems, *eg* car-like vehicles, and to further optimize Teddy. Considering for instance that the knowledge about the future behaviour is less reliable in the distant future, it could be interesting to monotonically decrease the influence of the obstacles with respect to time. Last but not least, Teddy remains to be integrated within a global navigation architecture and tested on an actual robotic system. It is planned to do so on the architecture and the vehicle presented in [2].

References

1. Brock, O., Khatib, O.: Elastic strips: a framework for motion generation in human environments. Int. Journal of Robotics Research 21(12) (December 2002)
2. Chen, G., Fraichard, T., Martinez-Gomez, L.: A real-time autonomous navigation architecture. In: Proc. of the IFAC Symp. on Intelligent Autonomous Vehicles, Toulouse (FR) (September 2007)
3. Delsart, V.: Autonomie du mouvement en environnement dynamique. Master report, Inst. Nat. Polytechnique, Grenoble (FR) (June 2007)
4. Delsart, V., Fraichard, T.: Reactive trajectory deformation. Research report, INRIA (in press)
5. Khatib, M., Jaouni, H., Chatila, R., Laumond, J.-P.: Dynamic path modification for car-like nonholonomic mobile robots. In: Proc. of the IEEE Int. Conf. on Robotics and Automation, Albuquerque, NM (US) (April 1997)
6. Kurniawati, H., Fraichard, T.: From path to trajectory deformation. In: Proc. of the IEEE-RSJ Int. Conf. on Intelligent Robots and Systems, San Diego, CA (US) (October 2007)
7. Lamiraux, F., Bonnafous, D., Lefebvre, O.: Reactive path deformation for nonholonomic mobile robots. IEEE Trans. on Robotics and Automation 20(6) (December 2004)
8. Lavalle, S.M.: Planning Algorithms. Cambridge University Press, Cambridge (2006)
9. Quinlan, S., Khatib, O.: Elastic bands: Connecting path planning and control. In: Proc. of the IEEE Int. Conf. on Robotics and Automation, Atlanta, GA (US) (May 1993)
10. Yang, Y., Brock, B.: Elastic roadmaps: Globally task-consistent motion for autonomous mobile manipulation. In: Proc. of the Int. Conf. Robotics: Science and Systems, Philadelphia PA (US) (August 2006)

Recovery in Autonomous Robot Swarms[*]

Damien Martin-Guillerez, Michel Banâtre, and Paul Couderc

INRIA Rennes - ACES Team, IRISA, Campus Universitaire de Beaulieu, F-35042 Rennes cedex, France
{dmartin,banatre,pcouderc}@irisa.fr

Summary. A swarm of robots is composed of several small and simple robots that can communicate to perform complex tasks. Those robots are subject to failures: battery outage, hardware destruction, or hardware failure. When a robot fails, its whole state is lost. The task realized by this robot needs to be restarted then the swarm might need some data from the failed robot. Even when this data is not needed for restarting the process, losing this data can entail a lot of cost for the swarm.

Ensuring state backup of the robots is, somehow, very useful. This can be done by a general wireless coverage. However, this coverage can be difficult and expensive to deploy. To address this issue, a collaborative backup system uses encounters of each robot as temporary backup points to save its internal state. In this paper, we present a collaborative backup system for swarm of robots.

Keywords: Swarm of robots, fault-tolerance, peer-to-peer, collaborative backup, mobile computing.

1 Introduction

Some researchers have envisioned the use of swarm of simple robots acting cooperatively to perform relatively complex tasks [3]. Swarm of robots can perform a wide range of tasks from cleaning and exploration to construction and surveillance.

Such robots are acting autonomously and sometimes in dangerous areas (chemical plants for instance). They generally relies on ad hoc communications for planning to avoid expensive global network usage. They can endure severe failures, like destruction, hardware failure, battery depletion or inability to act when falling into a hole, that can entail data loss. Data loss often means to restart a task and reacquire those data, which can be very expensive. The solution is to backup those data on a regular basis. If we only use a global server for data backup, we might need either to have a global wireless connection (often hard to deploy) or to often connect to the global server (time and energy consuming).

In this paper, we propose a collaborative backup system for swarms of robots. It uses neighbor robots and memories available in the physical space to backup data items that are the most expensive (in term of resources) to recover. This system is transparent

---

[*] This work is partially supported by the ROBOSWARM project (EU-IST-FP6-45255) and the ReSIST network (EU-IST-FP6-26764).

H. Bruyninckx et al. (Eds.): European Robotics Symposium 2008, STAR 44, pp. 243–252, 2008.
springerlink.com   © Springer-Verlag Berlin Heidelberg 2008

during the process and realizable without a global network nor constant reports to the backup server. It concentrates on the efficiency in term of resource usage for recovery (costs).

This paper is structured as follows: after some generalities in section 2 about collaborative backup and swarms of robots, section 3 introduces the architecture we want to support. A model for the tasks realized by the robots is then presented in section 4. Section 5 presents our solution for implementation of a collaborative backup system for swarm of robots. Finally, we conclude in section 6.

2   Background

A swarm of robots needs to plan the tasks to achieve. This planning is a complex process and, when a failure occurs, a *replanning* is needed [6]. A *plan repair* first aims at reducing the costs of the new plan by deleting tasks or task parts already performed. This *plan repair* needs the knowledge about past actions. Classical works like ALLIANCE [7] do this *replanning* in a collaborative way using wireless technologies that enable all robots to stay in contact with each other. Other works like [6] address the issue of *replanning* without investigating the problem of availability of data for the *plan repair*.

This knowledge about past actions is very easy to get when a global network is available and when all the entities of the system have a permanent access to this network (like in ALLIANCE). However, this network may be hard or impossible to set up. For example, 802.11 coverage on a large field needs for access points to relay informations between areas. The deployment of those access points can be expensive (time and money consuming). In dangerous area like exploration area (space, submarine) or fields on fire, this deployment is even impossible. Therefore, new techniques are needed.

A well known approach in increase data resilience is hardware replication. In network file systems, replication of data uses several data servers [9]. Peer-to-peer file systems have used replication to increase data availability and have paved the way for collaborative backups [1].

In our case, collaborative backup greatly helps when a robot fails. It can reduce the costs of a failure without assuming the availability of a global wireless infrastructure. The deployment of robots is then easier and faster. As the main aim of this system is to reduce costs of the *replanning*, this system must give priority to expensive task and data. This system can also decrease the swarm needs for a centralized entity (like a global server).

In a mobile context, several collaborative file systems have been developed [8, 2] as well as some backup systems [5]. However, all those systems suffer from the same limitations: they are designed for nearly static network where users stay in the same area for a long period of time. Those systems hardly exploit opportunistic replication to support high mobility. The MoSAIC project [4] has worked on a collaborative backup service for personal devices. They have addressed the specific issues for this kind of devices especially regarding security. However, security implies a loss of resources (memory especially) and prevents data to get aggregated.

3 Architecture

In this paper, we consider the architecture described in figure 1. Several mobile entities (swarm robots) can move and interact with the physical world.They may be specialized for some actions (e.g. some robots are vacuum machines while others are steam cleaner). They can communicate between them using high-bandwidth, short-range wireless communication facilities. The MEs (Mobile Entities) have an amount of available memory to run the swarm process and other applications, and a battery duration (Battery Life, BL). MEs are now considered to be fail-silent for now.

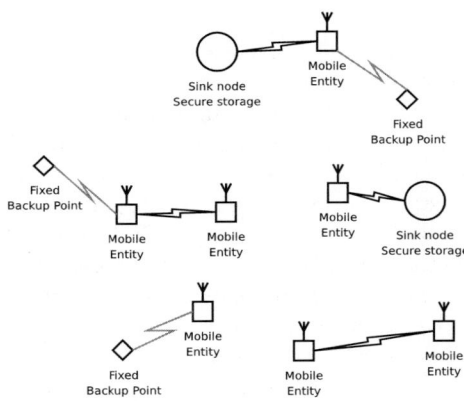

Fig. 1. Architecture

One or several wireless base stations act as data servers and task controllers, i.e. they provide data about the physical world and give tasks to the swarm robots. The planning can be completely done by the base stations or more cooperatively between swarm robots. We call Sink Nodes (SNs) those base stations. They are computers with the same kind of wireless capabilities than MEs. They are able to communicate with MEs when in range. The SNs have unlimited power and memory. They can perform evolved computation and have, at the beginning of the process of the MEs, all the initial information needed. The SNs are considered resilient. To ensure this assertion, usual techniques can be used like hardware replication (RAID disks) or network replication (a replicated file system).

Furthermore, several fixed memory points (like RFID tags) are positioned in the physical world. They have limited memory but unlimited battery (passive RFID tags). They are used to store information linked to the physical world. For instance, if a ME detects an object to move for the next ME it can put this context in a fixed memory point in the vicinity of this object. FMPs (Fixed Memory Points) are considered fail-silent. On some exceptions, FMPs can actually move along with the attached object (e.g. when moving a seat that has a tag attached). Those FMPs can be queried by MEs using specific wireless technology.

All entities (MEs, SNs and FMPs), with limited wireless range, are not able to communicate within all the field of action. Indeed, global connectivity requires expensive infrastructure and hard deployment. Using just one or two base stations along with cheap robots and RFID tags is fast to deploy and inexpensive (robots and RFID tags can be reused easily). The limited wireless connectivity implies the need of fault-tolerance mechanisms to recover from a failure outside connectivity.

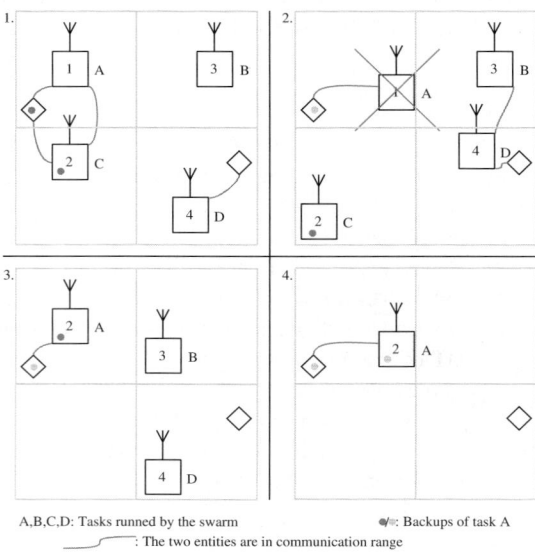

A,B,C,D: Tasks runned by the swarm   ◁▫: Backups of task A
                    ⌒ : The two entities are in communication range

Fig. 2. Example of the backup system in operation

1. Several MEs are performing a collaborative process (exploring area for example). The ME 1 is backuping its data on a FMP and on the ME 2.
2. ME 1 fails in the vicinity of a FMP.
3. ME 2 finishes its task and is reaffected to task A with the info saved on step 1.
4. ME 2 gets back the latest info from the FMP and restarts at the ME 1 failure position.

Figure 2 shows an example of the system running in an exploration process. The system saves the data of the current tasks on the encounters (FMPs and MEs) during the process. It tries to optimize the memory consumption and the recovery costs. When a ME fails, the planner (generally a sink node) reaffects failed tasks to available MEs. It uses the latest acquired information as a start. If a reaffected ME encounters an entity with more recent information on the task, then it tries to use these new data to reduce the cost and the duration of the task (and, by consequence, of the process). Depending on the situation, backup on FMP or on ME can be impossible. Our system adapts itself using only free available space on several entities. It also considers information locality when backuping on FMP.

To integrate this architecture to different swarms of robots, the numbers of SNs and FMPs can be modified. In section 4, we detail the tasks realized by those entities.

4   Mobile Entity Tasks

MEs are performing some tasks which, once planned, is a whole process. We can imagine a lot of different processes: waste removal and cleaning, field exploration, surveillance, construction, stock management, etc...

4.1   Atomic Tasks

The process is split in *atomic tasks* that can be executed by one ME. An atomic task can be very simple (like go to point P) to quite complicate (like go to area 1, clean the ground of area 1 and then come back). Atomic tasks can have requirements such as specific information or kind of ME.

We associate to each atomic task an estimated duration (EDT, Estimated Duration of Task) determined from the several components of this task. For instance, for waste collecting, the task will be to cover an area and remove a waste when finding it: the EDT of this task will be the EDT of the area coverage plus the EDT of one waste removal multiplied by the estimated number of waste.

We also associate to each atomic task, an ETTC (Estimated Time To Connection) which is the estimated time until we met a SN. It can be infinite when the task is not passing near a SN. It can be estimated using the known positions of all SNs.

4.2   Task Cost

Each task have several associated costs: time, consumable usage, energy, etc... Those costs can be combined into a generic one using factors for each (for example, the price of the corresponding consumable). These costs are determining the costs of losing the associated data.

These costs can be easily computed depending on the task. For example, a painting task would have three main costs: paint usage which can be converted to currency depending on the cost of the paint, energy consumption and time. A cleaning task would have the same kind of costs: time, energy and cleaning product.

Energy consumption can be directly converted to a currency. However, if there is a need to recharge the battery, time can be wasted and the corresponding cost must be included. Time is harder to convert to a general cost. The process is composed of parallel tasks, thus the global time taken by the process is not the sum of the time of each tasks. The process taking overtime could have very little consequences until it exceeds a certain duration (real-time constraints). For instance, if a cleaning process for a supermarket lasts after the opening then the supermarket might endure severe financial lost. The conversion might not be easy between duration and real cost. We consider for now that we do have a factor that let us convert the time to a general cost. The recovery process will then adjust to ensure a good tradeoff between cost usage and time taking.

5   The System

5.1   The Backup System

The backup system is composed as shown in figure 3. A memory management system keeps backups up to date and manages memory usage. A scheduler sends and receives backup depending on data priorities recomputed when needed. They are based on the costs of recovering each lost data items and on the probability of losing it, to ensure the best tradeoffs between backup costs and probable costs of a future failure.

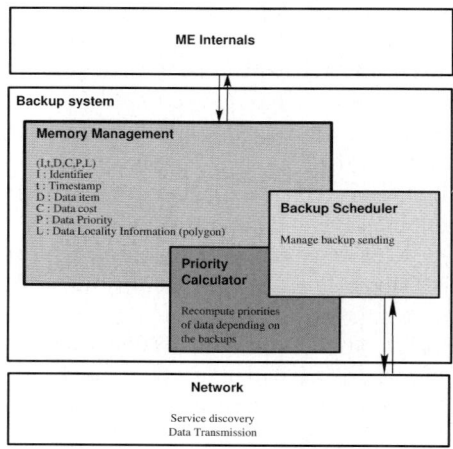

Fig. 3. Architecture of the backup system

Memory Management

The system should act transparently during the process without restraining other tasks. It means that backups should not take resources that are needed by ME activities. For memory, backups should be deleted if more memory is needed by a task. The more important data are, the more they are kept.

When backing-up a data item (D in figure 3), it is stored along a timestamp (t) and an identifier (I). The identifier is used to know how to recover the task from the data and the timestamp to know what to update when a new version of the data item is proposed (see section 5.1).

Tasks have costs, so do data items. The cost of a data item is the cost of recovering the process without a backup of this data item. For example, it can be the cost of painting to redo if the system forgot that it has painted a certain area. These costs should also contain the eventual costs of restarting the task if it is finished. It implies that the data of a finished task may cost more than the sum of the cost of each subtask. For example, if a construction task fails, we need to clean the site before we can restart the construction.

We also consider combination of data items together to reduce data size. For example, an exploration task just adds data to the already acquired ones (a polygon describing the covered area is increasing for instance). The cost of the reunited data items is then the cost of the new items added to the cost of the old ones.

Those costs are used to compute a priority (P) for each data item. These priorities are used to sort data items by importance. When more memory is needed for the process, a ME can delete data items with the lowest priority. Local data should always be preserved on the ME. For instance, a finished task data may be useless to the ME but needs to be conserved for process reconstruction. Hence we ensure that there is always at least one copy of each data.

To be efficient, the priorities include data locality, i.e. when doing a backup on an FMP, the data will be localized in a certain area. We want those data to be near the area of usage. Thus, we increase the data priority when backing-up on a FMP inside a locality zone (L in figure 3) and decrease it when backing-up on a FMP far from this zone. This disables the needs of travelling through the whole operational area to recover data.

Backup Scheduling

When two MEs meet, they initiate the backup process. It should not interfere with the standard process of the MEs to avoid useless costs. For example, we prefer to stop the backup process instead of stopping the robot or taking the communication channel used for the process.

The backup process is an exchange of data. First, each ME computes for each data item the priority P^1 if the item is saved on both ME and the priority P^2 if the item is removed from both MEs. Data items are ordered using P^1. Then the backup exchange can start. Each ME executes the same round until no more backup is needed, no more backup can be realized or the two MEs cannot communicate anymore (they are no more in communication range).

One round is executed as describe in figure 4. It uses the five kinds of message described next:

- $BACKUP\_REQ(I,t,S,C,P^1)$ asks to backup the data item identified by I and timestamped by t. This item has a size of S, a cost of C and a priority if saved of P^1.
- $BACKUP\_ACCEPT(S',t')$ replies for acceptance of the backup. S' and t' give, respectively, the size and the timestamp of the version of the same data item existing on the peer.
- $BACKUP\_DENY(S^1,S^2,P)$ replies for refusal of the backup.
- $BACKUP\_SEND(D_{|S',t'},L)$ sends the part of the data item needed to update the backup along with its locality.
- $BACKUP\_ACK(I,t)$ acknowledges the backup of the data item.

The backup acceptance phase replies whether the backup is accepted or not. The preparation phase extracts, depending on S' and t', $D_{|S',t'}$ which is a subpart of the data item D that permit the reconciliation of the backup. For example, if D is an incremental record then $D_{|S',t'}$ is the entries of D taken between time t and time t'. If $S' = S$ and $t' = t$, the acceptance phase goes directly to the next round as it means that the backup is already present on the peer. The storage phase just reconciliates the stored backup item with the new data and acknowledges the result. It eventually deletes some

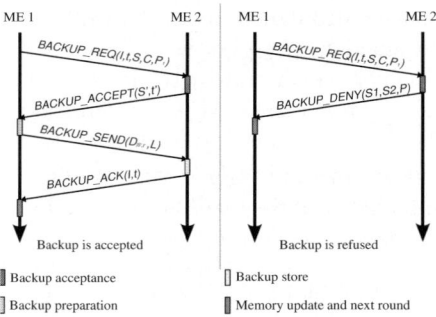

Fig. 4. Message exchanging for one round of the backup process

data items determined in the acceptance phase. Finally, the memory update phase updates the priority of D and searches for next data item to backup. If the reply is a $BACKUP\_DENY(C_1,C_2,P)$, data items which size is greater than $S^1 + S^2$ will not be proposed as well as data items which size is greater than S^1 and priority lower than P.

We call $D_1, D_2, ..., D_n$ the data items present in the ME memory ordered with ascending P^2. We logically use the same index than the data for t and P^2. S_i will be used for the size of the data item of index i. AMS will note the available memory space for backup.

To decide to accept or to refuse a backup, the ME must search for an existing version of the data item in its memory. S' and t' will be the size and the timestamp of the same item if present in the memory of the ME. Note that $S' = 0$ and $t' = 0$ if the item is not present. The ME can now decide whether to accept or refuse the item depending on the size it can free by the process described in algorithm 1.

Algorithm 1. Backup acceptance phase

$\quad$ ACCEPTANCE(S,S',t,t',P^1);
1 $\;$ **if** $(S' = S$ *and* $t' = t)$ *or* $S \leq S'$ **then**

2 $\quad\quad$ SEND$(BACKUP\_ACCEPT(S',t'))$
$\quad$ **else**

3 $\quad\quad j \leftarrow 1$;
4 $\quad\quad$ **while** $j < n$ *and* $P^1 \leq P_j^2$ *and* $(AMS + \sum_{i=1}^{j} S_i < S - S')$ **do**

5 $\quad\quad\quad$ $j \leftarrow j+1$;

6 **if** $(AMS + \sum_{i=1}^{j} S_i < S - S')$ **then**

7 $\quad\quad$ SEND$(BACKUP\_DENY(AMS, \sum_{i=1}^{j} S_i, P))$;
$\quad$ **else**

8 $\quad\quad$ $toDelete \leftarrow j$;
9 $\quad\quad$ SEND$(BACKUP\_ACCEPT(S',t'))$;

---

toDelete is here to know which data to delete. When receiving a *BACKUP_SEND* the ME should delete all data items $(D_i)_{1 \leq i \leq toDelete}$.

5.2   The Recovery Process

In this section, we present the recovery process triggered upon failure detections. We present how we reconstruct data and the corresponding tasks then we present the reassignment of the tasks to optimize the process.

Data Reconstruction

When a failure is detected, a SN initializes the reconstruction using the data items saved on SNs. It reconstructs each task using these items. Some tasks also need some preparation to restart.

We also need to add a data recovery phase where the robots goes to the FMP of the area. We add a task for the ME which is to go to the FMPs near the area to get the data and to update the task before restarting it. However, we should avoid useless cost consumption.

Useless cost consumption will happen if the cost of recovery the data is more important than the one of redoing the task. Given the time between the detection of the failure and the timestamp of the data item used to recover the task, we can compute the maximum cost C_{max} of the task part that could have been done by the robot before failure. If the time is long enough for the robot to perform the whole task then it means that no restart is needed and then the cost of the restart should be added to C_{max}. Then if the cost of getting back the data item from the FMPs is bigger than C_{max} the task should start immediately without the data recovery.

Process Reconstruction

Once each task has been reconstructed, a new plan needs to be created. The *plan repair* phase compares the resources needed for realizing the old task and the reconstructed one. It select the cheaper one for the *replanning*.

Besides, depending on the scheduling, the time of a task can change because of physical constraints. For example, if between two tasks a ME needs to move and if these tasks are far from each other then the second task takes longer.

If the system uses a very flexible scheduling system (i.e. MEs are very autonomous) then the SN needs to *flood* MEs with the update information. If the system uses a semi-flexible scheduling system (i.e. MEs can exchange tasks between them) then the scheduling might be a light scheduling allowing MEs to exchange tasks to optimize costs or process duration.

5.3   Failure Detection

A failure can be detected using several techniques: when a ME is not seen at a rendezvous point after a deadline, then we can assume it has failed when another ME arrives and detects that the preceding task has not been finished (missing information on a FMP, waste found for waste removal, etc...). A ME can also emits some kind of *red signal* on failure or when battery is running out. The *red signal* can be propagated to a SN using neighbors. Some other specific techniques might be used for failure detection (surveillance robot, etc...). [6] details many error detection mechanisms for autonomous systems.

6   Conclusion

In this paper, we have discussed about the implementation and the utility of a backup system for swarm robots. This scheme seems to be extendable to several mobile distributed system like mobile sensor networks. We have simulated this backup system for swarm robots on some applications (cleaning and exploration) and shown that it reduces costs in the case of failure without interference with the swarm process.

This system tends to improve the process even more in the case of less centralized systems (multiple tasks attributed to one entity with possibilities of dynamic task rescheduling). We plan on implementing this system on a real swarm of robots. We will make further analysis of data used in those swarm as well as how to reconstruct them.

Acknowledgement. We would like to thank our colleagues Fabien Allard and Ciaràn Bryce for their corrections.

References

1. Batten, C., Barr, K., Saraf, A., Treptin, S.: pStore: A Secure Peer-to-peer Backup System. Technical Report MIT-LCS-TM-632, MIT Laboratory for Computer Science (December 2001)
2. Boulkenafed, M., Issarny, V.: AdHocFS: Sharing Files in WLANS. In: The Second IEEE International Symposium on Network Computing and Applications (NCA 2003) (April 2003)
3. Dudek, G., Jenkin, M., Milios, E., Wilkes, D.: A Taxonomy for Swarm Robots. In: IEEE/RSJ International Conference on Intelligent Robots and Systems, vol. 1, pp. 441–447 (1993)
4. Killijian, M.-O., Powell, D., Banâtre, M., Couderc, P., Roudier, Y.: Collaborative Backup for Dependable Mobile Applications. In: ACM (ed.) The 2nd International Workshop on Middleware for Pervasive and Ad-Hoc Computing (October 2004)
5. Loo, B.T., LaMarca, A., Borriello, G.: Peer-To-Peer Backup for Personnal Area Networks. Technical Report IRS-TR-02-015, Intel Research Seattle - University of California at Berkeley (May 2003)
6. Lussier, B., Gallien, M., Guiochet, J., Ingrand, F., Killijian, M.-O., Powell, D.: In: IEEE/IFIP (ed.) IEEE/IFIP International Conference on Dependable Systems and Networks (DSN 2007) (June 2007)
7. Parker, L.E.: ALLIANCE: An Architecture for Fault Tolerant, Cooperative Control of Heterogeneous Mobile Robots. In: IEEE/RSJ/GI (ed.) Intelligent Robots and Systems (IROS 1994), vol. 2, pp. 776–783 (December 1994)
8. Ratner, D., Reiher, P., Pope1, G.J.: Roam: A Scalable Replication System for Mobile Computing. In: The Workshop on Mobile Databases and Distributed Systems (MDDS), pp. 96–104 (September 1999)
9. Satyanarayanan, M.: Scalable, Secure and Highly Available Distributed File Access. IEEE Computer 23(5), 9–21 (1990)

Robot Force/Position Tracking on a Surface of Unknown Orientation

Yiannis Karayiannidis and Zoe Doulgeri

Aristotle University of Thessaloniki, Department of Electrical and Computer Engineering,
Thessloniki, 54124 , Greece
yiankar@auth.gr, doulgeri@eng.auth.gr

Summary. The problem of robot force and position trajectory tracking is revisited in the case of an uncertain mapping of a surface into the robot space; then, although it is possible to define the desired trajectories with respect to the constraint surface, the lack of knowledge of the constraint direction in the robot space, means that the position and force control subspaces are uncertain. Such a case arises when for example the surface is misplaced. A novel adaptive controller is proposed using estimates of the constraint surface normal direction that converge to the actual value; the controller drives the actual force and position errors to zero given a persistently excited desired velocity on the surface. The performance of the proposed controller is demonstrated by a simulation example.

1 Introduction

Most adaptive schemes in robot control consider uncertainties in system dynamics and assume that system kinematics is known. For the force/position tracking of constrained robots in particular, parametric and structural dynamic uncertainties have been considered in [2, 1] and [7] respectively. In recent attempts, kinematic uncertainties have been incorporated in robot control problems. Such uncertainties refer to robot kinematic and Jacobian parameters [4, 3, 5]. In the case of constrained robots, the constraint Jacobian is however assumed known. This knowledge is based on the hypothesis that force measurements can be used to calculate the normal to the surface direction [3, 9, 5, 11]. However, the assumption of the measurement of the normal force does not hold in most practical cases due to the distortion of these measurements by the friction forces that arise during the end effector motion. The friction effect in the calculation of normal direction has been noticed in [10] where an algorithm to estimate the constraint direction is proposed that filters out force measurements along discrete position step directions. In a remark in [8] filtering out of the friction forces from force measurements is also proposed with the use of a projection matrix constructed from current tip velocities but this solution collapses each time the current tip velocity is zero. Another problem arises in position/force tracking controllers that use the constraint Jacobian derivative which requires force derivatives [9, 5, 10]. Force derivatives cannot however be calculated by numerical differentiating the noisy force measurements [8, 11].

This work proposes a position/force tracking controller based on on-line estimates of the unknown constraint Jacobian. As the normal to the surface direction and its complement are uncertain in the robot workspace the desired force and position trajectories

H. Bruyninckx et al. (Eds.): European Robotics Symposium 2008, STAR 44, pp. 253–262, 2008.
springerlink.com

cannot be correctly determined in this space. However, it is possible to define desired trajectories with respect to the surface; the desired force trajectory can be defined in the one-dimensional force subspace and the position trajectory in the two-dimensional space of the surface. We here consider that the mapping of these force and position subspaces into the respective robot force and position control subspace is uncertain. This situation arises when an object is misplaced in the robot's workspace and the robot is required to perform a specific contact task on one of its surfaces. A planar surface is considered in this work. The proposed control solution is of simple structure, uses joint motion variables and total force measurements, is designed to drive the estimated direction parameters to converge to their actual values and requires only the persistent excitation of the desired motion velocity on the surface.

2   Problem Description

Consider a n_q degrees of freedom robot with a rigid hemispherical tip of radius r in contact with a rigid surface (Fig. 1). Let $q \in \mathfrak{R}^{n_q}$ be the vector of the generalized joint variables and $\{B\}$ be inertia frame attached at the finger base. Let the surface frame $\{s\}$ be attached at some point on the surface; its position is denoted by p_s and its orientation by matrix $R_s = [n_s \ o_s \ a_s]$ such that $n_s \in \mathfrak{R}^3$ is the unit vector normal to the contact surface pointing inwards. Let the contact point be defined as the common point of the hemispherical tip and the surface. Let the frame $\{c\}$ be attached at the contact point on the surface and described by the position vector p_c and the orientation matrix $R_c = [n_c \ o_c \ a_c]$ such that the normal vector $n_c \in \mathfrak{R}^3$ at p_c points inwards the surface. Note that the hemispherical shape of the robot tip allows the rolling motion of the tip; as the robot tip may roll along the rigid surface, the contact point does not necessarily refer to the same point of the rigid tip. The spherical tip however allows simplified system kinematics by considering the position of the robot tip center $p_t \in \mathfrak{R}^3$ instead of the contact point p_c since the following relation holds:

$$p_t = p_c - n_c r \tag{1}$$

Consider also the frame $\{t\}$ attached at p_t with rotation matrix R_t that can be parameterized by the vector of the rotation angles $\varphi_t \in \mathfrak{R}^3$ around the inertia frame. Let the generalized tip position be $p = \begin{bmatrix} p_t^T & \varphi_t^T \end{bmatrix}^T \in \mathfrak{R}^6$. The generalized velocity $\dot{p} = \begin{bmatrix} \dot{p}_t^T & \omega^T \end{bmatrix}^T \in \mathfrak{R}^6$ is related to the joint velocity $\dot{q}$ through the robot tip Jacobian $J = \begin{bmatrix} J_v^T & J_\omega^T \end{bmatrix}^T \in \mathfrak{R}^{6 \times n_q}$ as follows:

$$\dot{p} = J(q)\dot{q} \tag{2}$$

In constrained motion the contact point must satisfy the surface equation $\psi(p_c) = 0$ that based on (1) can be translated to a constraint for the center of the robot tip $\psi'(p_t) = 0$; that is, the robot tip center constraint is parallel to the real surface but displaced by an offset equal to the tip radius. Notice that the direction normal to the surface n_c is given by $n_c = \left(\frac{\partial \psi(p_c)}{\partial p_c} \right)^T / \left\| \frac{\partial \psi(p_c)}{\partial p_c} \right\| = \left(\frac{\partial \psi'(p_t)}{\partial p_t} \right)^T / \left\| \frac{\partial \psi'(p_t)}{\partial p_t} \right\|$. For a flat surface, n_c is independent of p_c and hence $R_c = R_s$ ($n_c = n_s$) and $n_s^T p_c = n_s^T p_s$; thus, the contact constraint with respect to p_t can be expressed as follows:

$$n_s^T (p_t - p_s) = -r \tag{3}$$

Differentiating (3) and using (2) we get:

$$n^T J \dot{q} = 0 \tag{4}$$

where $n = [\, n_s^T \ 0_3^T \,]^T$ is the generalized normal vector.

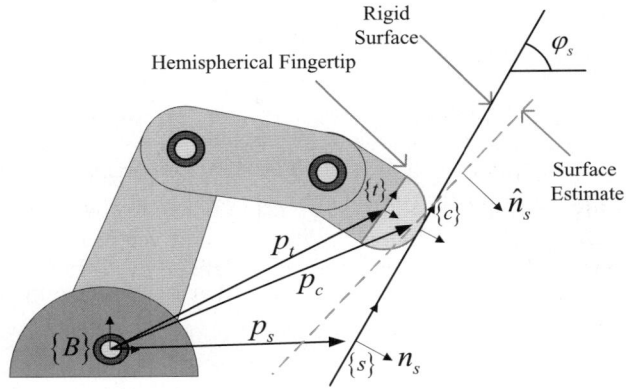

Fig. 1. A robotic finger with a rigid fingertip in contact with a rigid surface

The dynamic model of the robot can be written as follows:

$$L(q,\dot{q},\dot{q},\ddot{q}) + J^T F = u \tag{5}$$

where F is the interaction force between the robot tip and the environment that includes both normal force and frictional forces, u is the input control law and $L(q,\dot{q},\dot{q},\ddot{q})$ expresses the dynamic and gravity terms of the robot [1]:

$$L(q,\dot{q},\dot{q},\ddot{q}) = M(q)\ddot{q} + \left\{ \frac{1}{2}\dot{M}(q) + S(q,\dot{q}) \right\} \dot{q} + g(q)$$

where the second $\dot{q}$ in $L(q,\dot{q},\dot{q},\ddot{q})$ refers to the $\dot{q}$ outside the brackets of the equation above, $M(q) \in \Re^{n_q \times n_q}$ is the positive definite robot inertia matrix, $S(q,\dot{q}) \in \Re^{n_q \times n_q}$ is a skew symmetric matrix and $g(q) \in \Re^{n_q}$ denotes the gravity vector. For the general case of contact with friction, the interaction force can be decomposed into two orthogonal parts, $F_n = nn^T F$ that is normal to the surface and its complement $F_Q = QF$ where $Q = I_6 - nn^T$. The normal force can be written as $F_n = nf$ where f is the Lagrange multiplier associated with the constraint equation (4) while its complement arises owing to friction. We assume that joint positions and velocities are measured and that the robot tip Jacobian J is known; hence, p, $\dot{p}$ can be calculated by using the robot forward kinematic relationships.

The force and tip position desired trajectories are defined with respect to the surface as follows. Let $f_d(t) \in \Re^+$, be the desired trajectory of the normal force magnitude that should be tracked along the normal to the constraint surface direction; then the desired trajectory in the uncertain force control subspace is $\hat{n}_s f_d(t)$ (for a unit normal vector estimate) while the actual control target is $n_s f_d(t)$. Let $\phi_{td}(t) \in \Re^3$ and $\omega_{td} = \dot{\phi}_{td}(t) \in \Re^3$ be the desired orientation and angular velocity trajectories that are not affected by the surface uncertainty. Furthermore, let $d_s(t) \in \Re^2$ be the desired trajectory that should be tracked by the robot tip on the constraint planar surface and $\dot{d}_s(t) \in \Re^2$ the corresponding desired velocity trajectory. If $p_{ds} = \begin{bmatrix} 0 & d_s^T(t) \end{bmatrix}^T \in \Re^3$ is the desired trajectory expressed in the surface frame $\{s\}$ and $\hat{R}_s = \begin{bmatrix} \hat{n}_s & \hat{o}_s & \hat{a}_s \end{bmatrix}$ is the uncertain orientation of $\{s\}$ in the robot workspace then the desired tip position and velocity trajectories in the uncertain position control subspace are given by $p_{td} = p_t(0) + \hat{R}_s p_{ds}$ and $v_d = \hat{R}_s \dot{p}_{ds}$. Notice that the surface position is here taken as the initial tip position i.e. $p_s = p_t(0)$ and that $\dot{p}_{td}$ is not equal to v_d when the $\hat{R}_s$ is updated online. It is clear that although the desired force trajectory $\hat{n}_s f_d$ coincides with the actual target when the estimate of the normal to the surface direction converges to its actual value, p_{td} further requires the identification of one more axis of $\{s\}$ say for e.g axis o_s; we here assume that there is no relative rotation between the initial estimate and the actual surface frame around the n_s axis. Finally, the desired generalized position and velocity trajectories are concisely denoted by $p_d = \begin{bmatrix} p_{td}^T & \phi_{td}^T \end{bmatrix}^T$ and $V_d = \begin{bmatrix} v_d^T & \omega_{td}^T \end{bmatrix}^T$.

3   Controller Design

Let the estimates of the surface normal direction be denoted by $\hat{n}_s$ and its generalized expression by $\hat{n} = [\hat{n}_s^T \ 0_3^T]^T$ that are used to produce the projector $\hat{Q}$ on its orthogonal complement space, the force magnitude $\hat{f}$ and the surface rotation $\hat{R}_s$:

$$\hat{Q} = I_6 - \frac{\hat{n}\hat{n}^T}{\|\hat{n}\|^2} \tag{6}$$

$$\hat{f} = \frac{\hat{n}^T}{\|\hat{n}\|} F \tag{7}$$

$$\hat{R}_s = \begin{bmatrix} \dfrac{\hat{n}_s}{\|\hat{n}_s\|} & \hat{o}_s & \hat{a}_s \end{bmatrix} \tag{8}$$

where $\hat{o}_s$ and $\hat{a}_s$ can be defined from $\hat{n}_s$ and its update law (Appendix). Division of $\hat{n}$ by its norm $\|\hat{n}\|$ is required since $\hat{n}$ will be derived by an adaptation law and can take in general non-unit values. Notice that the adaptation law must be designed so that $\|\hat{n}\| \neq 0$.

Let us define the reference velocity vector $\dot{p}_r \in \Re^6$ in the robot tip space:

$$\dot{p}_r = \hat{Q}(V_d - \alpha \Delta p) - \beta \frac{\hat{n}}{\|\hat{n}\|} \Delta \hat{F} \tag{9}$$

where α, β are positive control gains, $\Delta p = p - p_d$ is the position error, $\Delta \hat{F} = \int_0^t \Delta \hat{f}(\tau) d\tau$ is the integral of the estimated force error $\Delta \hat{f} = \hat{f} - f_d$. We also define the error:

$$s_p = \dot{p} - \dot{p}_r \tag{10}$$

and using $\dot{p} = \hat{Q}\dot{p} - \frac{\hat{n}}{\|\hat{n}\|^2}\tilde{n}^T\dot{p}$ where $\tilde{n} = [\tilde{n}_s^T \ 0_3^T]^T$ with $\tilde{n}_s = n_s - \hat{n}_s$, we can express s_p as follows:

$$s_p = \hat{s}_Q + \hat{s}_n \tag{11}$$

where $\hat{s}_Q$, $\hat{s}_n$ are linearly independent quantities:

$$\hat{s}_Q = \hat{Q}\left[(\dot{p} - V_d) + \alpha \Delta p\right] \tag{12}$$

$$\hat{s}_n = \frac{\hat{n}}{\|\hat{n}\|}\left(\beta\Delta\hat{F} - \frac{\dot{p}^T}{\|\hat{n}\|}\tilde{n}\right) \tag{13}$$

On the other hand s_p can also be expressed as follows:

$$s_p = \dot{p} + \alpha\hat{Q}p + v(\hat{n}, V_d, p_d, \Delta\hat{F}) \tag{14}$$

where $v = \beta\frac{\hat{n}}{\|\hat{n}\|}\Delta\hat{F} - \hat{Q}(V_d + \alpha p_d)$. We can also define the reference joint velocity vector $\dot{q}_r$ and the error s as follows in the non-redundant case, :

$$\dot{q}_r = J^{-1}\dot{p}_r \tag{15}$$

$$s = \dot{q} - \dot{q}_r = J^{-1}s_p \tag{16}$$

Differentiating (15), the reference joint acceleration vector is given by:

$$\ddot{q}_r = -J^{-1}\dot{J}\dot{q}_r + J^{-1}\ddot{p}_r \tag{17}$$

The reference acceleration $\ddot{p}_r$ is found by taking the derivative of (9):

$$\ddot{p}_r = \frac{d}{dt}\left(\hat{Q}\right)(V_d - \alpha\Delta p) + \hat{Q}\left(\dot{V}_d - \alpha\Delta\dot{p}\right) - \frac{d}{dt}\left(\frac{\hat{n}}{\|\hat{n}\|}\right)\beta\Delta\hat{F} - \frac{\hat{n}}{\|\hat{n}\|}\beta\Delta\hat{f} \tag{18}$$

where $\frac{d}{dt}(\cdot)$ can be calculated using the estimates $\hat{n}$, $\dot{\hat{n}} \triangleq \left[\dot{\hat{n}}_s^T \ 0_3^T\right]^T$. The latter is given by an update law that will be defined through the subsequent stability analysis. Notice also that $\dot{V}_d$ and $\Delta\dot{p} = \dot{p} - \dot{p}_d$ involves the derivative of $\hat{R}_s$ (Appendix).

It is known that the dynamic model can be parameterized with respect to a set of constant parameters θ and hence we can write:

$$L(q, \dot{q}, \dot{q}_r, \ddot{q}_r) = Z(q, \dot{q}, \dot{q}_r, \ddot{q}_r)\theta \tag{19}$$

where Z is the dynamics regression matrix. The following model-based control law is proposed for force/position tracking under surface kinematic uncertainties:

$$u = J^T\frac{\hat{n}}{\|\hat{n}\|}(f_d - k_f\Delta\hat{f} - k_I\Delta\hat{F}) + J^T\hat{Q}F - Ds + Z(q, \dot{q}, \dot{q}_r, \ddot{q}_r)\hat{\theta} \tag{20}$$

where k_f, k_I are positive control gains and D is a bounded and positive definite matrix. The update laws for normal direction cosines and robot dynamic parameters are chosen as follows:

$$\dot{\hat{n}}_s = \mathscr{P}\left\{\Gamma_n[I_3\ O_{3\times3}]\|\hat{n}\|^{-1}\left(k_f'\Delta\hat{f} + k_I\Delta\hat{F}\right)\dot{p}\right\} \tag{21}$$

$$\dot{\hat{\theta}} = -\Gamma Z^T s \tag{22}$$

where $\Gamma_n = \mathrm{diag}\,[\gamma_{ni}]_{i=1}^{i=3}$, $\Gamma = \mathrm{diag}\,[\gamma_i]_{i=1}^{i=j}$ are diagonal matrices of positive parameter update gains and $\mathscr{P}$ is an appropriately designed projection operator [6] with respect to a convex set $\mathscr{S}$ so that $\|\hat{n}_s\| \neq 0$ and the absolute value of the angle formed between the current estimate and the actual normal vector defined by $\phi(t) = \arccos(\frac{n_s^T \hat{n}_s}{\|\hat{n}_s\|})$ is less than 90° at all times.

Substituting the input control law (20) into the robot dynamic model (5) we obtain the closed loop system:

$$M\dot{s} + \left(\tfrac{1}{2}\dot{M} + S\right)s + Ds + J^T\frac{\hat{n}}{\|\hat{n}\|}\left(k_f'\Delta\hat{f} + k_I\Delta\hat{F}\right) + Z\tilde{\theta} = 0 \tag{23}$$

where $k_f' = k_f + 1$. Taking the inner product of the closed loop system (23) with s we get:

$$\frac{d}{dt}\left\{\frac{1}{2}s^T M s + \frac{1}{2}\beta k_f'\Delta\hat{F}^2\right\} + \beta k_I\Delta\hat{F}^2 + s^T Ds$$
$$+ s^T Z\tilde{\theta} - \|\hat{n}\|^{-1}(k_f'\Delta f + k_I\Delta F)\dot{p}^T\tilde{n} = 0 \tag{24}$$

Using (21), (22) in (24) we get $\frac{dV}{dt} + W = 0$ where

$$V = \frac{1}{2}s^T M s + \frac{1}{2}\beta k_f'\Delta\hat{F}^2 + \frac{1}{2}\tilde{n}_s^T \Gamma_n^{-1}\tilde{n}_s + \frac{1}{2}\tilde{\theta}^T \Gamma^{-1}\tilde{\theta} \tag{25}$$

$$W = \beta k_I\Delta\hat{F}^2 + s^T Ds \tag{26}$$

Function V is positive definite with respect to s, $\Delta\hat{F}$ and parameter errors $\tilde{n}_s$, $\tilde{\theta}$ while function W is positive definite with respect to s, $\Delta\hat{F}$. Hence, function V has a negative derivative i.e. $\frac{dV}{dt} = -W \leq 0$ and can be regarded as a Lyapunov-like candidate function in order to prove the following theorem.

Theorem 1. *Given the persistent excitation of* $\|\dot{d}_s\|$, *the input control law (20) with the update laws (21), (22) applied in (5) achieves the boundedness of all signals and the convergence to zero of the force, position and velocity tracking errors as well as slope identification i.e.* $\Delta f \to 0$, $Q\Delta p \to 0$, $Q\Delta\dot{p} \to 0$ *and* $\frac{\hat{n}}{\|\hat{n}\|} \to n$.

Proof: $\frac{dV}{dt} = -W \leq 0$ implies $V(t) \leq V(0)$ and hence s $(s_p, \hat{s}_Q, \hat{s}_n)$, $\Delta\hat{F}$, $\tilde{n}_s$ $(\tilde{n})$, $\tilde{\theta}$ $\in \mathscr{L}_\infty$. Hence, given p_d, $\dot{p}_d$ are bounded trajectories, $v \in \mathscr{L}_\infty$ and consequently (14) implies $\dot{p} + \alpha\hat{Q}p \in \mathscr{L}_\infty$. Since $\hat{Q}$ is bounded and positive semi-definite (6), it can be easily proved that $\dot{p}, p \in \mathscr{L}_\infty$. If the robot moves away from singular positions, the boundedness of p, $\dot{p}$ implies that q, $\dot{q}$ are bounded. Moreover, the estimated force error can be written as $\bar{\mu}(q,\dot{q},\hat{n},\hat{\theta})\Delta\hat{f} = \mu(q,\dot{q},\hat{n},\hat{\theta})$, where $\bar{\mu}$, μ are state dependent scalar functions that are bounded if $q,\dot{q},\hat{n},\hat{\theta} \in \mathscr{L}_{\infty}$; hence, $\Delta\hat{f} \in \mathscr{L}_{\infty}$ given $\bar{\mu} \neq 0$ $\forall t \geq 0$. From (23), $\dot{s}$ can be expressed as a sum of bounded quantities and hence $\dot{s} \in \mathscr{L}_{\infty}$ and in turn $\ddot{q} \in \mathscr{L}_{\infty}$. From $\frac{dV}{dt} + W = 0$ and (26) $\Delta\hat{F}$, $s \in \mathscr{L}_2$. The boundedness of $\Delta\hat{F}$, s and their derivatives implies that $\Delta\hat{F}$, s are uniformly continuous and it follows from Desoer and Vidyasagar (1975) that $\Delta\hat{F} \to 0$, $s \to 0$ (s_p, $\hat{s}_n$, $\hat{s}_Q \to 0$). The boundedness of q, $\dot{q}$, $\hat{n}$, $\hat{\theta}$ and their derivatives implies that μ, $\bar{\mu}$ and consequently $\Delta\hat{f}$ are uniformly continuous. The uniform continuity of $\Delta\hat{f}$ in conjunction with the convergence of its integral to zero implies $\Delta\hat{f} \to 0$.

Furthermore, (21), (22) and the convergence of s (s_p, $\hat{s}_Q$, $\hat{s}_n$), $\Delta\hat{f}$ and $\Delta\hat{F}$ to zero imply that $\dot{\hat{\theta}}$, $\dot{\hat{n}}_s \to 0$ and consequently $\dot{\hat{Q}} \to 0$, $\dot{\hat{R}}_s \to 0$ and consequently $\dot{p}_{td} \to v_d$. In turn $\hat{s}_Q \to \frac{d}{dt}[\hat{Q}\Delta p] + \alpha\hat{Q}\Delta p \to 0$ and hence, $\hat{Q}\Delta p, \hat{Q}\Delta\dot{p} \to 0$. By projecting the estimated error $\hat{Q}(\dot{p} - \dot{p}_d)$ along the actual normal direction and using (6) and $n^T\dot{p} = 0$, $\forall t \geq 0$ we can find that $\frac{\hat{n}^T}{\|\hat{n}\|}\dot{p} \to -n_e^T V_d$ where $n_e = \frac{n}{\cos\phi} - \frac{\hat{n}}{\|\hat{n}\|}$. Notice, that the projection operator in the update law (21) must be designed in order to prevent $\cos\phi$ to take zero values. Vector n_e can be regarded as an indication of the error between the actual and estimated normal direction and is zero in the convex set $\mathscr{S}$ if and only if $\frac{\hat{n}}{\|\hat{n}\|} = n$. Decomposing the robot tip velocity $\dot{p}$ along the estimated orthogonal directions, we find that $\dot{p}$ converges to $(I - \frac{\hat{n}}{\cos\phi\|\hat{n}\|}n^T)V_d$. Using the velocity convergence result in (13) and the convergence of $\hat{s}_n$ and $\Delta\hat{F}$ to zero we find that $V_d^T n_e$ converges to zero. The desired velocity can be expressed as $v_d = \hat{e}_v(t)\|\dot{d}_s\|$ where $\hat{e}_v(t)$ is a unit direction vector lying on the estimated surface that satisfy $\hat{n}_s^T \hat{e}_v(t) = 0$. Hence, the convergence of $V_d^T n_e$ to zero implies that $n_s^T \hat{e}_v(t)\|\dot{d}_s\| \to 0$. Consequently $n_s^T \hat{e}_v(t) \to 0$ provided that $\|\dot{d}_s\|$ is a persistently excited signal, i.e. $\int_t^{t+T_0} \|\dot{d}_s\|^2 d\tau \geq \alpha_0 T_0$, $\forall t \geq 0$ for some α_0, $T_0 > 0$. The convergence of $n_s^T \hat{e}_v(t)$ means that $\hat{e}_v(t)$ that is updated on-line using the update of $\hat{n}_s$ becomes normal to n_s as time tends to infinity. This can only be achieved if the estimated normal direction converges to the actual direction, i.e. $\frac{\hat{n}}{\|\hat{n}\|} \to n$, that in turn implies that $\hat{f} \to f$, $\hat{Q} \to Q$ and in turn Δf, $Q\Delta p$, $Q\Delta\dot{p} \to 0$. Notice that despite the convergence of $\frac{\hat{n}}{\|\hat{n}\|}$ to n, $\hat{n}$ can converge to a constant vector or a vector of changing magnitude but in both cases its direction converges to the real one. Notice that the velocity persistent excitation condition is very easy to satisfy for a moving tip although the speed of the parameter convergence may depend on the motion speed. Furthermore, notice that the convergence of the dynamic parameters $\tilde{\theta}$ is not required.   □

4   Simulation Results

We consider a planar two-dof manipulator with revolute joints and link lengths $l_1 = 0.3$ m, $l_2 = 0.2$ m masses $m_1 = 0.8$ kg, $m_2 = 0.6$ kg and inertias $I_{z1} = 0.006$ kg·m^2, $I_{z2} = 0.002$ kg·m^2. The surface is modeled by a line with slope $\varphi_s = 45°$ with a normal vector $n = \begin{bmatrix} \sin\varphi_s & -\cos\varphi_s \end{bmatrix}^T$ and a tangent vector $o = \begin{bmatrix} \cos\varphi_s & \sin\varphi_s \end{bmatrix}^T$. The end-effector initial contact point position is $p_c(0) = \begin{bmatrix} 0.3515 & 0.1315 \end{bmatrix}^T$ (m). The control purpose is to exert a time-variant normal force with magnitude $f_d(t) = 10 + 4\cos(t)$ (N) and to track the desired position trajectory that is defined in a lower dimensional space as the

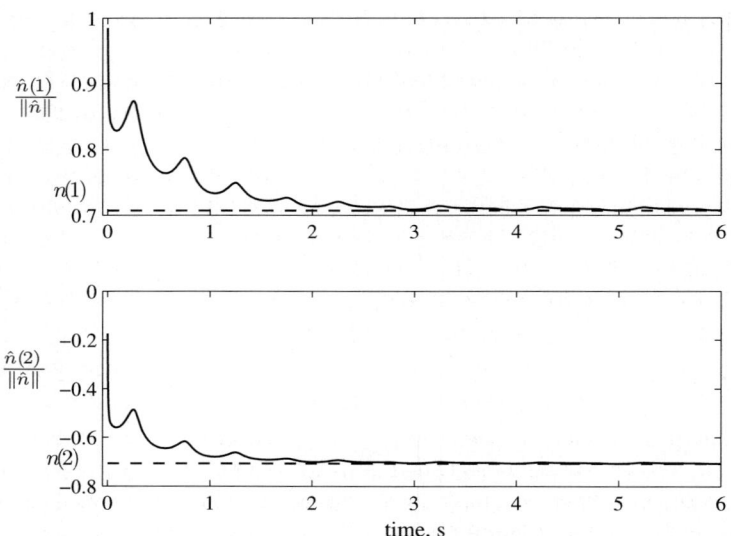

Fig. 2. Normal Vector Identification

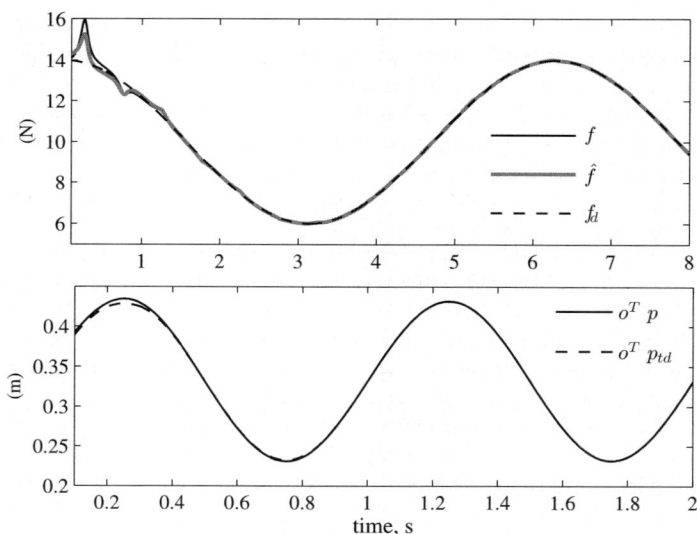

Fig. 3. Force and Position Tracking

distance $d_s(t) = -0.01 + 0.1\sin(2\pi t)$ (m) from the initial position. The initial estimate of the line slope that is taken equal to $80°$ and hence the initial line slope estimate corresponds to an initial angle error of $-35°$. The initial parameter estimates is $10\% - 30\%$

less than the actual. The gains of the controller are chosen as follows: $\alpha = 20$, $\beta = 0.8$, $k_I = 10$, $k_f = 3$, $D = 400 J^T J$, $\Gamma = \mathrm{diag}[0.01, 0.01, 0.01, 5, 5]$, $\Gamma_n = \mathrm{diag}[1, 2]$. Fig. 2 shows the convergence of the estimated normal vector direction coordinates to their actual values that is achieved in 3 s. Fig. 3 shows the estimated (gray solid line), the actual (black solid line) and the desired force trajectories (dashed line) as well as the desired feasible and the actual position trajectories. As the normal direction converges to its actual value the estimated force magnitude converges to the actual force magnitude and tracks the desired trajectory. The same is true for the convergence of the actual position trajectory to the desired feasible trajectory. Simulations have also shown that the proposed controller performs satisfactorily in case of smoothly curved surfaces.

5 Conclusions

This work proposes an adaptive controller for the problem of robot force/position tracking in the case of uncertain position and force control subspaces. The proposed control scheme is based on updated estimates of the surface normal that converge to the actual value achieving the control target provided that the desired velocity trajectory on the surface is a persistently excited signal.

References

1. Arimoto, S.: Control Theory of Non-linear Mechanical Systems, A Passivity-based and Circuit-theoretic Approach. Oxford University Press, Oxford (1996)
2. Arimoto, S., Liu, Y., Naniwa, T.: Model-based adaptive hybrid control for geometrically constrained robots. In: Proc. IEEE 1993 International Conference on Robotics and Automation, May 1993, pp. 618–623 (1993)
3. Cheah, C.C., Kawamura, S., Arimoto, S.: Stability of hybrid position and force control for robotic kinematics and dynamics uncertainties. Automatica 39, 847–855 (2003)
4. Cheah, C.C., Liu, C., Slotine, J.J.E.: Approximate Jacobian adaptive control for robot manipulators. In: Proc. IEEE 2004 International Conference on Robotics and Automation, pp. 3075–3080 (2004)
5. Cheah, C.C., Zhao, Y., Slotine, J.J.E.: Adaptive Jacobian motion and force tracking control for constrained robots with uncertainties. In: Proc.IEEE 2006 International Conference on Robotics and Automation, May 2006, pp. 2226–2231 (2006)
6. Ioannou, P.A., Sun, J.: Robust Adaptive Control. Upper Saddle River, Prentice Hall (1996)
7. Kwan, C.M., Yesildirek, A., Lewis, F.L.: Robust force/motion control of constrained robots using neural network. Journal of Robotic Systems 16(12), 697–714 (1999)
8. Namvar, M., Aghili, F.: Adaptive force-motion control of coordinated robot interacting with geometrically unknown environments. IEEE Transactions on Robotics 21(4), 678–694 (2005)
9. Xiao, D., Ghosh, B., Xi, N., Tarn, T.J.: Sensor-based hybrid position/force control of a robot manipulator in an uncalibrated environment. IEEE Transactions on Control System Technology 8(4), 635–645 (2000)
10. Yoshikawa, T., Sudou, A.: Dynamic hybrid position/force control of robot manipulators – online estimation of unknown constraint. IEEE Transactions on Robotics and Automation 9(2), 220–226 (1993)

11. Zhao, Y., Cheah, C.C., Slotine, J.J.E.: Adaptive vision and force tracking control for constrained robots. In: Proc.IEEE/RSJ 2006 International Conference on Intelligent Robots and Systems, October 2006, pp. 1484–1489 (2006)

Appendix

Let $\{\hat{s}_0\} = \{\hat{n}_{s0}, \hat{o}_{s0}, \hat{a}_{s0}\}$ be the initial estimate of frame $\{s\}$. Since we have assumed that there is no relative rotation between the $\{\hat{s}_0\}$, $\{s\}$ around the n_s axis either the actual axis o_s is parallel to the $\hat{n}_{s0}\hat{o}_{s0}$-plane or a_s is parallel to the $\hat{n}_{s0}\hat{a}_{s0}$-plane. Let us consider the first case; then current estimate $\hat{o}_s(t)$ is constrained to move on the $\hat{n}_{s0}\hat{o}_{s0}$-plane. Therefore:

$$\hat{a}_{s0}^T \dot{\hat{o}}_s = 0 \tag{27}$$

Moreover $\hat{o}_s(t)$ must preserve the unity of its magnitude and its orthogonality with the current estimate $\hat{n}_s$

$$\hat{o}_s^T \dot{\hat{o}}_s = 0 \tag{28}$$

$$\frac{\hat{n}_s^T}{\|\hat{n}_s\|} \dot{\hat{o}}_s = -\hat{o}_s^T \frac{d}{dt}\left(\frac{\hat{n}_s}{\|\hat{n}_s\|}\right) \tag{29}$$

We can find a unique $\dot{\hat{o}}_s$ that satisfy (27)-(29) if $\hat{a}_s$ and $\hat{a}_{s0}$ are not orthogonal i.e. $\hat{a}_s^T \hat{a}_{s0} \neq 0$; in particular the closed form solution is given by:

$$\dot{\hat{o}}_s = -\frac{\hat{o}_s \times \hat{a}_s(0)}{\hat{a}_s^T \hat{a}_{s0}} \hat{o}_s^T \frac{d}{dt}\left(\frac{\hat{n}_s}{\|\hat{n}_s\|}\right) \tag{30}$$

We can consequently find a unique solution of $\dot{\hat{a}}_s$ using the unity of magnitude and the orthogonality with $\hat{o}_s$ and $\hat{n}_s$:

$$\dot{\hat{a}}_s = -\frac{\hat{n}_s}{\|\hat{n}_s\|} \hat{a}_s^T \frac{d}{dt}\left(\frac{\hat{n}_s}{\|\hat{n}_s\|}\right) - \hat{o}_s \hat{a}_s^T \dot{\hat{o}}_s. \tag{31}$$

Scalable Operators for Feature Extraction on 3-D Data

Shanmugalingam Suganthan[1], Sonya Coleman[1], and Bryan Scotney[2]

[1] School of Computing and Intelligent Systems, University of Ulster, Northern Ireland
{S.Suganthan,SA.Coleman}@ulster.ac.uk
[2] School of Computing and Information Engineering, University of Ulster, Northern Ireland
BW.Scotney@ulster.ac.uk

Summary. Real-time extraction of features from range images can play an important role in robotic vision tasks such as localisation and navigation. Feature driven segmentation of range images has been primarily used for 3D object recognition, and hence the accuracy of the detected features is a prominent issue. Feature extraction on range data has proven to be a more complex problem than on intensity images due to both the irregular distribution of range images. This paper presents a general approach to the development of scalable derivative operators using a finite element framework that can be applied directly to processing regularly or irregularly distributed range image data. The gradient operators of varying scales are evaluated with respect to their performance on regular and irregular grids.

Keywords: 3D Range Data, Feature extraction, Gradient operators.

1 Introduction

The diverse world of mobile robotics has been in search of real-time electronic eyes that closely mimic the behaviour of the human eyes. Currently many machine vision techniques use range images to obtain useful descriptions of 3-D scenes [10, 25]. This is largely because range imagery can be used to obtain reliable descriptions of 3-D scenes; a range image contains distance measurements from a selected reference point or plane to surface points of objects within a scene [6], allowing more information about the scenes to be recovered [5]. Range image feature extraction and segmentation have been identified as means of scene representation and are used in applications such as mobile robot localization [19], object recognition [11,17], motion analysis [22], robot navigation [12], manufacturing process automation [26], automated visual inspection [20], and 3-D map reconstruction [13].

Range image data are acquired using range sensors; in an ideal situation, like intensity images, range data are uniformly distributed in the x- and y- directions. However, whilst a number of range image sensors are available [6][14], not all can sample the surface at equidistant x- and y- intervals; often the coordinates of the data points are dependent on the measured range of the point [9], as, for example, in the case of the commonly used ABW, K2T and Perceptron sensors [14]. Hence the data are often irregularly distributed.

A wide selection of feature detection methods are available for use on range image data; the well-known edge detection operators of Marr and Hildreth and Canny may be applied directly to range images, although they are not entirely appropriate as they cannot be readily applied to non-uniformly distributed data and may tend to misplace the detected feature [2]. Techniques designed specifically for feature extraction on range images include: scan line techniques [4][15][16][23]; mathematical morphology techniques [7][18]; and other methods including those of [5][17][24].

Whilst much research has been carried out to develop edge detection methods for range image data, little has focussed on the area of multiscale, or adaptive, edge detection methods. When features in an image that occur over a range of scales are extracted at only one scale, localisation error or false edges may be introduced. In order to successfully extract the various edge types found in range images, multiscale feature extraction algorithms are particularly pertinent for obtaining good feature localisation and reliability as smooth crease edges are low-frequency events and jump edges are high-frequency events. When features occur over a wide scale range, any method that extracts them at only one scale introduces either a localisation error or false edges. Multiscale boundary detection in range images has proven to be effective at dealing with discontinuities occurring at a variety of spatial scales, and one such example of a multiscale approach is that of [21] who fitted Legendre polynomials to one-dimensional windows of range data and varied the kernel size of the polynomial.

Due to the locational irregularity of range image data, multiscale feature detection on range images is a significantly different problem than that on intensity images. This paper presents a family of finite element-based shape-adaptive multi-scalable gradient operators that can be applied directly to range data without any data pre-processing for the purpose of detecting jump and crease edges. An overview of the range image representation, describing the finite element framework employed, and a brief overview of the scalable shape-adaptive gradient operator implementation is presented in Section 2. Section 3 presents a detailed evaluation of the operators using a set of images generated for a range of degrees of data irregularity. A summary of the work is provided in Section 4.

2   Operator Design Framework

We consider a range image to be represented by a spatially irregular sample of values of a continuous function $u(x, y)$ of depth value on a domain Ω. Our operator design is then based on the use of a quadrilateral mesh as illustrated in Figure 1 in which the nodes are the sample points. With each node i in the mesh is associated a piecewise bilinear basis function $\phi_i(x, y)$ which has the properties $\phi_i(x_j, y_j) = 1$ if $i = j$ and $\phi_i(x_j, y_j) = 0$ if $i \neq j$, where (x_j, y_j) are the co-ordinates of the nodal point j in the mesh. Thus $\phi_i(x, y)$ is a "tent-shaped" function with support restricted to a small neighbourhood centred on node i consisting of only those elements that have node i as a vertex. We then approximately represent the range image function u by a function $U(x, y) = \sum_{j=1}^{N} U_j \phi_j(x, y)$ in which the parameters $\{U_1, ..., U_N\}$ are mapped from the range image pixel values at the N irregularly located nodal points. Therefore, approximate image representation takes the form of a simple function (typically a low order polynomial) on each element and has the sampled range value U_j at node j.

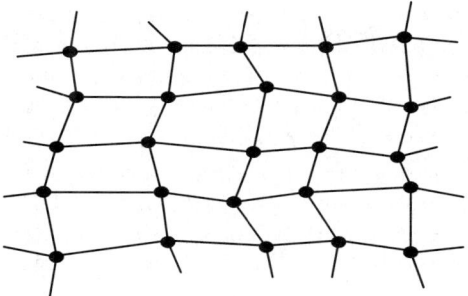

Fig. 1. Sample of the irregularly distributed range image

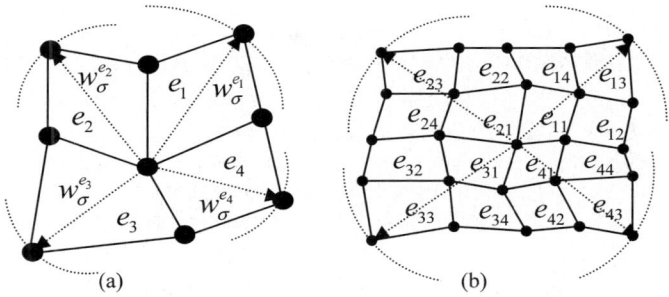

Fig. 2. Local (a) 3x3, (b) 5x5 operator neighbourhood

We describe the operator framework for the construction of both 3×3 and 5×5 irregular operators as illustrated in Figure 2. We formulate image operators that correspond to weak forms in the finite element method [3], in a similar manner as described in [8]. Corresponding to a first directional derivative $\partial u/\partial b \equiv \underline{b} \cdot \underline{\nabla} u$ we may use a test function $v \in H^1(\Omega)$ to define the weak form

$$E(u) = \int_{\Omega} \underline{b} \cdot \underline{\nabla} u v \, d\Omega \tag{1}$$

where $\underline{b} = (cos\theta, sin\theta)$ is the unit direction vector.

Since we are focussing on the development of operators that can explicitly embrace the concept of size and shape variability, our design procedure uses a finite-dimensional test space $T_\sigma^h \subset H^1$ that explicitly embodies a size parameter σ that is determined by the local data distribution. Using such test functions, the first order functional is defined as:

$$E_i^\sigma(U) = \int_{\Omega_i^\sigma} \underline{b}_i \cdot \underline{\nabla} U \psi_i^\sigma \, d\Omega_i \tag{2}$$

This generalisation allows sets of test functions $\psi_i^\sigma(x,y)$, $i=1,\ldots,N$, to be used when defining irregular derivative- based operators and the chosen test function is a Gaussian basis function.

$$\psi_i^\sigma(x,y) = \frac{1}{2\pi\sigma^2}e^{-\left(\frac{(x-x_i)^2+(y-y_i)^2}{2\sigma^2}\right)} \tag{3}$$

Hence, we naturally embody the scale parameter that supports the development of scalable operators, and also naturally builds in Gaussian smoothing. Within each neighbourhood, a different scale parameter is computed for each quadrant of the neighbourhood, enabling the Gaussian test function to adapt to the local area more accurately. As illustrated in Figure 2, W_σ^{em} is chosen as the diagonal of the neighbourhood from the operator centre (x_i,y_i), and in each case the quadrant scale parameter $\sigma_m = W_\sigma^{em}/1.96$ ensures that the diagonal of the quadrant through (x_i,y_i) encompasses 95% of the cross-section of the Gaussian.

On a neighbourhood Ω_i^σ we consider a locally constant unit vector $\underline{b}_i = (b_{i1},b_{i2})^T$ where $b_{i1}^2 + b_{i2}^2 = 1$. Substituting the image representation $U(x,y) = \sum_{j=1}^{N} U_j\phi_j(x,y)$ into the weak form $E_i^\sigma(U) = \int_{\Omega_i^\sigma} \underline{b}_i \cdot \nabla U \psi_i^\sigma \, d\Omega_i$ gives

$$E_i^\sigma(U) = b_{i1}\sum_{j=1}^{N} K_{ij}^\sigma U_j + b_{i2}\sum_{j=1}^{N} L_{ij}^\sigma U_j \tag{4}$$

where K_{ij}^σ and L_{ij}^σ are respectively entries in $N \times N$ global matrices K^σ and L^σ given by

$$K_{ij}^\sigma = \int_{\Omega_i^\sigma} \frac{\partial \phi_j}{\partial x} \psi_i^\sigma dxdy \qquad i,j = 1\ldots N \tag{5}$$

and

$$L_{ij}^\sigma = \int_{\Omega_i^\sigma} \frac{\partial \phi_j}{\partial y} \psi_i^\sigma dxdy \qquad i,j = 1\ldots N \tag{6}$$

These integrals need be computed only over the neighbourhood Ω_i^σ, rather than the entire image domain Ω, since ψ_i^σ has support restricted to Ω_i^σ. Each neighbourhood Ω_i^σ is composed of a set S_i^σ of elements. Hence, we may write K_{ij}^σ and L_{ij}^σ as the respective summations

$$K_{ij}^\sigma = \sum_{\{m|e_m \in S_i^\sigma\}} k_{ij}^{m,\sigma} \text{ and } L_{ij}^\sigma = \sum_{\{m|e_m \in S_i^\sigma\}} l_{ij}^{m,\sigma} \tag{7}$$

where $k_{ij}^{m,\sigma}$ and $l_{ij}^{m,\sigma}$ are the *element integrals*

$$k_{ij}^{m,\sigma} = \int_{e_m} \frac{\partial \phi_j}{\partial x} \psi_i^\sigma dxdy \text{ and } l_{ij}^{m,\sigma} = \int_{e_m} \frac{\partial \phi_j}{\partial y} \psi_i^\sigma dxdy. \tag{8}$$

The element integrals $k_{ij}^{m,\sigma}$ and $l_{ij}^{m,\sigma}$ are actually computed by mapping to the standard square element $\hat{e}$ in order to facilitate the integration of the Gaussian test functions

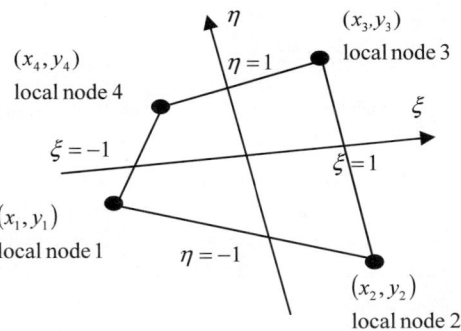

Fig. 3. 3x3 Operator, Quadrilateral Coordinates ans iso-parametric mappings

using simple quadrature rules. Figure [3] shows a typical quadrilateral element in which the nodes have locations in the x, y co-ordinates.

The local (x, y) co-ordinate reference system for element e_m is mapped to a co-ordinate system (ξ, η) with $-1 \le \xi \le 1$ and $-1 \le \eta \le 1$ where (ξ, η) is a rectangular co-ordinate system in the standard element $\hat{e}$. The co-ordinate transformation is defined as

$$x = \frac{1}{4}(x_1(1-\xi)(1-\eta) + x_2(1+\xi)(1-\eta) + x_3(1+\xi)(1+\eta) + x_4(1-\xi)(1+\eta)) \quad (9)$$

$$y = \frac{1}{4}(y_1(1-\xi)(1-\eta) + y_2(1+\xi)(1-\eta) + y_3(1+\xi)(1+\eta) + y_4(1-\xi)(1+\eta)) \tag{10}$$

Construction of the operators on an irregular quadrilateral grid differs from that of image processing operators on a typically regular grid in that it is no longer appropriate to build explicitly an entire operator, as each operator throughout an irregular mesh may be different with respect to the operator neighbourhood shape. When using an irregular grid, we work on an element-by-element basis, taking advantage of the flexibility offered by the finite element method as a means of adaptively changing the irregular operator shape to encompass the data available in any local neighbourhood.

3   Evaluation

In order to demonstrate the flexibility of the proposed multi-scale gradient operators for the purpose of 3D feature extraction over a range of irregularly distributed range images, we generate a set of synthetic test images for each edge type. As previously discussed, we concern ourselves with two main edge types: jump and crease. Crease edges can be further defined as convex roof, concave roof, convex crease, and concave crease edges. Each of the edge subtypes considered is illustrated in Figure 4 where the range images depth values, z, are defined as two planes, $S_1(x, y) = a_1 x + b_1 y + c_1$ and $S_2(x, y) = a_2 x + b_2 y + c_2$. Such plane equations are subsequently used to create a regularly distributed range image for each edge type.

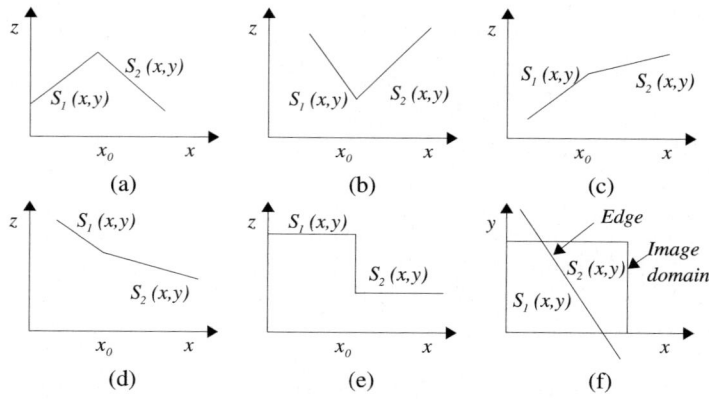

Fig. 4. Edge model: (a)-(e) edge model in x-z plane (a) Convex roof edge (b) Concave roof edge (c) Convex crease edge (d) concave crease edge (e) Jump edge (f) edge model in $x - y$ plane

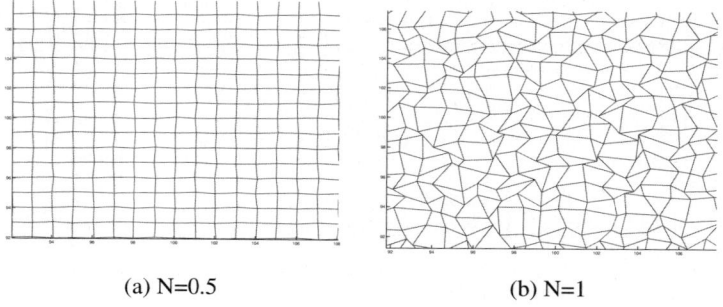

(a) N=0.5   (b) N=1

Fig. 5. Examples of irregular meshes

To generate the images with irregularly distributed data, we create a regularly distributed image and add varying degrees of random values to the x and y co-ordinates such that $x = x + r_x$, $y = y + r_y$ where $N \geq r_x \geq 0$ and $N \geq r_y \geq 0$ and N defines the degree of irregularity; examples of such image representation is illustrated in Figure 5.

For evaluation purposes, we use the Figure of Merit measure [1] and compare our proposed technique with that of the well-known scan-line approximation algorithm [15]. Pratt [1] considered three major areas of error associated with the determination of an edge: missing valid edge points; failure to localise edge points; classification of noise fluctuations as edge points. Pratt therefore introduced the Figure of Merit technique as one that balances these three types of error, defined as $R = \frac{1}{\max(I_A, I_I)} \sum_{i=1}^{I_A} \frac{1}{1 + \alpha d^2}$. Here I_A is the actual number of edge pixels detected, I_I is the ideal number of edge pixels, d is the separation distance of a detected edge point normal to a line of ideal edge points,

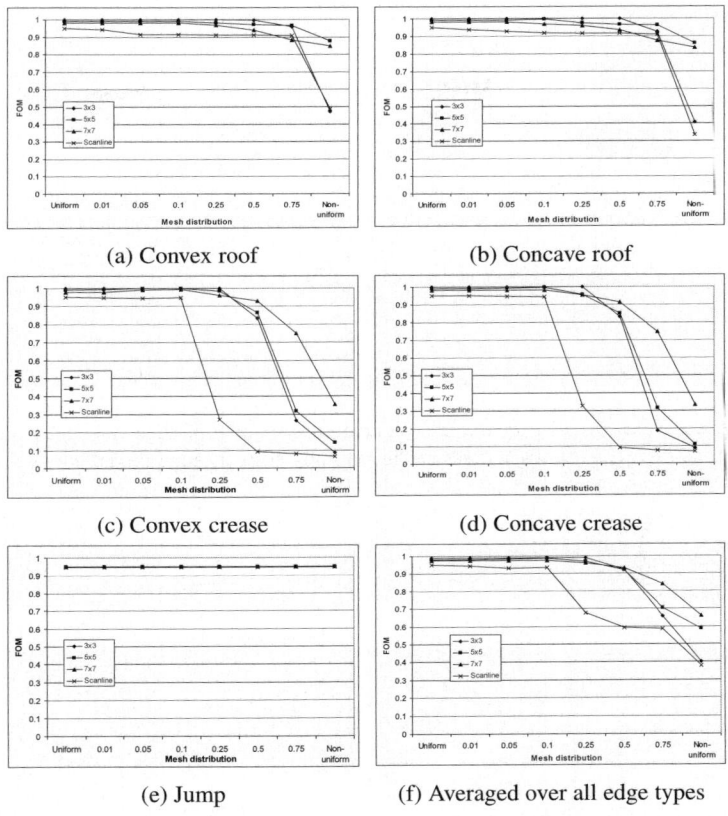

(a) Convex roof

(b) Concave roof

(c) Convex crease

(d) Concave crease

(e) Jump

(f) Averaged over all edge types

Fig. 6. Figure of Merit results for different edge types over a range of data irregularity

and α is a scaling factor, most commonly chosen to be 1/9, although this value may be adjusted to penalise edges that are localised but offset from the true edge position.

In Figure 6 we show results using the Figure of Merit evaluation technique for a vertical edge within a range image using varying degrees of irregularity and no noise. In each case the threshold value used provides the best Figure of Merit; similarly, it is computed over all possible parameter combinations for the scan line approach and again the optimal value selected. In additional, the results are computed on 5 randomly generated meshes, comprising five of each range edge type: Jump, convex roof, concave roof, convex crease, concave crease, and the Figure of Merit value is averaged for each. The patterns of behaviour of the $3 \times 3, 5 \times 5, 7 \times 7$ and scan-line methods are similar in the case of the roof edges although the proposed methods have slightly higher Figure of Merit values; however Figure 6 clearly demonstrates that the proposed technique becomes superior in the case of crease edges for higher degrees of irregularity than the scan line technique [15] with the proposed higher scale operators. For completeness, we present comparative edges maps for our proposed technique and for the scan line

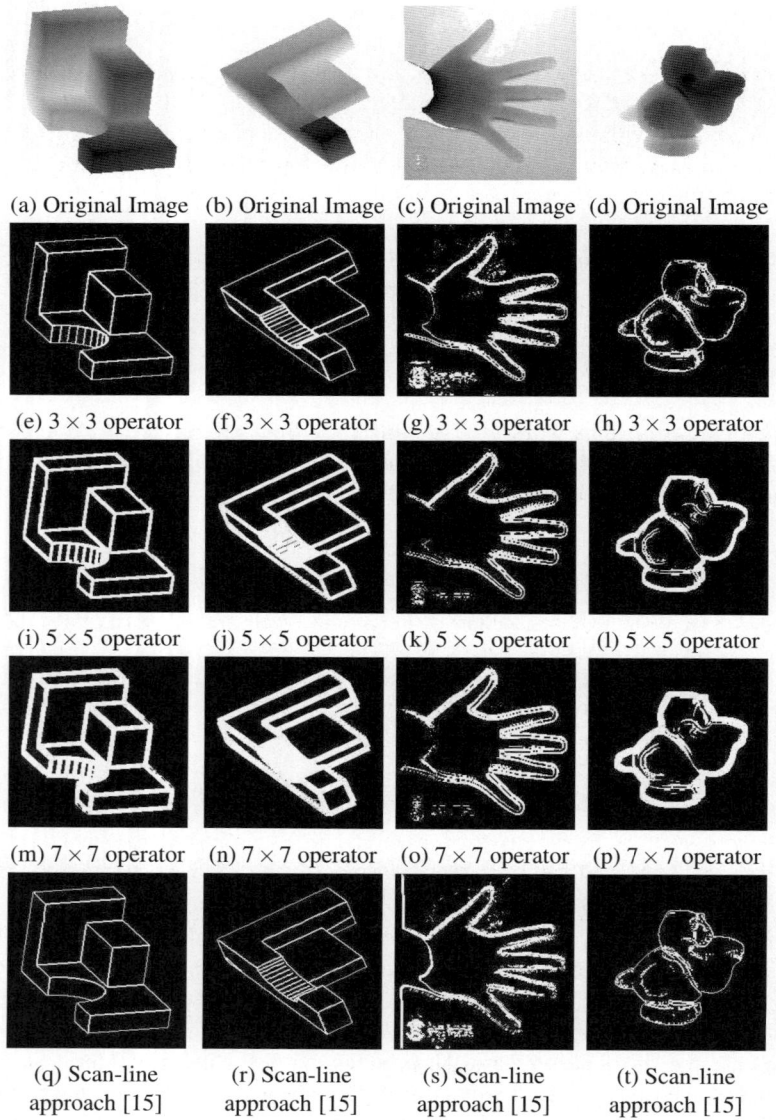

(a) Original Image (b) Original Image (c) Original Image (d) Original Image

(e) 3 × 3 operator (f) 3 × 3 operator (g) 3 × 3 operator (h) 3 × 3 operator

(i) 5 × 5 operator (j) 5 × 5 operator (k) 5 × 5 operator (l) 5 × 5 operator

(m) 7 × 7 operator (n) 7 × 7 operator (o) 7 × 7 operator (p) 7 × 7 operator

(q) Scan-line   (r) Scan-line   (s) Scan-line   (t) Scan-line
approach [15]   approach [15]   approach [15]   approach [15]

Fig. 7. Original range images from [27] and corresponding edge maps

approach in Figures 7. It should be noted that our proposed technique automatically finds all features whereas the technique in [15] does not automatically find the object boundary via the scan line approximation but instead, in all cases, assumes the boundary at the transition between data and no data in the range image.

4   Summary and Future Work

We have presented a design procedure within the finite element framework for the development of shape-adaptive scalable gradient operators that can be used directly on 3D range image data without the need for any image pre-processing. Through the use of the Figure of Merit evaluation measure, we have illustrated that our adaptive approach can accurately detect each edge sub-type over a varying degrees of data irregularity. We have compared performance with the scan-line approach of Jiang et al.[15] and found that the proposed approach is superior as the irregularity of the data increases, particularly in the case of crease edges. As the target application for this work is real-time robot vision, future work will involve additional comparison with other techniques for feature extraction and segmentation of range optimisation of the algorithms for real-time edge detection.

Acknowledgments. This work was supported by the U.K Research Council via EP-SRC. We would like to thank Professor Horst Bunke for providing us with the code for the scan line approximation algorithm in [15].

References

1. Abdou, I.E., Pratt, W.K.: Quantitative Design and Evaluation of Enhancement/ Threshold Edge Detectors. Proceedings of the IEEE 67(5) (1979)
2. Al-Hujazi, E., Sodd, A.: Range Image Segmentation with applications to Robot Bin-Picking Using Vacuum Gripper. IEEE Trans. Systems, Man, and Cybernetics 20(6) (1990)
3. Becker, E.B., Carey, G.F., Oden, J.T.: Finite Elements: An Introduction. Prentice Hall, London (1981)
4. Bellon, O.P., et al.: Edge Detection to Guide Range Image Segmentation by Clustering Techniques. In: IEEE Int. Conf. on Image Processing, Kobe, Japan (1999)
5. Bellon, O., Silva, L.: New Improvements on Range Image Segmentation by Edge Detection Techniques. In: Proceedings of the workshop on Artificial Intelligence and Computer Vision (2000)
6. Besl, P.J.: Active, optical range imaging sensors. Machine Vision and Apps 1, 127–152 (1988)
7. Cheng, J.-C., Don, H.-S.: Roof Edge Detection: A Morphological Skeleton Approach. In: Advances in Machine Vision: Strategies and Application, World Scientific, Singapore, pp. 171–191 (1992)
8. Coleman, S.A., Scotney, B.W., Suganthan, S.: Feature Extraction on Range Images - A New Approach. In: Coleman, S.A., Scotney, B.W., Suganthan, S. (eds.) Proceedings of IEEE International Conference on Robotics and Automation, Rome, pp. 1098–1103 (2007)
9. De Bakker, M.: The PSD chip, high speed acquisition of range images, PhD Thesis, Delft University of Technology (2000)
10. Dias, P., et al.: Combining Intensity and Range Images for 3D Modelling. In: Proceedings of the IEEE International Conference on Image Processing (ICIP 2003) (2003)
11. Flynn, P.J., Jain, A.K.: Three-dimensional object recognition. In: Handbook of Pattern Recognition and Image Processing: Computer Vision, pp. 497–541, Academic Press, San Diego (1994)
12. Franklin, D., Firby, R.J.: Integrating Range and Object Data for Robot Navigation. In: Proceedings of the first international conference on Autonomous agents,Marina del Rey, California, United States, pp. 185–192 (1997)

13. Huber, D., Carmichael, O., Hebert, M.: 3-D Map Reconstruction from Range Data, In:Proc. of the IEEE Inter. In: Conf. on Robotics & Automation, San Francisco, CA, pp. 891–897 (2000)
14. Jarvis, R.A.: Range Sensing for Computer Vision. In: Three-Dimensional Object Recognition Systems, pp. 17–56. Elsevier Science, Amsterdam (1993)
15. Jiang, X.Y., Bunke, H.: Edge detection in range image based on scan line approximation. Computer Vision ad Image Understanding 73(2), 183–199 (1999)
16. Jiang, X.Y., Bunke, H.: Fast Segmentation of Range Images into Planar Regions by Scan Line Grouping. Machine Vision and Applications 7(2), 115–122 (1994)
17. Kaveti, S., et al.: Second-Order Implicit Polynomials for segmentation of Range Images. Pattern Recognition 29(6), 937–949 (1996)
18. Krishnapuram, R., Gupta, S.: Morphological Methods for Detection and Classification for Edges in Range Images. Journal of Mathematical Imaging Vision, 351–375 (1992)
19. Neira, J., Tardos, J.D., Horn, J., Schmidt, G.: Fusing Range and Intensity Images for Mobile Robot Localization. IEEE Transactions on Robotics and Automation 15(1), 76–84 (1999)
20. Newman, T.S., Jain, A.K.: A system for 3D CAD-based inspection using range images. Pattern Recognition 28(10), 1555–1574 (1995)
21. Parvin, B., Medioni, G.: Adaptive Multiscale Feature Extraction From Range Data. Computer Vision Graphics, Image Understanding 45, 346–356 (1989)
22. Sabata, B., Aggarwal, J.K.: Surface correspondence and motion computation from a pair of range images. Computer Vision and Image Understanding 63, 232–250 (1996)
23. Sappa, A.D., Devy, M.: Fast Range Image Segmentation by an Edge Detection Strategy. In: Proc 3rd Int. Conference on 3D Digital Imaging and Modelling, Quebec, Canada, pp. 292–299 (2001)
24. Trucco, E., Fisher, R.B.: Experiments in Curvature-Based Segmentation of Range Data. IEEE Trans. Pattern Analysis and Machine Intelligence 17(2), 177–182 (1995)
25. Umeda, K., Arai, T.: Industrial Vision System by Fusing Range image and Intensity Image. Proceedings of the IEEE International Conference on Multisensor Fusion and Integration for Intelligent Systems, 337–344 (1994)
26. Zhao, D., Li, S.: A 3D image processing method for manufacturing process automation. In: Computer in Industry, vol. 56, pp. 975–985. Elsevier, Amsterdam (2005)
27. http://sampl.eng.ohio-state.edu/~sampl/data/3DDB/RID/-index.htm

Semi-autonomous Learning of an RFID Sensor Model for Mobile Robot Self-localization

Philipp Vorst and Andreas Zell

Department of Computer Science, University of Tübingen, Tübingen, Germany*
{philipp.vorst,andreas.zell}@uni-tuebingen.de

Summary. In this paper, we present a method of learning a probabilistic RFID reader model with a mobile robot in a semi-automatic fashion. RFID and position data, recorded during an exploration phase, are used to learn the probability of detecting an RFID tag, for which we investigate two non-parametric probability density estimation techniques. The trained model is finally used to localize the robot via a particle filter-based approach and optimized with respect to the resulting localization error. Experiments have shown that the learned models perform comparably well as a grid-based model learned from measurements in a stationary setup, but can be obtained easier.

1 Introduction

Radio frequency identification (RFID) is nowadays not only used for identification purposes in the industry, but also for navigation tasks in mobile robotics. The technology allows for the contactless identification of objects and landmarks which are marked with RFID tags (also called labels or transponders) by a reader device and its antennas via radio waves. Passive tags obtain the energy for operation and response from the radio field of the RFID reader, which makes them inexpensive and easily maintainable. In case of passive UHF technology as in this work, however, factors such as the relative position of a tag and nearby materials affect the readability of a tag. Hence, in practice detection rates can be poor and noisy, and whatever application is regarded, it will benefit from an accurate model of tag detection probabilities. For example, the modeled detection field may lead to an improvement in the placement of RFID readers in a plant. Moreover, such a model is the basis of probabilistic localization algorithms. If it is easy to derive, it can be adapted or rebuilt quickly if the setup of the RFID system changes.

In this paper, we present a method of learning a probabilistic RFID reader model with a mobile robot in a semi-automatic fashion. We have chosen a non-parametric approach, which means that we do not claim any specific functional form of the tag detection probability density. The approach should thus be applicable to other tasks, RFID standards, and hardware. A question which arises is how the quality of the learned model can be measured. We decided to plug it into the target application – the localization of the mobile robot in our case – and measure the resulting error there. Using RFID for self-localization is motivated by the idea that, as more and more goods are being labeled

---

* This work has been funded by the Landesstiftung Baden-Württemberg within the scope of the support program BW-FIT and the research cooperation AmbiSense.

H. Bruyninckx et al. (Eds.): European Robotics Symposium 2008, STAR 44, pp. 273–282, 2008.
springerlink.com

with RFID tags and RFID hardware is getting cheaper, mobile robots operated e.g. in trade scenarios will be able to exploit RFID as a lower-cost sensor for self-localization and navigation in general.

We proceed as follows: First, the robot explores its environment and records RFID and position measurements. Then, a probabilistic model of tag detection frequencies is built from the logged data by estimating the probability density for detecting a transponder, conditioned on its relative position to an RFID antenna. For this, we compare two techniques: a simple grid-based method and a k-nearest neighbors search. The estimation step is repeated with different parameter values in order to search systematically for the model which minimizes the localization error of the mobile robot. Each learning technique only depends on one parameter.

This procedure is new insofar as related approaches do not learn the RFID reader model during the navigation of the mobile robot. Moreover, we do not only present a method of learning the model with the robot, but also of validating and optimizing it. The only assumptions that we make are firstly that the robot is equipped with a quite accurate reference localization module (e.g. laser- or vision-based) and secondly that we provide the robot with a list of some tags and their positions in the global frame of reference. The preparation of this list is the reason why our method is only *semi-automatic*.

With regard to related work, Hähnel et al. were one of the first to gain a probabilistic RFID sensor model and use it for Monte Carlo localization [4]. They measured tag detection rates for a single passive UHF tag on a grid of fixed distances and angles. We detail their approach in the subsequent sections, since we follow their method of particle filter-based self-localization. Bajcsy et al. obtained an RFID reader model in a similar fashion, but for several tags spread on the floor and with the possibility to tilt the RFID antenna [1]. Kloos et al. learned a sensor model for RF-based localization via a parametric approach modeling signal strength and distances for active (battery-powered) sensor nodes [6]. In [5], Kantor and Singh presented localization and mapping with RF beacons which provided distance information. They determined probability densities over actual ranges for a discrete number of measured ranges. Djugash et al. also used RF beacons for self-localization [2]. Their likelihood function was explicitly represented by a 2D Gaussian with standard deviation estimated from the variance in range measurements. By contrast, a sensor model-free approach to RFID-based self-localization is described in [8], but its mapping phase can be time-consuming.

This paper is structured as follows: In Sect. 2 we clarify the background of our work and show how a mobile robot is able self-localize with RFID. Thereafter, we present our approach to learning an RFID reader model in Sect. 3. Numerical evidence for the quality of the different techniques is provided in Sect. 4, before we finally summarize and discuss our work in Sect. 5.

2   Monte Carlo Localization with RFID Sensors

For localizing the mobile robot via RFID, we pursue the particle filter-based approach by Hähnel et al. [4], which is Monte Carlo localization [3] with an observation model adapted to RFID sensors. In Monte Carlo localization, the robot pose $\mathbf{r}_t$ is represented by an arbitrary probability density function (pdf) over the space of locations. This pdf is approximated by a set of n particles. Each particle consists of a pose hypothesis (x^i, y^i, θ^i) and a weighting factor w^i, where (x^i, y^i) are the coordinates in a global frame of reference, θ^i is the global heading of the robot, and w^i states the importance of the i-th particle. The filtering algorithm itself iteratively performs three steps:

1. *Prediction*: The robot pose at time t is predicted by propagating all particle positions according to the latest odometry readings $\mathbf{o}_{t-1}$ and a motion model. Formally, one samples from the distribution $p(\mathbf{r}_t | \mathbf{o}_{t-1}, \mathbf{r}_{t-1})$.
2. *Correction*: Sensor data $\mathbf{z}_t$ are incorporated into the set of particles by correcting the particle weights according to some likelihood function $p(\mathbf{z}_t | \mathbf{r}_t)$:

$$w_t^i = \eta \cdot p(\mathbf{z}_t | \mathbf{r}_t)$$

Here, η is a normalizing constant which ensures that $\sum_{i=1}^{n} w_t^i = 1$.
3. *Resampling*: A new set of n particles with equal weights $1/n$ is obtained from the old one by drawing n samples from the old set of particles, where the probability of choosing particle i corresponds to its weight w_t^i. An option is to resample only if the estimate $\hat{n}_{eff} \approx 1 / \left(\sum_{i=1}^{n} (w_t^i)^2 \right)$ of the so-called effective sample size falls below some threshold, e.g. $n/2$.

Particle filtering has turned out to be a robust and versatile method for self-localization, even in presence of non-Gaussian noise and highly imprecise measurements, as it is the case also for self-localization with RFID.

To learn a sensor model of the RFID reader means in this paper to learn the likelihood function $p(\mathbf{z}_t | \mathbf{r}_t)$ required by step 2 of the algorithm. This function should represent the likelihood that the observed RFID measurements provide evidence for the current robot pose. More specifically, the measurements $\mathbf{z}_t$ at time t consist of two lists of detected RFID tags, $\mathbf{d}_t^l$ and $\mathbf{d}_t^r$, one for the left and one for the right RFID antenna of our robot (see Fig. 1 (a)). Assuming that both measurements are independent, we set $p(\mathbf{z}_t | \mathbf{r}_t) := p(\mathbf{d}_t^l | \mathbf{r}_t) p(\mathbf{d}_t^r | \mathbf{r}_t)$. The two $\mathbf{d}_t$ can be regarded as sequences $\mathbf{d}_t = (d_t^1, d_t^2, \ldots)$, where $d_t^i \in \{0, 1\}$ states whether $(d_t^i = 1)$ or not $(d_t^i = 0)$ transponder i was detected.

3 Learning an RFID Sensor Model

Gaining a model of an RFID reader means in our case to estimate the probability at which RFID tags can be detected from the perspective of a single antenna, conditioned on a number of parameters of the system. Hence, we first have to choose the types of parameters of the model and then decide on how the probabilities can be computed from the raw binary data. Recall that the information given by an RFID reader is only which transponders have been detected, not their direction or distance.

In this paper, we restrict the set of model parameters to the relative position $\mathbf{x} = (x, y)$ of an RFID tag to the antenna. As indicated in the introduction, this is a vast simplification, but works surprisingly well. Besides, parameters such as the materials in the vicinity of RFID tags are difficult to be taken into consideration. So, formally we wish to estimate the detection probability $q_l(\mathbf{x}_l) := p(d_l | \mathbf{x}_l)$ for some tag l, given its relative position $\mathbf{x}^l$ to an antenna. We further assume that q does not depend on the specific tag l, so we will simply write $q(\mathbf{x})$.

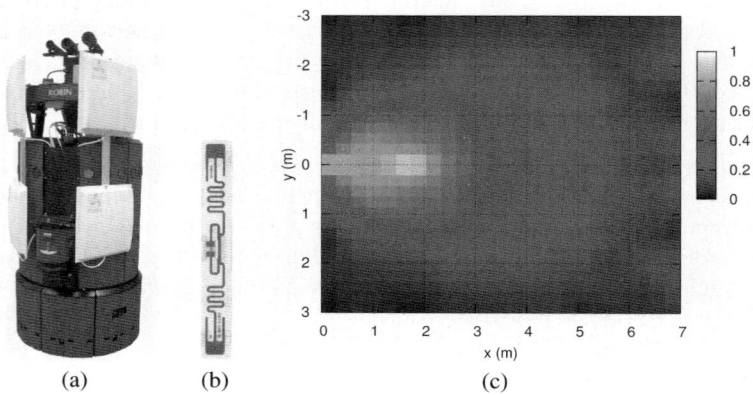

Fig. 1. (a) The RWI B21 robot used in our experiments, with its RFID antennas (white) and the front-mounted laser scanner (blue). (b) The type of tag ("squiggle"/"higgs" tag, by Alien Technology, approx. 10 cm × 1 cm) that we used for our studies. (c) The model obtained from fixed-setup measurements of a single tag. The x axis points to the opening direction of the RFID antenna, the y axis orthogonally, both parallel to the ground. The model was taken on 195 grid points and over 4 different heights and orientations with 100 measurements per configuration. Values between grid points are linearly interpolated. The recording took approx. 50 hours.

3.1   Fixed-Setup Recordings

In order to estimate $q(\mathbf{x})$, the approaches presented in related papers usually choose different fixed positions of the robot, equipped with RFID antennas, and some tag. For each such position $\mathbf{x}$ on a discrete grid, the numbers $n^+(\mathbf{x})$ and $n^-(\mathbf{x})$ of successful and failing detection attempts are counted in order to derive $q(\mathbf{x}) = n^+(\mathbf{x})/(n^+(\mathbf{x}) + n^-(\mathbf{x}))$. Hähnel et al., for instance, attached an RFID tag to a cardboard box and rotated the robot in front of it [4]. By repeating these measurements for various distances, they gained a mapping from a discrete number of relative distances and angles to the detection frequencies of a tag. We also recorded the detection frequencies on a grid of relative positions, depicted in Fig. 1 (c). There, the measurements were averaged over different heights and relative orientations of a single tag with respect to the RFID antenna.

The advantage of supervised grid-based measurements is that they are taken at precisely known positions of the grid, and the detection probability on a grid point is simply the rate of successful detection attempts measured there. If only one tag is taken into account, however, recording the RFID inquiries takes a lot of time. Even worse, the ideal case is investigated, claiming that the detection frequencies for one tag are independent of the presence of other tags.

In this paper, we pursue a slightly different approach. The idea is to exploit the fact that our ultimate goal is to localize a *mobile* robot. That is, provided that the robot is equipped with another, fairly accurate self-localization mechanism and that the robot knows at which positions RFID tags are fixed, it can explore its environment and continuously record tag detections. This procedure has a number of advantages: Firstly, much information is retrieved with every RFID inquiry (each known tag i contributes a response value $d_i = 1$ or $d_i = 0$ at a time), which makes the recording phase fast. Secondly, one does not have to worry about which grid points to choose. Of course, this comes at the expense that computations get slightly more complicated. But the learned models are supposed to be more realistic in presence of more than one tag.

3.2   Overview of the Learning Steps

In order to learn the RFID reader model in a semi-autonomous fashion, our solution comprises three phases: the *exploration phase* (during which RFID and position measurements are recorded), the *offline learning phase* (during which the RFID sensor model is actually computed and optimized), and the *localization and validation* phase (in which the robot is able to localize itself only via RFID and validates the accuracy of the resulting pose estimates).

Exploration phase. In the exploration phase, the robot logs RFID measurements and the poses where it takes these RFID measurements while traversing the environment. Note that therefore an additional positioning system (e.g. laser- or vision-based) is required which provides quite accurate position estimates. On the one hand, this seems to be a limitation of our approach. On the other hand, the models learned with our technique can later be used on other robots without the extra localization system. Moreover, the reference pose estimate enables us to assess the quality of the model with respect to self-localization during the last phase. And it is possible to combine the resulting RFID-based and the reference self-localization system, which has proven to be useful [4].

Offline learning phase. In this phase, the sensor model function is actually learned from the logged data of the previous phase. As additional input, it requires a map of RFID tags and their positions in the global coordinate frame. Note that we found that establishing a short list of RFID tag numbers and their positions can be done faster than the supervision of RFID measurements for a fixed-setup model. In the following, we detail the steps of the offline training phase.

1. Given a map of transponder positions in the environment, the robot transforms the global tag positions into coordinates relative to the RFID antennas for each time step t when an RFID inquiry was performed. In this, each tag l contributes a sample position $\mathbf{x}_t^l$ in relative coordinates and a response value $d_t^l \in \{0,1\}$, which states whether or not tag l was detected. From now on, we ignore the identity l of the tag and the time index t of the measurement. The result of this transformation is a set of samples $S := \{(\mathbf{x}_i, d_i)\}$ as depicted in Fig. 2 (a).

2. The raw samples are used to compute the conditional probability density $p(d|\mathbf{x}) = q(\mathbf{x})$ of detecting one tag, given its relative position to the antenna. Note that this function fully represents the target sensor model and is used as likelihood function in the correction step of the particle filter. For this step, we investigated two different techniques which are elaborated below.

3. The function $q(\mathbf{x})$ is evaluated at fixed positions of a fine-grained grid (with a resolution of 0.1 m) and stored as a look-up table. By this, $q(\mathbf{x})$ is only approximated, but the efficiency of evaluating the likelihood function for a potentially large number of particles is increased.

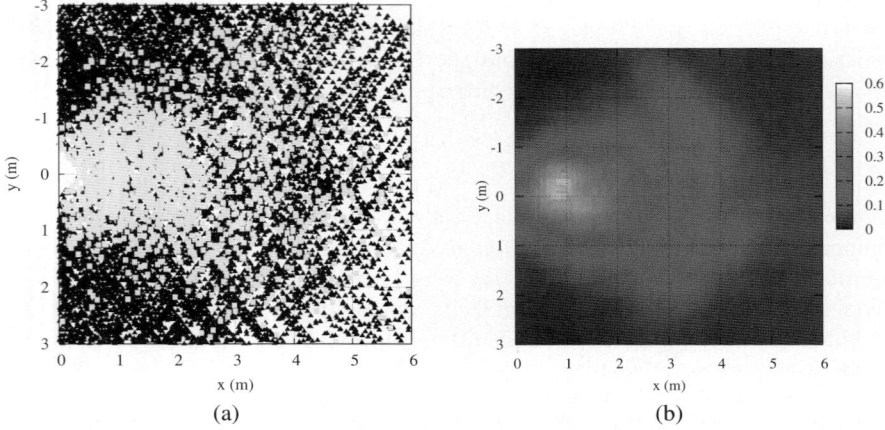

(a)   (b)

Fig. 2. (a) The aligned positive (green/gray) and negative (black) samples as obtained from RFID measurements while the robot was moving around. (b) The sensor model learned via the k-nearest neighbors approach ($k = 2000$) from 11,429 positive and 102,950 negative samples which were recorded in about 50 minutes.

Online localization and validation phase. The robot is finally able to localize itself, based on the method described in Sect. 2 and supplied with a map of RFID tags and the stored sensor model from the previous phase. Given the reference poses, it can validate and optimize the learned model.

3.3   Detection Probability Estimation

Grid-based Estimation. One of the simplest approaches to estimating the tag detection probability $q(\mathbf{x})$ is to divide the detection field (see Fig. 2 (a)) into a uniform grid with square bins of length λ and count the observed tag detections for the resulting bins. Optimizing a model of this kind then means to find the λ which minimizes the localization error. Note that this method resembles the manual recordings with a fixed setup, but samples are not forced to lie on discrete grid points only.
The advantage of the grid-based approach is that $q(\mathbf{x})$ can be stored as a look-up table, since it only depends on discrete bin indices. This makes it very efficient. A problem, however, is that the grid introduces discontinuities in the density function. Moreover, λ is fixed for the entire detection field. In parts where there are many samples available, a large value of λ will lead to over-smoothing, whereas a small value of λ will not be able to provide good estimates of $q(\mathbf{x})$ in areas for which only few (typically noisy) RFID measurements have been recorded. So, the optimal choice of λ may actually depend on the query position $\mathbf{x}$ in the detection field.

k-Nearest Neighbors (k-NN) Approach. The k-NN approach overcomes the limitation of the grid-based approach in that it takes the density of training samples around $\mathbf{x}$ into account. In this method we consider the k samples which are closest to $\mathbf{x}$ and compute the probability of detecting a tag as $q(\mathbf{x}) = n^{+}(\mathbf{x}, k)/k$, where

Table 1. Localization errors of the fixed-setup (manual) model and the two learned models, averaged over the 700 validation runs in the laboratory and 400 validation runs in the corridor, respectively

	Laboratory		Corridor	
	Validation 1	Validation 2	Validation 1	Validation 2
Manual model	0.5710 m	0.5101 m	0.4890 m	0.4803 m
Grid approach	0.5687 m	0.5190 m	0.4810 m	0.4342 m
k-NN approach	0.5641 m	0.5191 m	0.4864 m	0.4347 m

$n^+(\mathbf{x}, k)$ denotes the number of positive samples around $\mathbf{x}$ among the entire set of the k nearest samples in the vicinity of $\mathbf{x}$. In a sense, this approach is dual to the grid-based estimation technique: Instead of fixing the size of the bin inside which we count the response values d_i, we allow for arbitrary bin sizes and fix the number of samples from which we estimate the tag detection probability.

Optimizing a model of this kind then again means to find the value of k which minimizes the localization error. Analogously to the grid-based estimation technique, a too small value of k will not be robust to noisy data, but a too large value will lead to over-smoothing.

4   Experiments and Results

In order to evaluate and compare the different techniques of learning an RFID sensor model, we conducted a series of experiments with a B21 robot depicted in Fig. 1 (a). The robot is equipped with an Alien ALR-8780 RFID reader, two pairs of RFID antennas spanning an angle of $90°$, and a laser scanner. All experiments were conducted in a laboratory and an adjacent corridor. We installed 39 transponders in the corridor and 23 in the laboratory, spread over $75\,\text{m}^2$ and $50\,\text{m}^2$, respectively. The tags were attached to walls and desks at intervals of 1-2 m, roughly at the height of the upper antennas. We recorded 14 log files in the lab and 8 log files in the corridor, containing RFID data and the corresponding poses at which the robot performed the RFID inquiries. Ground truth was provided by laser-based Monte Carlo localization, for which we used the CARMEN toolkit [7]. The localization error can be assumed below 0.1 m on average. Within an average duration of 7 minutes the robot traveled distances of 32-155 m at varying speeds. RFID data arrived at approx. 2 Hz, resulting in 712 RFID inquiries on average per log file. The offline optimizations took less than 45 hours on a 3 GHz PC in total. They did not require human intervention, in opposition to recording the handcrafted model.

Each of the two sets of log files was randomly split into two halves for two-fold cross-validation. One half was once used to extract the raw RFID measurements and to optimize the models learned with the two techniques presented in Sect. 3, the other half was used to validate those models, and vice versa. Validation means that we played back the log files and measured the mean absolute localization error when the particle filter used the learned model. Because the mean accuracy in RFID-based positioning strongly

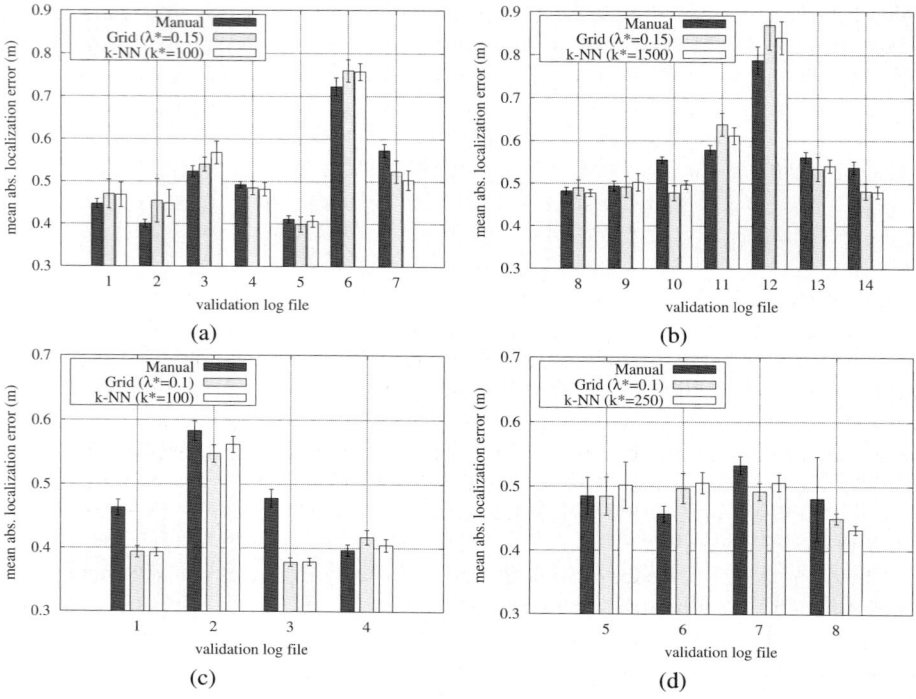

Fig. 3. Localization errors and standard deviations for the lab (a,b) and corridor (c,d) experiments. Each column represents one outcome of the 2-fold cross-validation: We first learned and optimized the models (the best choice of λ^* and k^*) on the one half of the log files. Thereafter, we measured the localization accuracy when using the optimized models on the respective other half of the log files.

depends on a good initialization, we localized with known initial pose (tracking case). For each probability estimation technique, we systematically searched over several parameter values ($k = 25, 50, 100, 250, 500, 1000, 1500$ and $\lambda = 0.1, 0.15, 0.2, 0.4, 0.6, 0.8$). Then we committed to the best parameter choices λ^* and k^* and investigated the localization error with the optimized model on the validation log files over 100 repeated experiments. The localization accuracy achieved by the hand-crafted sensor model served as a benchmark. The outcomes of the experiments are visualized in Fig. 3.

The results show that the learned models yield very similar results as the hand-crafted model. This is in spite of the small inaccuracies in the reference positions. On some log files, the learned models outperform the fixed-setup model, on others they are inferior. In both cases, the difference is in the range of few centimeters and therefore comparatively low with regard to the typical mean absolute localization error of approx. 0.5 m. Table 1 also shows that on average the learned models perform similarly well as or even better than the fixed-setup model. Moreover, the two employed estimation techniques provide similar accuracy. The grid-based approach performs surprisingly well, despite the arguments mentioned in Sect. 3.3. We observed, however, that the k-NN approach seems to be slightly more robust to changes of its parameter k. Firstly, it is consequently no contradiction that for the cross-validation experiments in the lab we obtained rather different best parameters $k^* = 100$ and $k^* = 1500$ (see Fig. 3). Secondly, we would therefore consider the k-NN approach the first choice if a preliminary reference sensor model is to be created quickly or whenever optimizations play a minor role.

5   Conclusion

In this paper, we have shown how to gain a probabilistic model of an RFID reader semi-automatically with a mobile robot. During an exploration phase RFID and position data are recorded. The logged data are then aligned and used to learn tag detection frequencies by means of simple bin-based averaging and a k-nearest neighbors search. By repeating the learning step with different parameter values, one can systematically search for the model which minimizes the localization error of the mobile robot. Note that the learned reader models were used to localize a mobile robot and benchmark the quality of the models, but in general, their utilization need not be restricted to self-localization.

The employed two model learning methods yield similar results in self-localization as a sensor model which was recorded by hand. None of the two estimation methods should be preferred to the other one, although the k-nearest neighbor approach appears to be slightly more robust to changes of its model parameter. The presented approach eases the creation of an RFID sensor model, because tedious measurements are replaced by automatic offline computations and optimization. This enables one to change the setup of an RFID system and re-learn the model quickly.

For the future, we plan to extend the learning to more model parameters, e.g. the height of a tag over ground or its orientation. And we strive to further automate the exploration phase.

References

1. Bajcsy, P., Kooper, R., Johnson, M., Soe, K.: Toward hazard aware spaces: Localization using passive RFID technology. Techn. Report ISDA06-002, National Center for Supercomputing Applications/Univ. of Illinois Urbana-Champaign (May 25, 2006)
2. Djugash, J., Singh, S., Corke, P.: Further results with localization and mapping using range from radio. In: International Conference on Field & Service Robotics (FSR 2005) (July 2005)
3. Fox, D., Burgard, W., Dellaert, F., Thrun, S.: Monte carlo localization: Efficient position estimation for mobile robots. In: Proceedings of the Sixteenth National Conference on Artificial Intelligence (AAAI 1999) (1999)
4. Hähnel, D., Burgard, W., Fox, D., Fishkin, K., Philipose, M.: Mapping and localization with RFID technology. In: Proceedings of the IEEE International Conference on Robotics and Automation (ICRA 2004) (2004)
5. Kantor, G., Singh, S.: Preliminary results in range-only localization and mapping. In: Proceedings of the IEEE Conference on Robotics and Automation (ICRA 2002) (May 2002)
6. Kloos, G., Guivant, J.E., Nebot, E.M., Masson, F.: Range based localisation using RF and the application to mining safety. In: Proceedings of the 2006 IEEE/RSJ International Conference on Intelligent Robots and Systems (IROS 2006), Beijing, China, pp. 1304–1311 (2006)

7. Montemerlo, M., Roy, N., Thrun, S.: Perspectives on standardization in mobile robot programming: The Carnegie Mellon Navigation (CARMEN) Toolkit. In: Proceedings of the IEEE/RSJ International Conference on Intelligent Robots and Systems (IROS 2003), vol. 3, pp. 2436–2441 (October 2003)
8. Schneegans, S., Vorst, P., Zell, A.: Using RFID snapshots for mobile robot self-localization. In: Proceedings of the 3rd European Conference on Mobile Robots (ECMR 2007), September 19-21, 2007, pp. 241–246 (2007)

A Simple Visual Navigation System with Convergence Property

Tomáš Krajník and Libor Přeučil

The Gerstner Laboratory for Intelligent Decision Making and Control
Department of Cybernetics, Faculty of Electrical Engineering
Czech Technical University in Prague
{tkrajnik,preucil}@labe.felk.cvut.cz

Summary. The aim of this paper is to present a convergence property of a simple vision based navigation system for a mobile robot. A robot equipped with a single camera is guided by a human operator along a path consisting of straight segments. During this guided tour, local image invariants are extracted from acquired frames and odometric data are collected. When navigating learned path, the vision is used to reckon direction to the start point of next straight segment. Odometric measurements are utilized to estimate distance to this point. A simple linear model of this navigation system is lined up and its properties are examined. We proclaim a theorem, which states, that for a limited odometric error and "reasonable" trajectory, the robot uncertainty in position estimation does not diverge. A formal proof of this theorem is given for regular polygonal trajectories. The proclaimed convergence theorem is also experimentally verified.

Keywords: visual navigation.

1 Introduction

1.1 Paper Structure

The paper is organized as follows: Introduction presents a very brief overview of current state of the art in vision-based mobile robot navigation. The next chapter describes proposed path learning and navigation algorithms. The following division presents the convergence theorem and its proof for cyclic trajectories. After that, the experiment setups and results are described. Conclusion discusses drawbacks of proposed method and possible solutions. Acknowledgments and references are placed at the end of this paper.

1.2 Monocular Navigation

In recent years, as the computational power of common systems increased and image processing became possible in real-time, the means of using vision to navigate mobile robots have been investigated. According to [5], the described system belongs to the "Map-building based" group. There have been several successful attempts to create such a "Map-building based" system, some [6] rely on stereo vision, while others [4] use single camera. Most systems extract invariant features from images [8] and build a

H. Bruyninckx et al. (Eds.): European Robotics Symposium 2008, STAR 44, pp. 283–292, 2008.
© Springer-Verlag Berlin Heidelberg 2008

threedimensional map of these. We present a system capable of autonomous navigation in known environment, which utilizes a single camera. Like in [3],[7], the system has to learn the environment during a teleoperated drive. Unlike in those cases, we use camera sensing only to correct small-scale errors in movement direction. Positions of significant locations, i.e. places where the robot changes its movement direction significantly, are estimated by odometric measurements. We explore convergence properties of such landmark navigation and state that for some trajectories, the camera readings can correct odometry imprecision without explicitly localizing the robot. While [2] describes convergence property by a vector field, we use a simple linear model.

2   Surf-Based Navigation System

The SURFNav system recognizes objects in the image taken by forward looking camera and corrects direction of robot movement. Data from compass and odometry are processed as well. The system works in two phases: learning and navigation. A brief explanation of object extraction from the image is given in subsection 2.1. The learning phase is described in subsection 2.2, the navigation phase is depicted in subsection 2.3.

2.1   Object Recognition

We have decided to use Speeded Up Robust Features [1] to identify landmarks in the image. This algorithm processes gray-scale image in two phases. At first, a local brightness extrema detector is applied to the image. In the next phase, a scale, rotation and skew invariant descriptor of detected extrema neighborhood is computed. Algorithm provides image coordinates of salient features together with their descriptor. To speed up computation time, the image is horizontally divided and both its parts can be processed paralelly by multiprocessor machine. Typically, image recognition duration is 300 ms while 250 features are detected. See processed image with highlighted feature positions on Figure 1.

2.2   Learning Phase

In the learning phase, the robot is guided through the environment on a polyline shaped trajectory. At the beginning of each segment, the robot resets its odometry counter, reads compass data and takes a serie of 15 images. Objects, which have been detected in 10 subsequent snapshots of this serie are considered to be stable. Stable objects with constant positions are regarded as stationary. Positions and descriptors of stored objects are saved. Afterwards, the robot starts to move forwards, obtains and processes images and records odometric data. When an object is detected for the first time, the algorithm saves its descriptor, image coordinates and robot distance from segment start. Saved objects are tracked over several pictures and their positions in image are assigned to current robot position within a segment. Tracking of an object is terminated after three subsequent unsuccessful attempts to detect it in the image. Its descriptor, image coordinates and odometric data in moments of the first and the last successful recognition

Fig. 1. Image and detected features

are inserted into the dataset describing the traversed segment. Segment learning is terminated by an operator, which stops the robot (segment length is saved) and turns it in the direction of next movement. After that, the learning algorithm either runs for next segment or quits.

2.3   Navigation

When navigation mode is started, the robot loads description of relevant segment and turns itself to the indicated direction. After that, the odometry counter is reset, forward movement and picture scanning are initiated. Objects, which are expected to occur in the image, are selected from learned set. These are the objects with the first and the last detection distance greater, respectively lower than the current robot distance from segment start. Expected image coordinates in current camera image are calculated by linear interpolation using aforementioned distances. Selected objects are rated by a number of frames which they have been detected in and 50 best-rated objects are chosen as suitable for navigation. For each candidate, the most similar object is searched in the set of actually detected ones. The similarity is calculated from an Euclidean distance of descriptors of both compared objects. A difference in horizontal image coordinates is computed for each such couple. A modus estimate of those differences is then converted to a correction value of movement direction. After the robot travels distance greater or equal to the length of given segment, the next segment description is loaded and the algorithm is repeated.

An important aspect of this navigation algorithm is its functionality without the need to localize the robot or to create a three-dimensional map of detected objects. Even though the camera readings are utilized only to correct the direction and the distance is measured by imprecise odometry, it is shown, that if the robot changes direction often enough, it will keep close to learned trajectory.

3   Convergence Property of Navigation

To defend the last statement of the previous chapter, we first need to create a model of robot movement. We will explore the properties of this model and give a formal proof of aforementioned statement for certain trajectories.

Theorem 3.1 (Convergence theorem). *If the robot uses navigation described in chapter 2 to travel a regular polygonal trajectory, its position uncertainty is bound for any polygon size and bounded odometric error.*

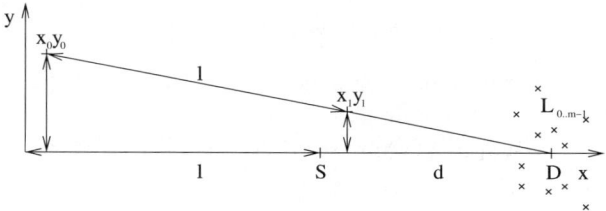

Fig. 2. Navigation model for one segment

3.1   Movement Model

Let us suppose, that the learned trajectory starts at coordinate origin, leads in direction of x axis and consists of one segment of length l with endpoint S, see figure 2. Let the robot has observed and recorded a landmark set $L_{0...m-1}$ during path learning phase. Let the robot is placed at x_0, y_0 and headed in direction of segment endpoint. Let us assume, that the robot has been switched to navigate the segment. Because its camera is heading forwards, detected landmarks are not distributed along the way, but are rather shifted in current segment direction. As a result, the robot does not head directly to segment end S, but rather behind it, to the point D. After it travels distance l, it gets to x_1, y_1. Assuming previous conditions have been fulfilled, we can compute x_1, y_1 (denoted as $\mathbf{x_1}$) as follows:

$$\mathbf{x_1} = \frac{D - \mathbf{x_0}}{\|D, \mathbf{x_0}\|} l + \mathbf{x_0}. \tag{1}$$

If we assume that $\|x_0\| < l$, we can introduce a linear representation of (1):

$$\mathbf{x_1} = \begin{pmatrix} 1 & 0 \\ 0 & \frac{d}{d+l} \end{pmatrix} \mathbf{x_0} + S. \tag{2}$$

This model assumes precise odometry, so we choose to model odometric imperfection as a multiplicative error υ with normal distribution, giving us movement model (3).

$$\mathbf{x_1} = \begin{pmatrix} 1 & 0 \\ 0 & \frac{d}{d+l} \end{pmatrix} \mathbf{x_0} + S + \upsilon = \mathbf{M_1}\mathbf{x_0} + S + \upsilon. \tag{3}$$

Equation (3) holds for a segment starting at coordinate origin and ending at a point on x axis. Let us have a path of n segments numbered $0 \ldots n-1$ and denote starting point of segment k as $\widehat{\mathbf{x}}_\mathbf{k}$. We designate the robot position at the start of kth segment as $\mathbf{x_k}$ and mark $\widetilde{\mathbf{x}}_\mathbf{k} = \mathbf{x_k} - \widehat{\mathbf{x}}_\mathbf{k}$. When we want to apply our simple movement model to segment k, we first compute a rotation matrix R_k to align k^{th} segment with x axis, then we apply linear model (3) and odometric noise and rotate the result back by applying $\mathbf{R_k^T}$. This is expressed by next relation:

$$\mathbf{x_{k+1}} = \widehat{\mathbf{x}}_\mathbf{k+1} + \mathbf{R_{k+1}^T} \left(\mathbf{M_{k+1}} \mathbf{R_{k+1}} (\mathbf{x_k} - \widehat{\mathbf{x}}_\mathbf{k}) + \upsilon \right) \tag{4}$$

Since $\mathbf{x_k} = \widehat{\mathbf{x}}_\mathbf{k} + \widetilde{\mathbf{x}}_\mathbf{k}$, then

$$\widetilde{\mathbf{x}}_\mathbf{k+1} = \mathbf{R_{k+1}^T} \left(\mathbf{M_{k+1}} \mathbf{R_{k+1}} \widetilde{\mathbf{x}}_\mathbf{k} + \upsilon \right) \tag{5}$$

The covariance matrix of position uncertainty $\widetilde{\mathbf{x}}_k$ is then calculated by

$$\widetilde{\mathbf{x}}_\mathbf{k+1}\widetilde{\mathbf{x}}_\mathbf{k+1}^T = \left(\mathbf{R_{k+1}^T} \left(\mathbf{M_{k+1}} \mathbf{R_{k+1}} \widetilde{\mathbf{x}}_\mathbf{k} + \upsilon \right) \right) \left(\mathbf{R_{k+1}^T} \left(\mathbf{M_{k+1}} \mathbf{R_{k+1}} \widetilde{\mathbf{x}}_\mathbf{k} + \upsilon \right) \right)^T \tag{6}$$

since $\int \upsilon \widetilde{\mathbf{x}}_\mathbf{k}^T = 0$, then

$$\widetilde{\mathbf{x}}_\mathbf{k+1}\widetilde{\mathbf{x}}_\mathbf{k+1}^T = \mathbf{R_{k+1}^T}\mathbf{M_{k+1}}\mathbf{R_{k+1}}\widetilde{\mathbf{x}}_\mathbf{k}\widetilde{\mathbf{x}}_\mathbf{k}^T\mathbf{R_{k+1}^T}\mathbf{M_{k+1}^T}\mathbf{R_{k+1}} + \mathbf{R_{k+1}^T}\upsilon\upsilon^T\mathbf{R_{k+1}} \tag{7}$$

$$\mathbf{R_{k+1}}\widetilde{\mathbf{x}}_\mathbf{k+1}\widetilde{\mathbf{x}}_\mathbf{k+1}^T\mathbf{R_{k+1}^T} = \mathbf{M_{k+1}}\mathbf{R_{k+1}}\widetilde{\mathbf{x}}_\mathbf{k}\widetilde{\mathbf{x}}_\mathbf{k}^T\mathbf{R_{k+1}^T}\mathbf{M_{k+1}^T} + \upsilon\upsilon^T \tag{8}$$

Proof (Convergence theorem). We assume, that the robot moves on a regular polygon with n edges with length l. Then $\mathbf{M_k} = \mathbf{M_n}$, $\mathbf{R_k} = \mathbf{R_n^k}$, where

$$\mathbf{R_n} = \begin{pmatrix} \cos\left(\frac{\pi}{2} - \frac{\pi}{n}\right) & -\sin\left(\frac{\pi}{2} - \frac{\pi}{n}\right) \\ \sin\left(\frac{\pi}{2} - \frac{\pi}{n}\right) & \cos\left(\frac{\pi}{2} - \frac{\pi}{n}\right) \end{pmatrix} \mathbf{M_n} = \begin{pmatrix} 1 & 0 \\ 0 & \frac{d}{d+l} \end{pmatrix} \tag{9}$$

if we denote $\mathbf{R_n^k}\widetilde{\mathbf{x}}_\mathbf{k} = \widecheck{\mathbf{x}}_\mathbf{k}$,

$$\widecheck{\mathbf{x}}_\mathbf{k+1}\widecheck{\mathbf{x}}_\mathbf{k+1}^T = \mathbf{M_n}\mathbf{R_n}\widecheck{\mathbf{x}}_\mathbf{k}\widecheck{\mathbf{x}}_\mathbf{k}^T\mathbf{R_n^T}\mathbf{M_n^T} + \upsilon\upsilon^T \tag{10}$$

and $\mathbf{R_n^k}\mathbf{M} = \widecheck{\mathbf{M}}_\mathbf{n}$,

$$\widecheck{\mathbf{x}}_\mathbf{k+1}\widecheck{\mathbf{x}}_\mathbf{k+1}^T = \widecheck{\mathbf{M}}_\mathbf{n}\widecheck{\mathbf{x}}_\mathbf{k}\widecheck{\mathbf{x}}_\mathbf{k}^T\widecheck{\mathbf{M}}_\mathbf{n}^T + \upsilon\upsilon^T \tag{11}$$

Since left side of (11) represents a covariance matrix, (11) is a discrete Lyapunov equation. Since $\widecheck{\mathbf{M}}_\mathbf{n}$ is stable, a finite solution to (11) always exists.   □

Thus, if the robot traverses a regular polygon of more than two edges with length l, its position uncertainty is bound for bounded odometric error υ and landmark distance d.

4   Experiments

During experiments, we first compared linear (3) and nonlinear (1) system models to check whether linear model is not too crude. After that, the real-world experiments were conducted to verify whether the theoretical assumptions correspond to real world properties.

4.1   Simulations

Simulations were conducted for regular polygons with vertices on a circle of 10 m radius. Landmark distance d was chosen to be 5 m. To simulate nonlinear model, particle filters were utilized. First, 10^5 positions estimating normal distribution with mean at trajectory initial point and unit variance were generated. Equation (1) was then applied 100 times to each generated position and covariance matrix was computed after each step.

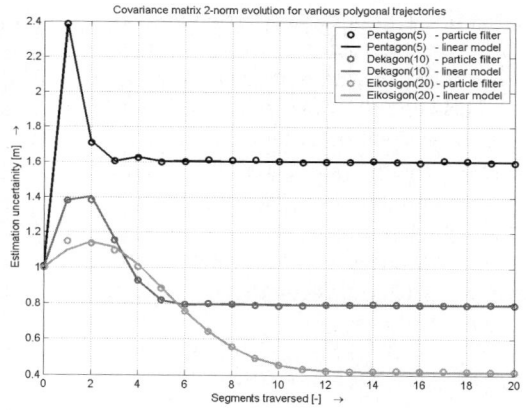

Fig. 3. Comparison of linear and nonlinear model

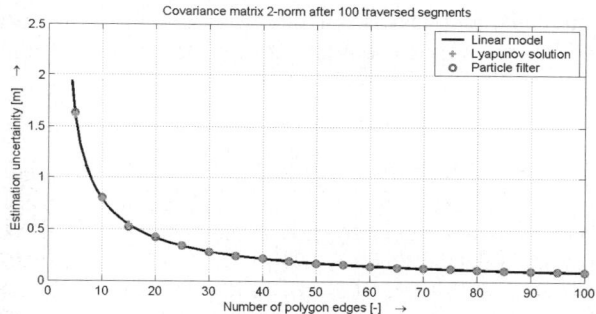

Fig. 4. Comparision of linear and nonlinear model

Resulting matrix was then compared with the one obtained by (3). Figure 3 compares evolution of covariance matrices 2-norm computed by particle filter and by linear model for polygons of 5, 10 and 20 segments. Dependency of 2-norm covariance matrix after 100 computation steps on the number of egdes of the polygon is shown on figure 4.

4.2 Real World Experiments

Experiment Setup

Experiments were performed by Pioneer 3AT robotic platform with TCM2 compass. Robot was equipped with Fire i-400 camera providing 15 images per second at 640x480 pixel resolution. A wide angle objective with focus length 2.1 mm was used. Images were processed in real time by Intel Core 2 Duo notebook. Only the upper half of the picture was processed in order to use more distant objects as landmarks.

The robot was learnt a closed trajectory first. Then it was placed on the trajectory start point and switched to navigate the learned path five times. Every time it completed a loop and started the next one, its position relative to the trajectory start point was measured. The robot was then placed 1 m away from the start point in direction perpendicular to the first segment, and navigated the loop five more times while measurements were taken. The same position set was collected for another initial position, which was 1 m away from learned trajectory trailhead in direction parallel to the first path segment. These measurements were taken for two trajectories, one being a straight line and second of triangular shape.

When navigating a straight line trajectory, the robot should be able to correct position deviations perpendicular to traversed segment. Deviations in direction of line trajectory can not be corrected. Robot traversing triangular trajectory was expected to be able to correct deviations in either direction.

Indoor Environment Setup

Indoor experiment was performed in a corridor of CTU FEE. Since this environment is small and detected landmarks were close to segment endpoints, robot was quickly converging to original initial position. Convergence speed was also fortifed by small odometric error on planar and smooth corridor floor. The first trajectory was a straight line of 5 m length. The second path was an equilateral triangle with 4 m long sides.

Table 1. Indoor test results

Loop num.	Position difference to learned trajectory start point [m]					
	Line trajectory			Triangular trajectory		
0	0.00, 0.00	-1.00, 0.00	0.00, 1.00	0.00, 0.00	1.00, 0.00	0.00, -1.00
1	-0.05, 0.07	-0.95, 0.30	-0.03, 0.03	0.08, -0.08	0.47, 0.14	0.02, -0.47
2	-0.07, 0.09	-0.93, 0.38	-0.05, 0.05	0.09, -0.10	0.26, 0.07	0.18, -0.19
3	-0.10, 0.10	-0.93, 0.47	-0.07, 0.07	-0.05, 0.05	0.18, -0.08	0.03, -0.10
4	-0.13, 0.12	-0.92, 0.47	-0.14, 0.14	-0.05, 0.05	0.08, -0.02	0.01, -0.07

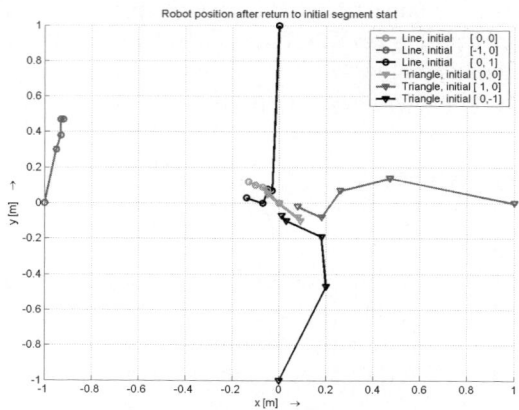

Fig. 5. Indoor test results

Indoor Experiment Results

Triangular trajectory was stable and the robot could correct deviations in both directions. In the case of line trajectory, the robot could correct position deviation perpendicular to traversed segment if its position estimation within a segment was precise.
However, odometric errors acumulated and each time a robot completed the loop, its
distance from learned trajectory origin increased. Since learned landmark positions are
bound to position of the robot within a segment, its course and distance perpendicular
to the linear trajectory deteriorated as well.

Outdoor Enviroment Setup

Outdoor experiments were perfomed at Charles square in Prague. This environment
was large (est. average landmark distance from segment end was 20 m), the surface
was rugged and pedestrians generating noisy readings were abundant. Therefore, the
convergence speed was not expected to be fast. As in indoor case, the first learned
trajectory was a straight line of 5 m length. Triangular path was a bit larger than indoors,
the triangle side was 5 m long.

Table 2. Outdoor test results

Loop num.	Position difference to learned trajectory start point [m]					
	Line trajectory			Triangular trajectory		
0	0.00, 0.00	-1.00, 0.00	0.00, -1.00	0.00, 0.00	1.00, 0.00	0.00, 1.00
1	0.09, 0.08	-0.98, 0.00	-0.02, -0.76	0.12, -0.15	0.92, 0.33	0.32, 0.62
2	-0.23, 0.20	-1.16, -0.16	0.03, -0.62	0.23, -0.16	0.59, 0.35	0.14, 0.22
3	-0.18, 0.26	-1.25, -0.32	0.07, -0.35	-0.15, -0.03	0.35, -0.12	0.05, 0.21
4	-0.11, 0.29	-1.31, -0.32	0.15, -0.59	-0.10, 0.03	0.23, -0.03	-0.12, 0.10

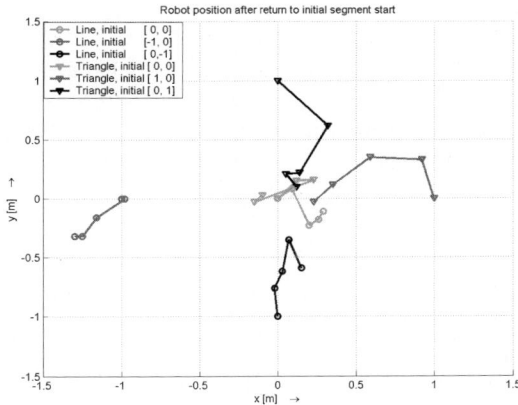

Fig. 6. Outdoor test results

Outdoor Experiment Results

Robot behaviour was similar as in the indoor environment, but the convergence was slower and less precise.

Proposed system was also tested during the RoboTour07[1] outdoor contest. The robot was able to travel approximatelly 150 m faultlessly, until reaching a wide area, where its position estimation dropped below required precision. While moving through this area, it left the pathway and had to be stopped.

5   Conclusion

A simple navigation system, where movement direction is calculated from visual information and movement distance is based on odometry measurent was presented. In this paper, we have stated, that if the robot changes directions often enough, position estimation errors resulting from odometry imperfection can be corrected by more precise direction assesment. To formalise such a statement, we formulated a "convergence theorem", which claims that for closed trajectories and finite odometric error robot position estimation error is bounded. Linear model of proposed navigation system was devised and a formal proof of the aforementioned convergence property for regular polygon trajectories was outlined. Experimental results supporting the convergence theorem were also presented.

Future work will focus on modifying the proposed system in order be able to follow a wider set of trajectories. We will try to extend presented proof to trajectories different from regular polygon. A framework to combine this navigation system with existing visual-based collision avoidance algorithms will be implemented.

---

[1] http://robotika.cz/competitions/robotour2007/en

Acknowledgements

I would like to thank my colleagues for valuable remarks, my friends for help with outdoor tests and language corrections. This work was supported by a research grant CTU0706113 of Czech Technical University in Prague and the Research program funded by the Ministry of Education of the Czech Republic No. MSM 684077038.

References

1. Bay, H., Tuytelaars, T., Gool, L.: Surf: Speeded up robust features. In: Proceedings of the ninth European Conference on Computer Vision (2006)
2. Bianco, G.M., Zelinsky, A.: The convergence property of goal based visual navigation. In: Proceedings of International Conference on Intelligent Robots and Systems EFPL Lausanne, Switzerland, pp. 649–654 (2002)
3. Blanc, G., Mezouar, Y., Martinet, P.: Indoor navigation of a wheeled mobile robot along visual routes. In: Proceedings of International Conference on Robotics and Automation (2005)
4. Blanc, G., Mezouar, Y., Martinet, P.: A visual landmark recognition system for autonomous robot navigation. In: Proceedings of CIMCA-IAWTIC 2006 (2006)
5. DeSouza, G.N., Kak, A.C.: Vision for mobile robot navigation: A survey. IEEE Trans. Pattern Anal. Mach. Intell. 24(2), 237–267 (2002)
6. Kidono, K., Miura, J., Shirai, Y.: Autonomous visual navigation of a mobile robot using a human-guided experience. In: Proceedings of 6th Int. Conf. on Intelligent Autonomous Systems, pp. 620–627 (2000)
7. Matsumoto, Y., Inaba, M., Inoue, H.: Visual navigation using view-sequenced route representation. In: Proceedings of the International Conference on Robotics and Automation, Minneapolis, USA, pp. 83–88 (1996)
8. Se, S., Lowe, D., Little, J.: Vision-based mobile robot localization and mapping using scale-invariant features. In: Proceedings of the IEEE International Conference on Robotics and Automation (ICRA), Seoul, Korea, May 2001, pp. 2051–2058 (2001)

Stability of On-Line and On-Board Evolving of Adaptive Collective Behavior

L. König, K. Jebens, S. Kernbach, and P. Levi

Institute of Parallel and Distributed Systems, University of Stuttgart, Universitätsstr. 38,
D-70569 Stuttgart, Germany
lukas-koenig@gmx.net,
{jebenskf,korniesi,levipl}@uni-stuttgart.de

Summary. This work focuses on evolving purposeful collective behavior in a swarm of Jasmine micro-robots. We investigate the stability of the on-line and on-board evolutionary approaches, where mutation, crossover as well as fitness calculation are performed only by interacting micro-robots without using any centralized resources. In this work it is demonstrated that the environment-adaptive collective behavior can be obtained, where the evolving fitness and behavior are partially stable. To increase stability of the approach, some reduction methodology of the search space is proposed.

1 Introduction

An important challenge in modern network researching and swarm robotics is a design of purposeful collective behavior [4]. Such a behavior should be technically useful, adaptive to environmental changes and scalable in size and functional metrics [1]. One possible paradigm here is to use evolutionary approaches for evolving desired collective behavior [7]. Using an evolutionary approach, robots can start with simple behavior primitives and gradually increase their cooperative complexity until the collective behavior satisfies some imposed fitness.

The application of evolutionary approaches in swarm robotics is known. Some essential references can be given in works of evolving control [2], evolving shapes [8], evolving communication [12], [9] and others. However, essential obstacles for successful application of evolutionary approaches are "on-line" and "on-board" requirements imposed on the calculation of fitness and execution of evolutionary operators. The "on-line" requirement means that all evolutionary results should be obtained during the life-cycle of a robot, "on-board" means that only available on-board sensors, computational and communication resources can be used. Both requirements originate from the practical robotic field.

The present work focuses on the problems of stability in on-line and on-board evolving of collective behavior. More exactly, we assume that there exists some collective behavior with relatively high fitness. The questions are whether on-line and on-board evolutionary processes: (a) will preserve this originally effective behavior? (b) will replace the original behavior with a better one? (c) will destroy the original behavior without creating a better one? Answering these questions, we intend to acquire more insight about possibilities of performing on-line and on-board artificial evolution.

H. Bruyninckx et al. (Eds.): European Robotics Symposium 2008, STAR 44, pp. 293–302, 2008.
springerlink.com

In this work, the robot controller is represented as a generating hierarchy of *genome* $\xrightarrow{generator}$ *phenome* $\xrightarrow{interpretor}$ *behavioral automaton*. The behavioral automaton is based on a finite Moore automaton [3]. At each state an "atomic action" (e.g. *move*, *stop*, *turn left*) is produced. For encoding the Moore automaton a symbolic string, called *phenome* is used. The phenome contains the same information as the actual automaton and can be used to copy a behavior between different robots using local robot-robot communication. The genome contains generating rules for the phenome sequence. To simplify treating of on-board and on-line issues, we apply evolutionary operators only to *phenome strings* from one or different robots. The fitness is selected as a measure of collision, so that the final goal is to evolve an effective collision avoidance behavior. Since evolving the behavior is only possible when many robots interact, collision avoidance represents a result of collective behavior. Experiments are performed using Jasmine IIIp micro-robots.

The rest of the paper is organized as follows. The framework of evolutionary experiments is described in Sec. 2, evolutionary operators in Sec. 3 and experiments in Sec. 4. In Sec. 5 we discuss and conclude this work.

2   Hardware and Software Framework

In this section we briefly describe the used hardware and software framework. Experiments are performed with the Jasmine IIIp micro-robots, see Fig. 1(a). Each micro-robot has two Atmel AVR-microcontrollers with 8 MIPS each and totally 2 kB RAM, 24 kB Flash memory and 1 kB non-volatile memory. For the evolutionary approach only 6.5 kB Flash and 700 bytes RAM are available, the phenome strings are written in 512 bytes non-volatile memory. Each robot has a local 360° IR-based communication with an effective communication radius of about 2-10 cm. When two robots meet within this radius, they establish a bi-directional communication channel and can exchange phenome strings at 500 bytes/sec. The communication system can also be used for proximity sensing (see more on www.swarmrobot.org).

The software framework consists of a low-level BIOS (interface to hardware) and an operational system [5], and the high-level genetic framework. For the genetic framework there are several design alternatives, e. g. to use classical robot control, to use paradigms from the evolutionary community or to use bio-inspired ideas. We have chosen the bio-inspired way, firstly, due to interest of exploring alternatives to well-known solutions, secondly, because this framework is used in the projects [11] and [10], which are related to bio-inspired artificial evolution.

The structure of the genome framework is shown in Fig. 1(b). The behavior of the robot is controlled by a behavioral automaton (Petri-nets in the overall framework [5]) and is influenced by the environment. We call this mutual influence the ***phenotype*** of the robot. Behavioral automata represent an explicit description of the phenotype. We call this the ***phenome***. This is a contradiction to biological systems, where the phenome is not directly available. More exactly, the phenome is a symbolic string which generates the automata. This symbolic string contains direct low-level as well as high-level (from libraries) behavioral commands, therefore can be thought of as a functional descriptor. In turn, the phenome is generated by the ***genome*** of a robot. The genome is

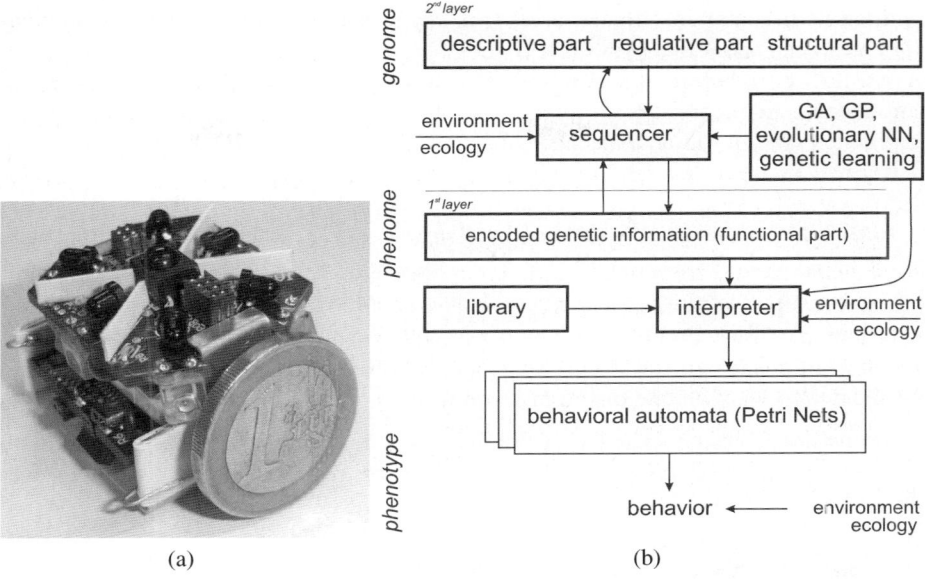

Fig. 1. (a) Micro-robot Jasmine IIIp. **(b)** Structure of the high-level genome framework.

also a symbolic string and consists of descriptive, structural and regulative parts and usually does not contain any direct behavioral commands. By analogy, the genome is a structural descriptor of the system. In this framework, both genome and phenome can underly evolutionary operators and can also be influenced by the environment.

The two-layer control structure *genome-phenome* is very effective for reconfigurable multi-robot organisms, where the robot system can change its own structure and therefore the functionality [6]. However, in this work, where we do not change the structure of robots, evolutionary operators will be applied to the phenome. In the following we describe the theoretical approach for evolving the phenome and show experimental results.

3   Implementation of Evolutionary Framework

This section describes the theoretical model for our evolutionary approach. We use a finite automaton based model to describe robot behavior [3] and a genetic programming approach [7] for evolving new behaviors.

3.1   The Automaton Model

We denote a set of byte values and a set of positive byte values as $B = \{0, ..., 255\}$ and $B^+ = \{1, ..., 255\}$. The behavior of a robot depends on the sensor data. We assume a set H of n sensor variables $H = \{h_1, ..., h_n\}$. The sensor variables stand as placeholders for sensor data from real or virtual sensors (i. e. any internal variables of the robot). Every variable h_i can be set to a byte value.

As mentioned before, the main behavior of a robot is controlled by a finite Moore automaton[1]. At each state, an output is produced, which is interpreted as an *atomic instruction* (which can be a mechanical action like *move* or a whole C-program) to be performed by the robot. The transitions between states depend on the values of the sensor variables $h_1, ..., h_n$.

States. The set of states is denoted by $Q = \{q_0, ..., q_m\}$, where q_0 is the initial state of the automaton. A state contains the following information:

- an identification number $N \in B^+$,
- an atomic instruction $I \in B^+$ to be performed,
- an additional parameter $P \in B^+$ to gain more information about the instruction, and
- the definition of all outgoing transitions from that state.

Transitions. We associated a condition to each transition, which has to evaluate to *true* for the transition to be taken. We define for this purpose a set of conditions over the sensor variables as follows:

Definition 3.1. *Conditions*
A condition is an element of the set, defined by:

$$c ::= true \mid false \mid z_1 \lhd z_2 \mid (c_1 \circ c_2),$$

$z_1, z_2 \in B^+ \cup H$,
$\lhd \in \{<, >, \leq, \geq, =, \neq, \approx, \not\approx\}$, *where $\approx$ means the range of ± 5.*
$\circ \in \{AND, OR\}$,
c_1, c_2 *are conditions themselves.*

The set of all conditions is denoted by C.

Valid conditions are e. g. $h_1 < h_2$, $((h_1 \not\approx h_2 \ AND \ h_3 = 104) \ OR \ h_5 \geq h_7)$, (*true AND false*). The result of a condition (*"true"* or *"false"*) is calculated in the obvious way, by feeding the variables with actual sensor values and evaluating the comparisons and logical operations. A transition is taken, if its corresponding condition evaluates to *true*. At this point two special cases have to be considered:

1. A state can have no condition that evaluates to *true*.
2. A state can have more than one condition that evaluates to *true*.

In case 1, we define an implicit transition to the initial state to be taken. In case 2, the "first" outgoing transition of that state is taken (the corresponding order is not important, as long as the transition to be taken does not vary from case to case; we took the order in which the transitions were generated).

Moore automaton for robot behavior. A Moore automaton for robot behavior is built by states and transitions as mentioned above. Each state carries identification N, instruction I, parameter P and its outgoing transitions, which are defined by a condition $c \in C$ and a following state $f \in B^+$. The automaton is defined as follows:

---

[1] Moore automaton with an underlying operational system and BIOS, which perform e.g. sensor data acquisition or interruption handling.

Definition 3.2. *Finite Moore automaton for robot behavior*
A finite Moore automaton for robot behavior A is defined as follows:

$$A = (Q, \Sigma, \Omega, \delta, \lambda, q_0, F),$$

where:

- *The set of states $Q = B^+ \times B^+ \times B^+ \times (C \times B^+)^*$, C being the set of conditions.*
  Let $\forall q \in Q$:

$$q = \left(N^q, I^q, P^q, \left(c_1^q, f_1^q \right), ..., \left(c_{|q|}^q, f_{|q|}^q \right) \right),$$

  where $|q|$ denotes the number of transitions of state q.
- *The input alphabet $\Sigma = H = (B^+)^n$.*
- *The output alphabet $\Omega = (B^+)^2$.*
- *The transition function (for $q \in Q$, $h \in H$): $\delta : Q \times H \to Q$:*

$$\delta(q,h) = \begin{cases} q', & \text{if } \exists\, k \in \{1,...,|q|\} : f_k^q = q' \text{ and } c_k^q \text{ evaluates to } true \text{ under } h \\ & \text{and } \forall\, j \in \{1,...,|q|\}, \text{ where } c_j^q \text{ evaluates to } true \text{ under } h, \\ & \text{it holds: } k \leq j, \\ q_0 & \text{otherwise} \end{cases}$$

- *The output function $\lambda : Q \to (B^+)^2 : \lambda(q) = (I^q, P^q)$ (for $q \in Q$).*
- *The initial state q_0.*
- *The empty set of final states: $F = \emptyset$.*

3.2   Evolutionary Operators

Mutation. We developed a mutation operator, which is *complete* and *smooth*, which means that every part of the search space is reachable and that **every single mutation causes only a small step in the search space**. The mutation operator consists of 10 atomic mutations, one of which is randomly chosen, when the operator is used. The atomic mutations are:

1. Toggle inactive transitions:
   a) Remove a random transition, associated with the condition *false*.
   b) Add a random transition, associated with the condition *false*.
2. Remove a state:
   a) Without incoming transitions.
   b) With all outgoing transitions being associated with the condition *false* and the state being associated with the instruction *IDLE*.
3. Add a new state with:
   - no incoming transitions, and
   - no outgoing transitions, and
   - arbitrary action, and
   - parameter $\leq k$ ($k \in \{1,...,255\}$ being a constant).

4. Change a condition: Let $a, b \in \{1, ..., 255, h_1, ..., h_{|H|}\}$, $A \in C$ a condition. Every part of a condition that matches the following patterns can be mutated:

a) $false \leftrightarrow a = b \leftrightarrow a \approx b \leftrightarrow \begin{array}{c} a \leq b \leftrightarrow a < b \\ a \geq b \leftrightarrow a > b \end{array} \leftrightarrow a \not\approx b \leftrightarrow a \neq b \leftrightarrow true$

b) One of the following:

$$
\begin{array}{ccccc}
(A \ AND \ true) & \leftrightarrow & A & \leftrightarrow & (true \ AND \ A) \\
(A \ OR \ true) & \rightarrow & true & \leftarrow & (true \ OR \ A) \\
(A \ AND \ false) & \rightarrow & false & \leftarrow & (false \ AND \ A) \\
(A \ OR \ false) & \leftrightarrow & A & \leftrightarrow & (false \ OR \ A)
\end{array}
$$

c) Let i be a number in a condition. Let $i' = i + rand[-k, k]$, $k \in \{1, ..., 255\}$ being a constant parameter. Mutate:

$$
i \rightarrow \begin{cases} i', & \text{if } 1 \leq i' \leq 255 \\ 1, & \text{if } i' < 1 \\ 255, & \text{if } i' > 255 \end{cases}.
$$

d) Let h_i be a sensor variable in a condition. Mutate:

$$
h_i \rightarrow h_{rand[1, ..., |H|]}.
$$

5. Change a state: Let $I_1, I_2, ..., I_l$ be the set of instructions, $(N, I, P, *Trans*)$ a state. Mutate: $(N, I, P, *Trans*) \rightarrow (N, J, (P + m) \mod 255 + 1, *Trans*)$, where $m = rand[-k, k]$, $k \in B^+$ a constant parameter,

$$
J = \begin{cases} I, & \text{if } P + m > 1 \\ I_{rand[1, l]} & \text{otherwise} \end{cases}.
$$

Crossover. We used a simple crossover operator, where two parental phenomes produce one child phenome, which is a clone of the better parent phenome. A usual crossover operator has not been implemented due to insufficient RAM memory resources (1 kB only).

3.3   Fitness Function

For each behavior to evolve, a separate fitness function has to be defined. Concerning the on-board, on-line approach, some problems arise:

1. The fitness cannot be calculated exactly from the phenome, since the environmental influence cannot be calculated.
2. Global fitness cannot be calculated because there is no central instance, which collects information about the whole population.
3. There exists a delayed fitness, i.e. fitness can be first approximated after a certain time period.
4. There may be unknown anomalies of a robot's and an environment's physical properties, which have influences on the fitness function.

For the experiments, we have developed a fitness function, which measures the goodness of a collision avoidance behavior.

```
void CalcFitness_CollAvoid(void) {
    fitness += 2;
    if (LastAction != MOVE_FLAG) fitness -= 1;
    for (int i = 1; i < 7; i++) {
        if (Sensor(i) > 100) {
            fitness -= 1;
            break;
        }
    }
    if (time_in_s % 30 == 0) fitness /= 2;
}
```

MOVE_FLAG indicates, whether the last action was *move*; the function $Sensor(i)$ returns the value of the i-th sensor. The fitness value is changed each time the fitness function is executed (5 times per second), depending on the last performed instruction and the distance to the nearest obstacle. The idea is to keep in motion, but away from obstacles. Assuming such a capability can evolve evolutionary, it should be stable and therefore not disappear. Once each 30 seconds the fitness is divided by 2 so that the fitness points from earlier behaviors cannot be accumulated. Initially, the fitness is equal to zero, but within the first minute of the experiment, it approximates the current behavior.

4   Experiments

We have conducted six experiments to check whether the selective process is stable enough to keep a once developed behavior through the whole experiment. Each experiment was performed with 20 robots, which initially had a collision avoidance automaton. The arena size was $70 \times 115 \ cm^2$ of rectangular geometry, mutation was performed each 15 seconds, crossover was performed when robots meet each other, duration of each experiment was 25 min.

In Fig. 2(a)-(c) we show three different states of the fifth experiment: 0, 13 and 25 minutes from the beginning. There were several commonalities in all experiments: several robots showed a behavior which can be described as collision avoidance, but different than with the initial automaton. Most robots, however, clustered in corners and were principally dead for the experiment, see marked spots in upper right corner in Figs. 2(b)-(c). Some robots showed an extremely robust wall following behavior (which is related to collision avoidance). Although wall following has been implemented in the past several times manually, we were never able to implement it in such a robust and fluent way.

Fig. 2(d) shows the fitness averaged through all experiments. The robots can be divided into two groups: moving ones or ones caught in clusters. The clustered robots, shown by the light spot in Figs. 2(b),(c), are dissociated from the evolutionary process. The dotted curve shows the fitness for *all robots*. This curve decreases averagely through all six experiments. The solid curve shows the average fitness only over *robots,*

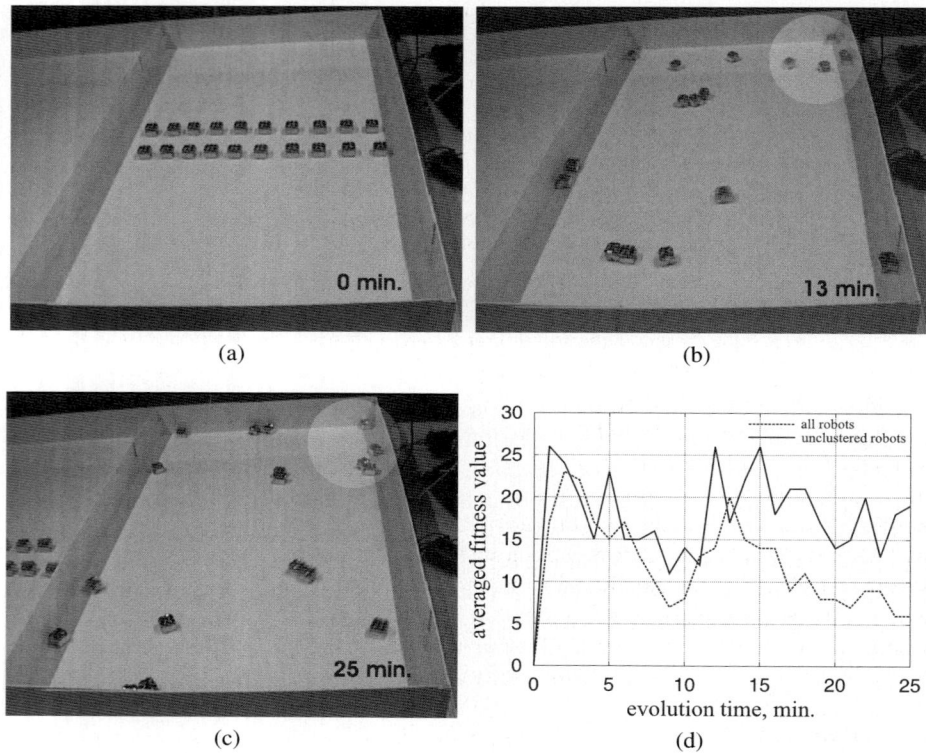

Fig. 2. Course of the experiment: **(a)** Start; **(b)** 13 minutes; **(c)** Finish; **(d)**; The light spot in (b) and (c) shows clustered robots. Changes of the averaged fitness value during 25 min. of experiments.

which were not stuck in clusters and, therefore, were still participating in the evolutionary process. The average fitness of these robots stays at a similar level until the end of the experiments.

In the end of the experiments, more than 75% of the unclustered robots had a positive fitness. More than 33% showed an observable behavior close to collision avoidance. However, other robots also had automata that were close to a sort of collision avoidance, but were unable to show this behavior because of hardware or environmental anomalies.

These experiments indicate several important properties:

1. It appears, that the selective pressure is not sufficient, since a majority of robots is lost in corners. Most likely the reason is an insufficient selection process since the communication between robots has hardware difficulties in large clusters of robots. This anomaly, effecting the search space, is also influencing the fitness function as these robots change the environment of the other robots by blocking them. Other anomalies are also thinkable: e. g. one robot might shift another robot "blindly" and make its movement instructions ineffective.

2. Though the number of experiments is statistically low and the population size is small, the appearance of several very good solutions could point to a limited search space.

5 Discussions

The experiments have shown, that an on-board and on-line evolution is partially capable of keeping a once developed behavior. Problems lie in the clustering of robots, which cannot move and communicate. These robots dramatically decrease collective fitness. The reason for clustering is diverse: robots hook each other with docking connectors, anomalies in sensor data, the delayed fitness, limited communication capability in clusters and so on. Additional experiments need to be performed to fix hardware and communication anomalies.

However, it is demonstrated that experiments with robots cannot be performed in a statistically significant manner. It is hardly possible to perform even 100 experiments in a large population of micro-robots. It seems that there is a need of finding new approaches by combining evolutionary paradigms with other non-evolutionary techniques. For example, the fitness measurement can be improved by using several plausibility checks, e. g. for a collision avoidance the robot could first check if there even exists a *move* instruction in its automaton and otherwise continue to mutate. Such a virtual measurement combined with the actual measurement could even more dramatically reduce the search space. This seems to be a realistic alternative for evolutionary approaches in robotics.

The last issue to be mentioned is the possibility to perform a robotic evolution first in simulation and then to copy the solution to the real robots. The problem is that the fitness calculation is embedded into the environment. Since we are unable to reproduce the complexity of the real environment in the simulative one, it could be expected that the evolved behavior cannot achieve the same qualities as in the real environment. However, the following experiment is thinkable: the real sensor data from robots can be transferred via wireless communication into an evolutionary simulation. It could be expected that two scenarios appear: either the evolution will develop similar solutions in real and virtual environments or even a small inaccuracy in simulation can create an essential change in the quality of the evolved behavior.

The next work will deal with a more sophisticated crossover operator, which will be applied when some spatial task is achieved successfully. In this way a more collective form of behavior can be evolved, moreover spatial statistics can be collected.

References

1. Constantinescu, C., Kornienko, S., Kornienko, O., Heinkel, U.: An agent-based approach to support the scalability of change propagation. In: Proc. of ISCA 2004, pp. 157–164 (2004)
2. Floreano, D., Husbands, P., Nolfi, S.: Evolutionary Robotics. In: Handbook of Robotics, Springer, Berlin (2008)
3. Fogel, L.J., Angeline, P.J., Fogel, D.B.: An Evolutionary Programming Approach to Self-Adaptation on Finite State Machines. In: Evolutionary Programming, pp. 355–365 (1995)

4. Kornienko, S., Kornienko, O., Levi, P.: About nature of emergent behavior in micro-systems. In: Proc. of the Int. Conf. on Informatics in Control, Automation and Robotics (ICINCO 2004), Setubal, Portugal, pp. 33–40 (2004)
5. Kornienko, S., Kornienko, O., Levi, P.: IR-based communication and perception in micro-robotic swarms. In: Proc. of the IROS 2005, Edmonton, Canada (2005)
6. Kornienko, S., Kornienko, O., Nagarathinam, A., Levi, P.: From real robot swarm to evolutionary multi-robot organism. In: Proc. of the CEC 2007, Singapore (2007)
7. Koza, J.: Genetic programming: on the programming of computers by means of natural selection. MIT Press, Cambridge, Massacgusetts, London, England (1992)
8. Lipson, H., Pollack, J.B.: Automatic design and Manufacture of Robotic Lifeforms. Nature 406, 974–978 (2000)
9. Marocco, D., Nolfi, S.: Origins of Communication in Evolving Robots. In: Şahin, E., Spears, W.M., Winfield, A.F.T. (eds.) SAB 2006 Ws 2007. LNCS, vol. 4433, pp. 789–803. Springer, Heidelberg (2007)
10. REPLICATOR. 2008-2012. REPLICATOR: Robotic Evolutionary Self-Programming and Self-Assembling Organisms, 7th Framework Programme Project No FP7-ICT-2007.2.1. European Communities.
11. SYMBRION. 2008-2012. SYMBRION: Symbiotic Evolutionary Robot Organisms, 7th Framework Programme Project No FP7-ICT-2007.8.2. European Communities.
12. Wischmann, S., Pasemann, F.: The Emergence of Communication by Evolving Dynamical Systems. In: Nolfi, S., Baldassarre, G., Calabretta, R., Hallam, J., Marocco, D., Meyer, J.-A., Parisi, D. (eds.) From animals to animats 9: Proceedings of the Ninth International Conference on Simulation of Adaptive Behaviour. LNCS (LNAI), pp. 777–788. Springer, Heidelberg (2006)

A Unified Framework for Whole-Body Humanoid Robot Control with Multiple Constraints and Contacts

Oussama Khatib, Luis Sentis, and Jae-Heung Park

Computer Science Department, Stanford University, Stanford, California 94305, USA
{ok,lsentis,park73}@cs.stanford.edu

Summary. Physical interactivity is a major challenge in humanoid robot-ics. To allow robots to operate in human environments there is a pressing need for the development of control architectures that provide the advanced capabilities and interactive skills needed to effectively interact with the environment and/or the human partner while performing useful manipulation and locomotion tasks. Such architectures must address the robot whole-body control problem in its most general form: task and whole body motion coordination with active force control at contacts, under various constraints, self collision, and dynamic obstacles. In this paper we present a framework that addresses in a unified fashion the whole-body control problem in the context of multi-point multi-link contacts, constraints, and obstacles. The effectiveness of this novel formulation is illustrated through extensive robot dynamic simulations conducted in SAI, and the experimental validation of the framework is currently underway on the ASIMO platform.

1 Introduction

Robotics is rapidly expanding into human environments and vigorously engaged in its new emerging challenges. Interacting, exploring, and working with humans, the new generation of robots will increasingly touch people and their lives. The successful introduction of robots in human environments will rely on the development of competent and practical systems that are dependable, safe, and easy to use. Physical interactivity is a key characteristic for providing these robots with the ability to perform and interact with the surrounding world. While much progress has been made in robot locomotion and free-space motion behaviors, the physical interaction ability has remained very limited. Over the past ten years, our effort in humanoid robotics was aimed at addressing the various aspects of humanoid robot control (motion, contacts, constraints, and obstacles) in an integrated coherent fashion. An important milestone in this direction was reached in 2004 [1], where full integration of free-space task and posture control was established. This result was based on the concept of task-consistent posture Jacobian and the development of models of the robot dynamic behavior in the posture space. These models were the basis for the development of the whole-body free-space motion control structure, which we have demonstrated on various platforms. In subsequent development, our effort has also addressed constraints [2] and multiple contacts [3]. This paper finally brings the integration of these components in a unified fashion. The new framework addresses the essential capabilities for task-oriented whole-body control in

H. Bruyninckx et al. (Eds.): European Robotics Symposium 2008, STAR 44, pp. 303–312, 2008.
springerlink.com

the context of integrated manipulation and locomotion under constraints, multiple contacts, and multiple moving obstacles.

In this paper. We will first summarize our previous results on the dynamic decomposition of task and posture behaviors. We will then propose a novel representation for under-actuated systems in contact with the environment, which will serve as the basis for the development of the unified whole-body control framework.

Several research groups in academic and private institutions have developed and implemented whole-body control methods for humanoid systems, with leading research by Honda Motor Corporation [4] and the National Institute of Advanced Industrial Science and Technology in Japan [5]. Their platforms, based on inverse kinematic control techniques, are designed for position actuated robots. In contrast, the unified control framework discussed here relies on torque control capabilities which are typically not available in most humanoid robots. Addressing this limitation in actuation, we have developed the Torque to Position Transformation Methodology [6] that allows to implement torque control commands on position controlled robots.

2   Whole-Body Control Framework

To create complex behaviors, humanoids need to simultaneously govern various aspects of their internal and external motion. For instance, locomotion, manipulation, posture, and contact stability are some important tasks that need to be carefully controlled to synthesize effective behaviors (see Figure 1).

As part of a methodological framework, we will develop here kinematic, dynamic, and control representations for multi-task control of humanoid systems in contact with their environment.

2.1   Task and Posture Control Decomposition

The task of a human-like robot generally involves descriptions of various parts of the multi-body mechanism, each represented by an operational point $x_{t(i)}$. The full task is represented as an $m \times 1$ vector, x_t, formed by vertically concatenating the coordinates of all operational points. The Jacobian associated with this task is denoted as J_t. The derivation of the operational space formulation begins with the joint space dynamics of the robot [7]

$$A\ddot{q} + b + g = \Gamma \tag{1}$$

where q is the the vector of n generalized coordinates of the articulated system, A is the $n \times n$ kinetic energy matrix, b is the vector of centrifugal and Coriolis generalized forces, g is the vector of gravity forces, and Γ is the vector of generalized control forces.

Task dynamic behavior is obtained by projecting (1) into the space associated with the task, which can be done with the following operation

$$\bar{J}_t^T [A\ddot{q} + b + g = \Gamma] \Longrightarrow \Lambda_t \ddot{x}_t + \mu_t + p_t = \bar{J}_t^T \Gamma. \tag{2}$$

Here, $\bar{J}_t^T$ is the dynamically-consistent generalized inverse of the J_t [7], Λ_t is the $m \times m$ kinetic energy matrix associated with the task and μ_t and p_t are the associated centrifugal/Coriolis and gravity force vectors.

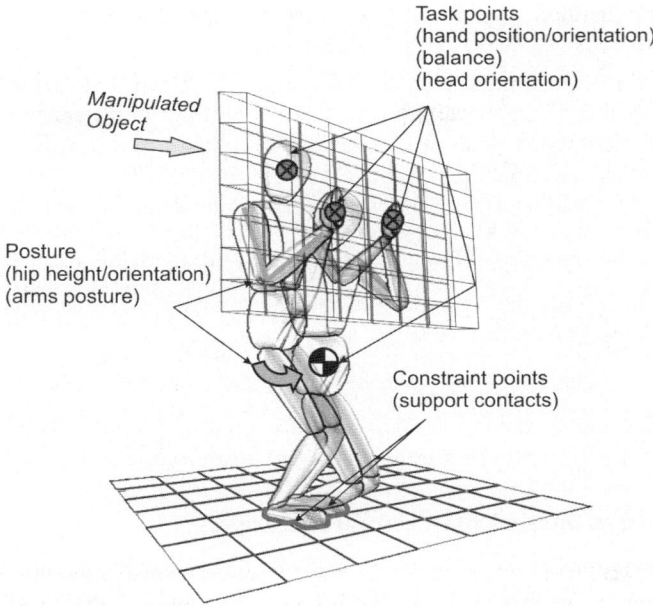

Fig. 1. Task decomposition: We depict here a decomposition into low-level tasks of a whole-body multi-contact behavior. Each low-level task needs to be instantiated and controlled individually as part of a whole-body behavior.

The above equations of motion provide the structure for the decomposition of behaviors into separate torque vectors: the torque that corresponds to desired task behavior and the torque that only affects postural behavior in the tasks's null space, i.e.

$$\Gamma = \Gamma_{\text{task}} + \Gamma_{\text{posture}}. \tag{3}$$

The following task control torque yields linear control of task forces and accelerations [8]

$$\Gamma_{\text{task}} = J_t^T F_t, \tag{4}$$

providing a flexible platform to implement motion or force control strategies for various task points in the robot's body. The second vector, Γ_{posture}, provides the means for posture control. The general form of these term is [8]

$$\Gamma_{\text{posture}} = N_t^T \Gamma_p \tag{5}$$

where Γ_p is a control vector assigned to control desired posture behavior and N_t^T is the dynamically-consistent null-space matrix associated with J_t [8].

Given a desired posture coordinate representation x_p with Jacobian J_p, in [1] we introduced a novel representation called task-consistent posture Jacobian defined by the product

$$J_{p|t} \triangleq J_p N_t. \tag{6}$$

Therefore the equation

$$\dot{x}_{p|t} = J_{p|t}\dot{q} \tag{7}$$

defines a vector of postural velocities that comply with task constraints, i.e. that do not involve motion of task coordinates [9]. Here, the subscript $p|t$ denotes that postural behavior that is consistent with task behavior. With this projections in mind, postural dynamics can be derived by multiplying (1) by the dynamically-consistent generalized inverse of $J_{p|t}$ [1], leading to the following equation of motion in postural space

$$\bar{J}_{p|t}^T[A\ddot{q} + b + g = \Gamma_{task} + \Gamma_{posture}] \Longrightarrow \Lambda_{p|t}\ddot{x}_{p|t} + \mu_{p|t} + p_{p|t} = F_{p|t} + \bar{J}_{p|t}^T\Gamma_{task}, \tag{8}$$

where $\Lambda_{p|t}$ is the inertia matrix in posture space, $\mu_{p|t}$ and $p_{p|t}$ are the associated Coriolis/centrifugal and gravity force vectors, and

$$F_{p|t} \triangleq \bar{J}_{p|t}^T\Gamma_{posture} \tag{9}$$

is a vector of desired control forces that only affect postural motion. In [9] we explored different methods to control postural behavior by means of the above equations.

2.2   Handling of Internal and External Constraints

Humanoids are aimed at executing realtime manipulation and locomotion tasks in complex environments, possibly with a high degree of autonomy. Operating in these environments entails responding to dynamic events such as moving obstacles and contact events without interrupting the global task.

Realtime response to motion constraints has been extensively addressed as a secondary process. In contrast, our approach consists on handling motion constraints as priority processes and executing operational tasks in the null space of constrained tasks [9].

To illustrate our approach, let us consider the control example shown in Figure 2, where the robot's end-effector is commanded to move towards a target point. When no constraints are active, the robot's right hand behavior is controlled using the decompositions presented in the previous section, i.e.

$$\Gamma = J_t^T F_t + J_{p|t}^T F_{p|t}. \tag{10}$$

When the obstacle enters an activation zone, we project the task and posture control structures in a constraint-consistent motion manifold, decoupling the task from the constraint (see [9]). At the same time, an artificial attraction potential is implement to keep a minimum safety distance between robot and obstacle. The simultaneous control of constraints and operational tasks is expressed as [9]

$$\Gamma = J_c^T F_c + J_{t|c}^T F_{t|c} + J_{p|t|c}^T F_{p|t|c} \tag{11}$$

where J_c and F_c are the Jacobian and control forces associated with the constrained parts of the robot (e.g. the closest point on the robot to upcoming obstacles in Figure 2), and

$$J_{t|c} \triangleq J_t N_c, \tag{12}$$

$$J_{p|t|c} \triangleq J_p N_t N_c \tag{13}$$

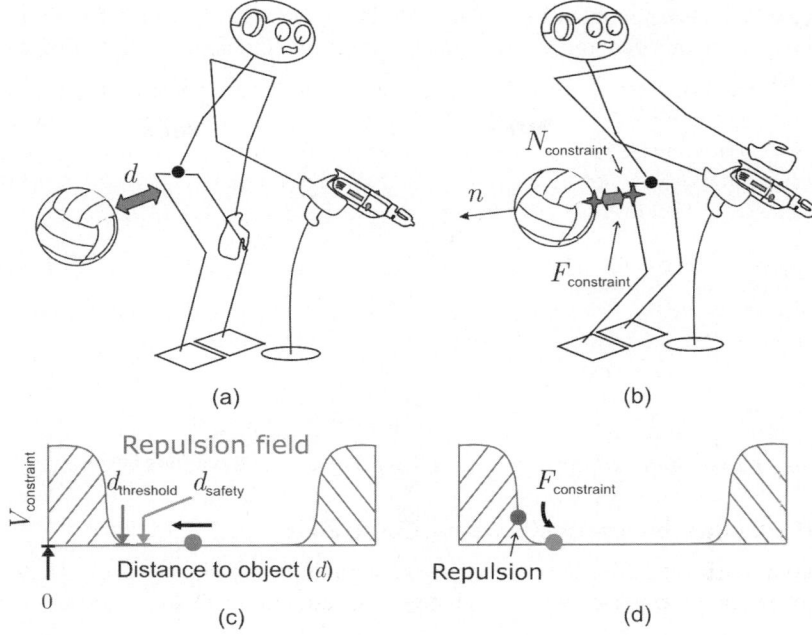

Fig. 2. Illustration of obstacle avoidance: When an incoming obstacle approaches the robot's body a repulsion field is applied to the closest point on the robot's body. As a result, a safety distance can be enforced to avoid the obstacle.

are constrained projections of task and postural Jacobians. Here the subscript $t|c$ indicates that the task point is consistent with the acting constraints and the subscript $p|t|c$ indicates that postural motion is consistent with both task motion control and physical constraints [9].

2.3   Unified Whole-Body Control Structure

To support the synthesis of whole-body behaviors, we define here unified control structures meant to serve as the main primitives of motion for interfacing with high-level execution commands.

In view of Equation (11) a global control primitive is sought to abstract the motion behavior of the entire robot. This primitive is expressed in the following form similar in structure to the original operational space formulation [8]

$$\Gamma = J_{\otimes}^{T} F_{\otimes} \tag{14}$$

where

$$J_{\otimes} \triangleq \begin{bmatrix} J_c \\ J_{t|c} \\ J_{p|t|c} \end{bmatrix}, \quad F_{\otimes} \triangleq \begin{bmatrix} F_c \\ F_{t|c} \\ F_{p|t|c} \end{bmatrix}. \tag{15}$$

Here $J_\otimes$ is a unified Jacobian that relates the behavior of all aspects of motion with respect to the robot's generalized coordinates and $F_\otimes$ is the associated unified control force vector.

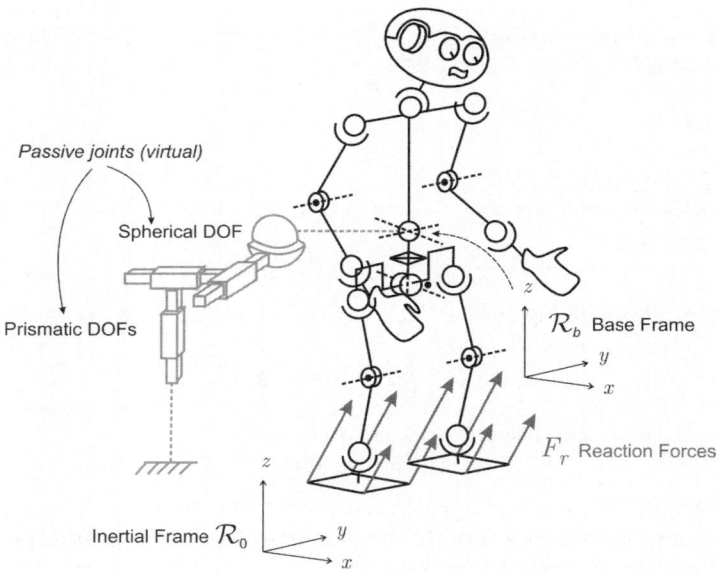

Fig. 3. Kinematic representation of a humanoid robot: The free moving base is represented as a virtual spherical joint in series with three prismatic virtual joints. Reaction forces appear at the contact points due to gravity forces pushing the body against the ground as well as due to COG accelerations.

2.4   Whole-Body Dynamics and Control Under Supporting Contacts

To create kinematic and dynamic representations under supporting constraints we represent multi-legged robots as free floating systems in contact with the ground and analyze the impact of the associated constraints and the resulting reaction forces. Reaction forces appear on the supporting surfaces due to gravity forces and center of gravity (COG) accelerations (see Figure 3). These reaction forces or contact constraints provide the means for stability, locomotion, and postural stance. Using Lagrangian formalism and expressing the system's kinetic energy in terms of the individual link kinetic and potential energies we can derive the following equation of motion describing robot dynamics under supporting contacts

$$A\ddot{q} + b + g + J_s^T F_s = \Gamma. \tag{16}$$

Here the term $J_s^T F_s$ corresponds to the projection of reaction forces acting on the feet into forces acting in passive and actuated DOFs and J_s corresponds to the Jacobian associated with all supporting links. Supporting contacts at the feet, and in general in any other place in the robot's body provide the support to realize advanced locomotion and manipulation behaviors. Therefore, they affect the robot's motion at the kinematic, dynamic, and control levels.

With the premise that stable balance is maintained and that internal forces are controlled to keep the feet flat against the ground, no relative movement occurs between contact points and the supporting ground. Therefore, relative velocities and accelerations at the contact points are equal to zero, i.e.

$$\dot{x}_s = 0, \quad \ddot{x}_s = 0. \tag{17}$$

By right-multiplying (16) by the term $J_s A^{-1}$ and considering the equality $\ddot{x}_s = J_s \ddot{q} + \dot{J}_s \dot{q}$, we solve for F_s using the the the above constraints leading to the following estimation of supporting forces [10]

$$F_s = \bar{J}_s^T \Gamma - \mu_s - p_s. \tag{18}$$

which leads to the following more elaborate expression of (16) [10]

$$A\ddot{q} + b + g - J_s^T \mu_s - J_s^T p_s = N_s^T \Gamma. \tag{19}$$

Here

$$N_s \triangleq I - \bar{J}_s J_s \tag{20}$$

is the dynamically-consistent null-space matrix associated with J_s.

The following constrained expression determines the mapping between arbitrary base and joint velocities to task velocities [9]

$$J_{\otimes|s} \triangleq J_\otimes N_s. \tag{21}$$

Here we use the subscript $\otimes|s$ to indicate that the unified Jacobian $J_\otimes$ is projected in the space consistent with all supporting constraints.

The tasks' equation of motion can be obtained by left multiplying (19) by the transpose of the dynamically consistent generalized inverse of the constrained Jacobian, $\bar{J}_{\otimes|s}$ yielding the following task space equation of motion

$$\Lambda_{\otimes|s} \dot{\vartheta}_\otimes + \mu_{\otimes|s} + p_{\otimes|s} = \bar{J}_{\otimes|s}^T N_s^T \Gamma \tag{22}$$

Because we represent humanoids as holonomic systems with n actuated articulations and 6 passive DOFs describing the position and orientation of its base, the vector of generalized torques Γ contains 6 zeros in its upper part and n actuated values in its lower part corresponding to control values. This is reflected in the following expression

$$\Gamma = U^T \Gamma_a, \tag{23}$$

where $U \triangleq [0_{n\times6} I_{n\times n}]$ is a selection matrix that projects actuated torques, Γ_a into the above underactuated equation of motion. To enforce linear control of whole-body accelerations and forces, we impose the following equality on the RHS of Equation (22)

$$\bar{J}_{\otimes|s}^T N_s^T U^T \Gamma_a = F_{\otimes|s}, \tag{24}$$

where $F_{\otimes|s}$ is an arbitrary vector of whole-body control forces. Although there are several solutions for the above equation, our choice is the following

$$\Gamma_a = \overline{[UN_s]}^T J_{\otimes|s}^T F_{\otimes|s}, \tag{25}$$

where the term $\overline{[UN_s]}$ is a dynamically-consistent generalized inverse of UN_s as described in [9]. The above solution reflects the kinematic dependency between task and joint velocities (also shown in [9]), i.e.

$$\dot{x}_{\otimes|s} = UN_s J_{\otimes|s} \dot{q}_a, \tag{26}$$

where $\dot{q}_a$ is the vector of actuated joint velocities.

3   Results

We conduct an experiment on a simulated model of a humanoid shown in Figure 4. This model measures 1.70 m in height and weights 66kg. Its body is similar in proportions to the body of an average healthy human. Its articulated body consists of 29

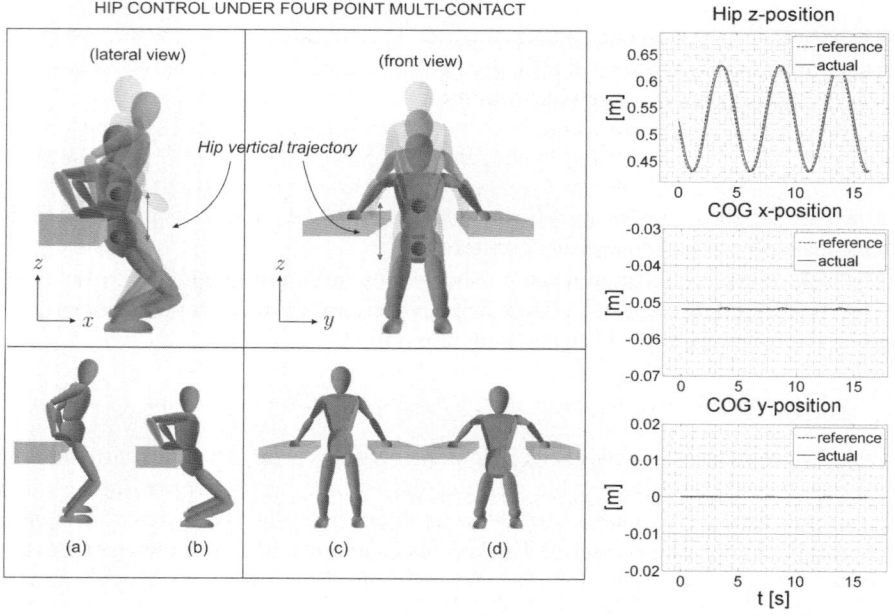

Fig. 4. Hip control under four point multi-contact: In this experiment, a virtual point located between the robot's hips is commanded to track a vertical sinusoidal trajectory while maintaining stable control of all contact points on its hands and feet. The two top sequences correspond to the same experiment seen from lateral and front view perspectives. The second row of images (a), (b), (c), and (d) correspond to snapshots of the resulting behavior. The data graph correspond to hip and center of gravity trajectories.

joints, with 6 joints for each leg, 7 for each arm (3 shoulder, 1 elbow, 3 wrist), 2 for waist movement, 1 for chest rotations, and 2 for head pitch and yaw movements. The masses and inertias of each link have been calculated to approximate those of real humans. A dynamic simulation environment and a contact solver based on propagation of forces and impacts are used to conduct the experiment [11]. The whole-body controller described in Equations (23) and (25) is implemented. For this experiment, the tasks being controlled are the center of gravity, the hip height, the hip saggital position, and a posture that resembles that of a human standing up (see [9] for details on multi-task and posture control). Initially, the robot stands up with its hands away from the side strips. When the multi-contact behavior is commanded, the hands reach towards the contact goals. When the hands make contact with the strips, a multi-contact behavior with four contact points is initiated. This behavior involves projecting task controllers in the contact-consistent null-space matrix as shown in Equation (25), while commanding the hip vertical position to track the sinusoidal trajectory shown in Figure 4. The simulated experiment reveals excellent tracking of hip trajectories and COG positions. Despite the fast commanded movements, the maximum error on both the COG and hip trajectories is around 3 mm.

4   Conclusion

In this paper, we have presented a unified whole-body control framework that integrates manipulation, locomotion, and diverse dynamic constraints such as multi-contact interactions, obstacle avoidance, and joint limits. The proposed methodology provides the basic structures to synthesize whole-body behaviors in realtime and allows human-like robots to fully interact with their dynamic environments. This framework is currently being implemented in the Honda humanoid robot Asimo, providing a platform to explore advanced manipulation and locomotion behaviors.

Beyond robotics, this framework is being applied in related fields: synthesis of human motion (biomechanics), optimization and design of spaces where humans operate (ergonomics), and synthesis of realistic interactions in computer-simulated environments (interactive worlds).

References

1. Khatib, O., Sentis, L., Park, J.H., Warren, J.: Whole body dynamic behavior and control of human-like robots. International Journal of Humanoid Robotics 1(1), 29–43 (2004)
2. Sentis, L., Khatib, O.: Synthesis of whole-body behaviors through hierarchical control of behavioral primitives. International Journal of Humanoid Robotics 2(4), 505–518 (2005)
3. Park, J., Khatib, O.: A haptic teleoperation approach based on contact force control. International Journal of Robotics Research 25(5), 575–591 (2006)
4. Hirai, K., Hirose, M., Haikawa, Y., Takenaka, T.: The development of Honda humanoid robot. In: Proceedings of the IEEE International Conference on Robotics and Automation, Leuven, Belgium, vol. 2, pp. 1321–1326 (1998)
5. Kajita, S., Kanehiro, F., Kaneko, K., Fujiwara, K., Harada, K., Yokoi, K., Hirukawa, H.: Resolved momentum control: Humanoid motion planning based on the linear and angular momentum. In: Proceedings of the IEEE/RSJ International Conference on Intelligent Robots and Systems, Las Vegas, USA, October 2003, pp. 1644–1650 (2003)

6. Khatib, O., Thaulaud, P., Park, J.: Torque-position transformer for task control of position controlled robots. Patent, Patent Number: 20060250101 (2006)
7. Khatib, O.: Advanced Robotic Manipulation. Stanford University, Stanford, USA, Class Notes (2004)
8. Khatib, O.: A unified approach for motion and force control of robot manipulators: The operational space formulation. International Journal of Robotics Research 3(1), 43–53 (1987)
9. Sentis. Synthesis and Control of Whole-Body Behaviors in Humanoid Systems. PhD thesis, Stanford University, Stanford, USA (2007)
10. Park, J.: Control Strategies For Robot. In: Contact. PhD thesis, Stanford University, Stanford, USA (2006)
11. Chang, K.C., Khatib, O.: Operational space dynamics: Efficient algorithms for modeling and control of branching mechanisms. In: Proceedings of the IEEE International Conference on Robotics and Automation (April 2000)

Visual Approaches for Handle Recognition

E. Jauregi, E. Lazkano, J.M. Martínez-Otzeta, and B. Sierra

Robotics and Autonomous Systems Group, University of Basque Country, Donostia
e.lazkano@ehu.es

Summary. Objects can be identified in images extracting local image descriptors for interesting regions. In this paper, instead of making the handle identification process rely in the keypoint detection/matching process only, we present a method that first extracts from the image a region of interest (ROI) that with high probability contains the handle. This subimage is then processed by the keypoint detection/matching algorithm. Two methods for extracting the ROI are compared, Circle Hough Transform (CHT) and blobs, and combined with three descriptor extraction methods: SIFT, SURF and USURF.

1 Introduction

Door recognition is a key problem to be solved during mobile robot navigation. Many navigation tasks can be fulfilled by point to point navigation, door identification and door crossing [13]. Indoor semi-structured environments are full of corridors that connect different offices and laboratories where doors give access to many of those locations that are defined as goals for the robot. Hence, endowing the robot with the door identification ability would undoubtedly increase its navigating capabilities.

Several references can be found that tackle the problem of door identification. Muñoz-Salinas et al. [16] present a visual door detection system that is based on the Canny edge detector and Hough transform to extract line segments from images. Then, features of those segments are used by a genetically tuned fuzzy system that analyzes the existence of a door. On the other hand, a vision-based system for detection and traversal of doors is presented in [4]. Door structures are extracted from images using a parallel line based filtering method, and an active tracking of detected door line segments is used to drive the robot through the door.

A different proposal for vision-based door traversing behavior can be found in [19]. Here, the PCA (Principal Component Analysis) pattern finding method is applied to the images obtained from the camera for door recognition. A door identification and crossing approach is also presented in [15]; there, a neural network based classification method was used for both, the recognition and crossing steps. More recently, in [11] a Bayesian Network based classifier is used to perform the door crossing task.

The proposal in [10] differs in the sense that doors are located in a map and do not need to be recognized, but rectangular handles are searched for manipulation purposes.

H. Bruyninckx et al. (Eds.): European Robotics Symposium 2008, STAR 44, pp. 313–322, 2008.
springerlink.com   © Springer-Verlag Berlin Heidelberg 2008

The handles are identified using cue integration by consensus: for each pixel the probability of being part of a handle is calculated by combining the gradient and the intensity of every pixel in order to have the degree of membership. The template model is used to obtain the consensus over a region.

But navigating in narrow corridors makes it difficult to identify doors by line extraction due to the inappropriate viewpoint restrictions imposed by the limited distance to the walls. On the other hand, door handles are at the same height of robot's camera and small in size. Moreover, door blades are rarely cluttered surfaces and handles are commonly the unique element doors have on them. Previously, we tackled the problem using CHT for handle identification [9] but there, the circle information was combined with color segmentation inside and around the circle. This approach showed to be too specific to robot's particular environment and not easily generalizable.

Objects can be identified in images extracting local image descriptors for interesting regions. These descriptors should be distinctive and invariant to image transformations. SIFT and SURF (together with its upright version USURF) [2] are well known methods to extract these kind of descriptors. In this paper, instead of making the handle identification process rely in the keypoint detection/matching process only, we present a method that first extracts from the image a region of interest (ROI) that with high probability contains the handle. This subimage is then processed by the keypoint detection/matching algorithm. Two methods for extracting the ROI are compared and combined with the three descriptor extraction methods previously mentioned.

2   SIFT

SIFT (*Scale Invariant Feature Transform*) is a method to extract features invariant to image scaling and rotation, and partially invariant to change in illumination and 3D camera viewpoint. Those properties make it suitable for being used in robotics applications, where changes in robot viewpoint distort the images taken from a conventional camera. SIFT can be used for different goals as object recognition in images [14], image retrieval [12], mobile robot localization [21] [7] and SLAM [18].

After keypoints are localized, for each keypoint a descriptor is computed by calculating an histogram of local oriented gradients around the interest point and storing the bins in a 128 dimensional vector. These descriptors can be then somehow compared with stored ones for object recognition purposes.

3   SURF/USURF

SURF [2] is another detector-descriptor algorithm developed with the aim of speeding up the keypoint localization step without losing discriminative capabilities. Instead of using a different measure for selecting the location and the scale of the keypoints, it relies on the determinant of the Hessian for both. The second order Gaussian derivatives needed to compute the Hessian are approximated by Box filters that are evaluated very fast using integral images. Such filters of any size can be applied at exactly the same speed directly on the original image. Therefore, the scale-space is analyzed by up-scaling the filter size instead of by iteratively reducing the image size as occurs in the SIFT approach.

SURF descriptors are computed in two steps. First, a reproducible orientation is found based on the information of a circular region around the interest point. This is performed using Haar-wavelet responses in the x and y directions at the scale the interest point was detected. The dominant orientation then is estimated by calculating the sum of all responses within a sliding window. Next, the region is split up into smaller square subregions and some simple features are computed (weighted Haar wavelet responses in both directions, sum of the absolute values of the responses). This yields a descriptor of length 64, half size of the original SIFT descriptor and hence, offers a less computationally expensive matching process.

The upright version of SURF, named USURF, skips the first step of the descriptor computation process, resulting in a faster version. USURF is proposed for those cases for which rotation invariance is not mandatory.

4   Extracting the Interesting Region of the Image

Instead of computing the invariant features of the whole image, the approach presented here aims to take advantage of the properties of the robot environment and reduce the size of the image to be processed by extracting the portion of the image that, with high probability, will contain most of the features. In this way, and considering that door blades in the robot environment are featureless surfaces, almost every keypoint is located at the door handle and only a few of them appear at the handle surroundings (see Figure 1). Therefore, there is no reason to process the whole image, only the handle must be selected as the region of interest (ROI) and processed afterwards.

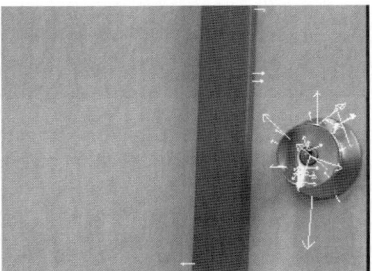

Fig. 1. SIFT Keypoints in a 320×240 image

4.1   Hough Transform for Circle Identification

Most handles in our environment are round in shape. Hence, the ROI can be located by finding circles and taking the subimage associated to the most probable circle. Although many circle extraction methods have been developed, probably the most well-known algorithm is the Circle Hough Transform (CHT) [3]. Yuen et al. [22] investigated five circle detection methods which are based on variations of the Hough transform. One of those methods is the *Two stage Hough Transform* and it is the one implemented in OpenCV vision library[1] experiments described below.

Unfortunately, the existence of a circle on the picture does not guarantee the existence of a handle. A deeper analysis of the function indicates that the circle has to have a limited radius in order to the figure be candidate of containing a door handle. In this manner only the identified circumferences with a radius that lies within the known range would be considered as handle candidates.

4.2   Extracting Blobs

The Hough transform limits the approach to circular handles. Robot's viewpoint affects the shape of the handle in the sense that the circular handle distorts and often does not look like a circle but more like an ellipsoid.

Instead of looking for circles, and again taking profit of the textureless property of the door blades, the image can be scanned for continuous connected regions or *blobs*. Blob extraction, also known as region labeling or extracting, is an image segmentation technique that categorizes the pixels in an image as belonging to one of many discrete regions. The image is scanned and every pixel is individually labeled with an identifier which signifies the region to which it belongs (see [8] for more details). Blob extraction is generally performed on the resulting binary image from a thresholding step. Instead, we apply the SUSAN (Smallest Univalue Segment Assimilating Nucleus) edge detector [20], a more stable and faster operator.

Again, some size restrictions have been imposed to the detected blobs in order to ensure that the selected one corresponds, with high probability, to the door handle. Using the blob information, a squared subimage is obtained based on the center of the blob and scaled to a fixed size. The OpenCV blobs library[2] offers the basic functionalities needed for blob extraction and analysis.

5   Off-Line Performance of the Approach

Once we get the circle, a squared subimage is obtained based on the radius and center of the circle or the blob and scaled to a fixed size. Then, the keypoint descriptor extraction procedure (SIFT/SURF/USURF) is only applied to this obtained ROI instead of applying it to the whole image.

In order to measure the performance of the developed approach, experiments were carried out with a database of about 3000 entries. All the images (references and testing DB) were taken while the robot followed the corridors using its local navigation strategies and therefore, they were taken at distances at which the robot is allowed to approach the walls. This database contained positive and negative cases. It must be mentioned that the test cases were collected in a different environment from where reference cases were taken. Therefore, the test database did not contain images of the handles in the reference database.

---

[1] http://www.intel.com/research/mrl/research/opencv
[2] http://opencvlibrary.sourceforge.net/cvBlobsLib

Figure 2 compares the results obtained with the different approaches. Plot on figure 2(a) shows the classification accuracies. But accuracy is considered a fairly crude score that does not give much information about the performance of a categorizer. Instead, F1 measure is employed as the main evaluation metric as it combines both precision and recall into a single metric and favor a balanced performance of the two metrics (see figure 2(b)).

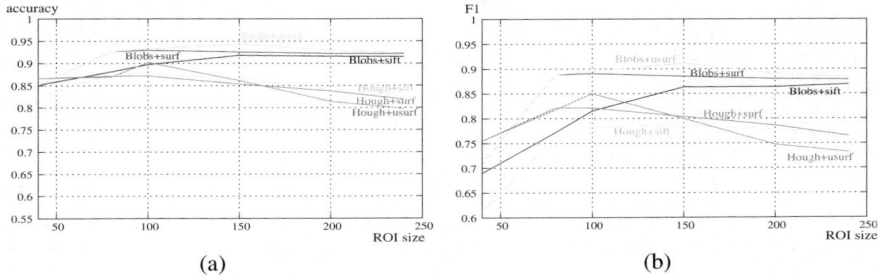

Fig. 2. Results: a) accuracy. b) F1 measure.

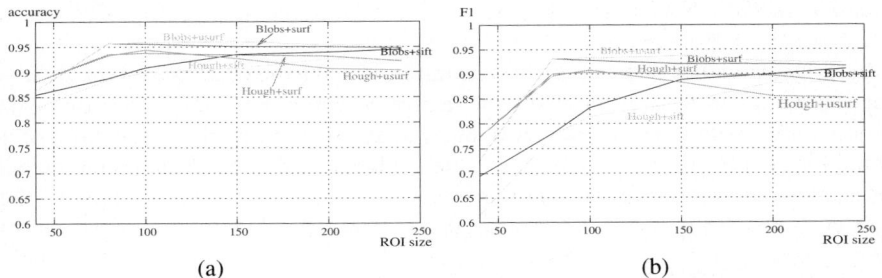

Fig. 3. Morphological restriction: a) accuracy. b) F1 measure.

As mentioned in [17], vision turns out to be much easier when the agent interacts with its environment. Taking into account the robot's morphology and the environmental niche, more specifically the height at which the camera is mounted on the robot, and the height at which the handles are located on the doors, these handles should always appear at a specific height on the image. This information, if used, could help to eliminate false positive candidates proposed by the Hough transform or the blob identification and not discarded by the keypoint extraction processes. The improvement introduced by this *morphological restriction* (MR) is also plotted in figure 3.

Table 1 shows the outstanding results obtained with each approach and the corresponding ROI size. Table 1(a) shows the results obtained just limiting the circle radius or the blob area, while Table 1(b) introduces the improvement obtained by the morphological restriction. These results clearly outperform results obtained when no ROI is extracted (see table 1(c)).

Analyzing the two ROI extraction procedures, blob information gives better subimages than the CHT does. Although the resulting subimage after the Blob based scaling process offers a smaller amount of keypoints (a 90% in average of the number of keypoints obtained after the Hough based scaling), the repeatability of the keypoints is higher according to the obtained results. On the other hand, among the keypoint extraction algorithms tested, the USURF seems to outperform both, SIFT and SURF with respect to classification accuracy and F1 measure. Although the biclassifier Blob+USURF requires smaller ROIs, when using the blobs approach for ROI extraction the behavior of all the three keypoint extracting processes shows to remain stable when increasing the ROI size. On the contrary, the performance of the pair Hough+USURF degrades highly for big ROI sizes.

Table 1. Outstanding accuracy results for each approach

| | Hough | | Blobs | | | Hough | | Blobs | | | | |
	acc.	size	acc.	size		acc.	size	acc.	size		acc	F1
Sift	87.59	100	91.82	150	Sift	92.55	150	94.48	240	Sift	62.39	59.25
Surf	87.16	100	92.94	100	Surf	93.73	100	95.70	80	Surf	72.5	69.17
Usurf	90.18	100	94.35	150	Usurf	94.35	150	96.09	150	Usurf	72.5	69.19
	a) No MR					b) MR					c) No ROI	

In many real-world domains, errors may differ in significance and may have different consequences. The system should predict in a way to minimize unwanted side effects, namely costs. Cost-sensitive classification systems aim to minimize the total cost acquired by the prediction process. Since conventional predictive accuracy metric does not include cost information, it is possible for a less accurate classification model to be more cost-effective in reality. This means that to obtain the minimal cost, cost-sensitive learning systems may need to trade off some of the predictive accuracy and are subject to make more mistakes in quantity [5]. A common measure used in cost-sensitive systems evaluation is the *Total Cost Ratio* (TCR). Following the cues in [5] ($\lambda = 7$, and allowing a maximum global error of 0.02), we obtained the TCR values in figure 4.

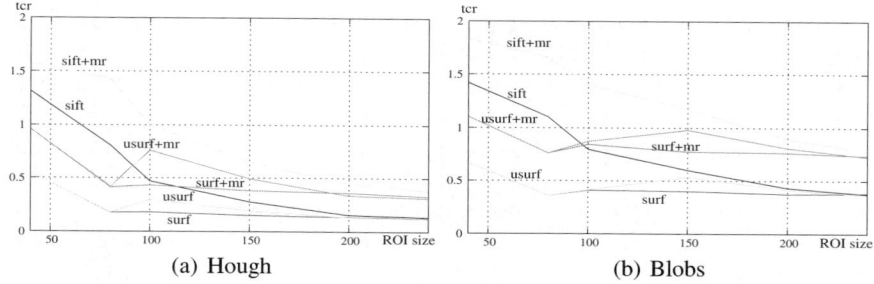

(a) Hough   (b) Blobs

Fig. 4. Total cost ratios

6   Door Knocking Behavior

Tartalo is a PeopleBot robot from MobileRobots. This robot is provided with a Canon VCC5 monocular PTZ vision system, a Sick Laser, several sonars and bumpers and some other sensors. All the computation is carried out in its on-board Pentium (1.6GHz). Player-Stage [6] is used to communicate with the different devices and the

Table 2. Computational payload (s)

Method	40	80	100	150	200	No ROI
Hough+Sift	0.093	0.209	0.243	0.488	0.777	0.307
Blobs+Sift	0.068	0.184	0.218	0.463	0.752	0.307

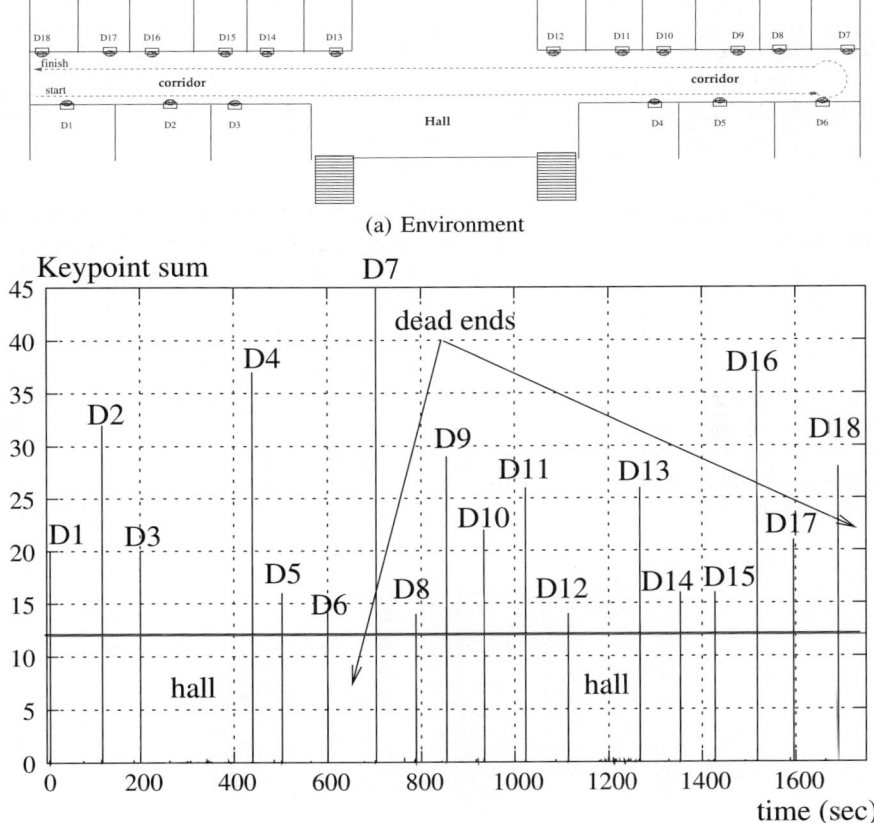

(a) Environment

(b) Keypoint sum over time

Fig. 5. Door handle identification during navigation

software to implement the control architecture is SORGIN [1], a specially designed framework that facilitates behavior definition. To evaluate the robustness of the handle identification system developed, it has been integrated in a behavior-based control architecture that allows the robot to travel across corridors without bumping into obstacles. When the robot finds a door, it stops, turns to face the door and knocks it with its bumpers a couple of times asking for the door to be opened and waiting for someone to open it. If after a certain time the door is still closed, Tartalo turns again to face the corridor and continues looking for a new handle. On the contrary, if someone opens the door the robot detects the opening with its laser and activates a door crossing behavior module that allows it to enter the room.

Experiments were performed in a different environment from where reference images were acquired (see figure 5(a)). The reference DB contained the same 40 images used for the off-line experimental phase previously described.

Although the off-line experimental step showed a degraded accuracy for the ROI of size 40 extracted using the blobs approach and SIFT with MR, the short time needed to compute the keypoints and perform the matching process (see table 2), together with the lowest percentage of false positives confirmed by the TCR, makes it more appealing for the real time problem stated in this paper.

To make the behavior more robust, instead of relying on a single image classification, the robot will base its decision upon the sum of the descriptor matches accumulated for the last five consecutive images.

Figure 5(b) shows the evolution of the sum of the matching keypoints over time. The horizontal line represents the value at which the threshold was fixed. The 18 doors present in the environment were properly identified and no false positives occurred.

7   Conclusions and Further Work

The experiments here described showed an attempt to use scale invariant image features to identify door handles during robot navigation. Taking advantage of the featureless property of the door blades, the area corresponding to the handle is extracted. ROI extraction improves handle identification procedure and depending on the ROI size, the computational time to classify an image can be considerably reduced. Blob based

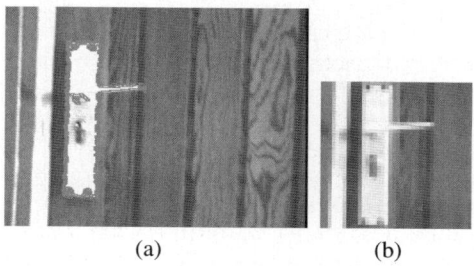

(a)   (b)

Fig. 6. Blob extraction in a non circular handle

scaling outperforms Hough, makes the approach more general and applicable to other type of handles (see figure 6) and requires less computational effort.

The developed system outperforms the performances obtained without extracting the ROIs and experiments carried out in a real robot-environment system show the adequateness of the approach. The system showed a very low tendency to give false positives while providing a robust handle identification.

Optimizing the contents of the reference DB would be desirable and could be achieve by using a genetic algorithm based search. The keypoint matching criteria also has to be analyzed more deeply. More sophisticated and efficient algorithms (SVN, AdaBoost) remain to be tested and the performance of different distance measures still needs to be studied.

Our interest now focuses on extending the approach to non circular handles in order to generalize the behavior to cross every door in the environment. It is worth mentioning that the approach could also be applied to different tasks like face recognition, taking profit of the performance of the descriptor extraction approaches at a lower computational payload.

Acknowldegments. This work has been supported by SAIOTEK (S-PE06UN16) and by the Gipuzkoako Foru Aldundia (OF 0105/2006).

References

1. Astigarraga, A., Lazkano, E., Rañó, I., Sierra, B., Zarautz, I.: SORGIN: a software framework for behavior control implementation. In: CSCS14, vol. 1, pp. 243–248 (2003)
2. Bay, H., Tuytelaars, T., Vam Gool, L.: SURF: Speeded up robust features. In: Proceedings of the 9th European Conference on Computer Vision (2006)
3. Duda, R., Hart, P.E.: Use of Hough transform to detect lines and curves in pictures. Communications of the ACM 15(1), 11–15 (1972)
4. Eberset, C., Andersson, M., Christensen, H.I.: Vision-based door-traversal for autonomous mobile robots. In: Proceedings of the IEEE/RSJ International Conference on Intelligent Robots and Systems, pp. 620–625 (2000)
5. Elkan, C.: The foundations of cost-sensitive learning. In: IJCAI, pp. 973–978 (2001)
6. Gerkey, B.P., Vaughan, R.T., Howard, A.: The Player/Stage project: tools for multi-robot and distributed sensor systems. In: Proc. of the International Conference on Advanced Robotics (ICAR), pp. 317–323 (2003)
7. Gil, A., Reinoso, O., Vicente, A., Fernández, C., Payá, L.: Monte Carlo localization using SIFT features. In: Marques, J.S., Pérez de la Blanca, N., Pina, P. (eds.) IbPRIA 2005. LNCS, vol. 3522, pp. 623–630. Springer, Heidelberg (2005)
8. Horn, B.K.P.: Robot Vision. MIT Press, Cambridge (1986)
9. Jauregi, E., Martínez-Otzeta, J.M., Sierra, B., Lazkano, E.: Handle identification: a three-stage approach. In: Intelligent Autonomous Vehicles (september 2007)
10. Kragic, D., Petersson, L., Christensen, H.I.: Visually guided manipulation tasks. Robotics and Autonomous Systems 40(2-3), 193–203 (2002)
11. Lazkano, E., Sierra, B., Astigarraga, A., Martínez-Otzeta, J.M.: On the use of bayesian networks to develop behavior for mobile robots. In: Robotics and Autonomous Systems (2006), doi:10.1016/j.robot.2006.08.003
12. Ledwich, L., Williams, S.: Reduced sift features for image retrieval and indoor localisation. In: Australian Conference on Robotics and Automation (2004)

13. Li, W., Christensen, H.I., Orebäck, A.: An architecture for indoor navigation. In: Proceedings of the IEEE International Conference on Robotics and Automation, pp. 1783–1788 (2004)
14. Lowe, D.G.: Object recognition from local scale-invariant features. In: International Conference on Computer Vision, Corfu, Greece, pp. 1150–1157 (1999)
15. Monasterio, I., Lazkano, E., Rañó, I., Sierra, B.: Learning to traverse doors using visual information. Mathematics and Computers in Simulation 60, 347–356 (2002)
16. Muñoz-Salinas, R., Aguirre, E., García-Silvente, M.: Detection of doors using a genetic visual fuzzy system for mobile robots. Technical report, University of Granada, 2005.
17. Pfeifer, R., Bongard, J.: How the body shapes the way we think. A new view of intelligence. MIT Press, Cambridge (2006)
18. Se, S., Lowe, D.G., Little, J.: Mobile robot localization and mapping with uncertainty using scale-invariant visual landmarks. International Journal of Robotics Research 21(9), 735–758 (2002)
19. Seo, M.W., Kim, Y.J., Lim, M.T.: Door Traversing for a Vision Based Mobile Robot using PCA. LNCS (LNAI), pp. 525–531. Springer, Heidelberg (2005)
20. Smith, S.M., Brady, J.M.: Susan: a new approach for low level image processing. International Journal of Computer Vision 23(1), 45–78 (1997)
21. Tamimi, H., Halawani, A., Burkhardt, H., Zell, A.: Appearance-based localization of mobile robots using local integral invariants. In: Proc. of the 9th International Conference on Intelligent Autonomous Systems (IAS-9), Tokyo, Japan, pp. 181–188 (2006)
22. Yuen, H.K., Princen, J., Illingworth, J., Kittler, J.: Comparative study of Hough transform methods for circle finding. Image and Vision Computing 8(1), 71–77 (1990)

Visual Top-Down Attention Framework for Robots in Dynamic Environments

Ulrich Kaufmann and Guenther Palm

University of Ulm, Department of Neural Information Processing, 89069 Ulm, Germany
ulrich.kaufmann@uni-ulm.de

Summary. In this paper a framework for flexible top-down visual attention for robots is introduced. On development time it is often not clear which objects should be in the focus of attention. On the other hand it is usually not enough computing time available to compute all possible region of interests (ROI) for each frame from the camera. Therefore we describe here a framework, allows the application client to steer the attention and compute only the necessary image processes at the time. Two possible application scenarios, RoboCup and a Servicerobot, are shown.

1 Introduction

Image processing is not to be excluded today by the mobile robots. Robots recognize objects for manipulation or recognize obstacles with the inserted cameras. By fast robot translations it is necessary to notice objects in the environment in time correctly, at least 30 FPS must be processed. Often robots have even several cameras, whose data will processes in short time. That is called even at least 30 or 60 frames per second are to be worked on, to find potentially interesting parts in the images. If you use a complex imaging processing, not only color blobdetection (often used in robotics), the computing time becomes very fast scarcely on a normal PC. Now if we focus on investigations for the actual occurrence, we can save much of our computing time in image processing and object recognition. For example a robot stands in front of a table and has the task to manipulate a glass on the table, it is not necessary in this task to recognize chairs or humans behind the table. On the other hand if the robot drives with full speed, it has to recognize obstacles standing on its path. In order to ensure this the image processing information needs information about the current tasks and if possible information of the current environment. So the vision system can automatically adapt itself to the task. This is the point of this paper, only aimed compute, what is needed for the moment and you reach this by a task specific visual attention control. This Framework we describe here is a part of the video image processing framework " (VIP [11]) based on Miro [12]. Miro is a complete middleware for autonomous mobile robots developed at the University of Ulm. It includes up to date technologies and follows modern approaches like object oriented programming, a modular architecture, and XML parametrization. The software modules communicate by CORBA, so it is very simple to plug parts together from different locations. A publisher/subscriber model is implemented to deal with asynchronous events.

H. Bruyninckx et al. (Eds.): European Robotics Symposium 2008, STAR 44, pp. 323–332, 2008.
springerlink.com

First the VIP is described and the technology we call visual attention control. Afterwards the two most important Methods of the attention frameworks are described, followed by sample applications to clarify that implementation.

2 Video Image Processing

The video image processing framework (VIP) based on Miro (Middleware of Robots, University of Ulm). The goal of VIP was to build a flexible working environment for the image processing for mobile robots. The framework should be flexible by using different cams and different ranges of application. In the following sections you can see the necessary points for such a framework:

Parametrization

Most image operators need to be parametrized in order to tune the quality of their results. Examples are the width of a Gaussian filter or the number of buckets for an color or orientation histogram. The calibration and optimization of parameters is an important part of the development process. Also, the configuration of the flow of different image processing steps has to be altered frequently during development, which can be significantly facilitated by a flexible and configurable development environment for robot vision systems.

Fixed Frame Rate Image Streams

In most applications new images usually arrive at a fixed frame rate form a camera. As the value of the obtained information rapidly decreases in a dynamic environment, a sensor-triggered evaluation model is required.

Timeliness Constraints

Robots are situated in a physical world. For tasks like obstacle detection and object tracking the image processing operations must be calculated many times per second and preferably at full frame rate, which is typically 30Hz or 60 Hz. The system designer needs to repeatedly assess the performance of the vision system and to ensure its efficiency. Whenever possible, image processing operations should be executed in parallel in order to fully exploit the available resources, such as dual-CPU boards and hyperthreading and multicore processor technologies. More complex image operations, which need not be applied at full frame-rate, should be executed asynchronously in order to ensure that the performance of other image evaluations is not jeopardized. Adequate processing models are required to support such designs.

Efficient Organization of Control and Data Flow

Video image processing on a mobile robot is usually sensor triggered and is started as soon as a new image is available to the robot as an image taken one second before does not necessarily resemble anymore the actual situation in a dynamic environment. At the same time, the performed processing needs to be demand driven, to not misspend the available computational resources.

Parallel and Asynchronous Evaluation

Multiple image sources allow for interleaving processing, and the true parallelism of the advanced hardware features stay unused by single-threaded applications. The actual challenge however, lies in the proper synchronization between different image processing tasks for the fusion of their results.

3   Related Work

The largely known vision-related architectures, systems, and associated relevant literature can be divided in three different categories: subroutine libraries, command languages and visual programming languages.

The mostly used ones are the subroutine libraries. They mainly deal with a efficient implementation of image operations. Classical examples are e.g. SPIDER system [9] and NAG's IPAL package [6] written in C or Fortran. More recent libraries include the LTI-Lib [3] or VXL [4], which are open-source, written in C++, and provide a wide range of image operation, ranging from image processing methods, visualization tools and I/O functions. The commercial Intel Performance Primitives [2] are an example for highly (MMX and SSE) optimized processing routines with a normal C-API. What they all have in common is there lack of support for some kind of flow control support.

The second block are the command languages. In case of the imlib3d package [1], the image processing operators can be called from the Unix command line, the CVIP-tools [10] are delivered with an extended tcl command language. Both packages have the ability to include conditional and looping facilities. But the programmer has not a flexible way of complete control over the system, but also the full liability over the processing cycle.

The most sophisticated solutions are the visual programming languages. They allow the user to connect a flow-chart of the intended processing pipeline using the mouse. They combine the expressiveness and the flexibility of both above groups. Often they contain not only a real mass of image processing functions and statistical tools, but also a complete integrated development environment. One of the most advanced one is VisiQuest (formerly known as Khoros/Cantata).

To the best of our knowledge, there is no image processing framework, that combines all of our above described features like processing on demand of complete parts of the filter tree in a flexible yet powerful way, making the system suitable for a wider range of image processing tasks, like e.g. active vision problems on autonomous mobile robots.

4   Basic of Visual Attention

In our framework we call every processing step in a processing tree filter. A filter is only one operation in the tree. It can be a filtering function (e.g. Gaussian filters), linear transformation (e.g. Fourier), color conversions or morphological transformation. A filter can also be a neural Network for object recognition. Mostly a filter gets one image and returns a image. But it is possible to have multiple inputs e. g. a Canny operation [5]. Additional a filter can receive and return meta information like histograms or region

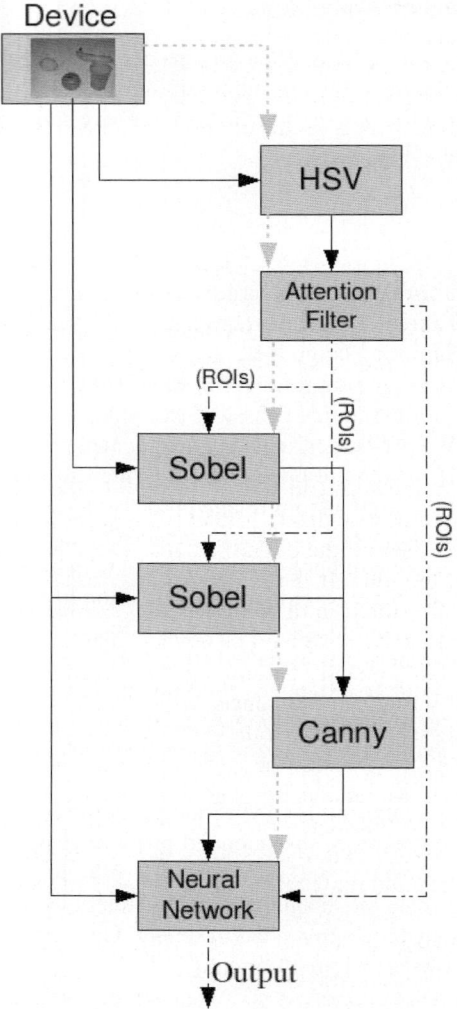

Fig. 1. A simple example for a image processing tree with an attention filter and transmission of vectors of ROIs

of interests. In our robotic image processing system we often split the image processing job in three parts (attention, feature calculation, and classification).

The first step is to decide which parts of the image are interesting. We call filter with region of interest (ROI) output attention filters. To save computing time every ROI is a rectangle. So it is very easy and fast to calculate only in this ROI in later filters.

The second step calculates only features for the classification in the image parts corresponding to the ROIs.

In the last step a neural network [7] (e.g. MLP, RBF) classifies the calculated features and returns the classification results.

In figure 1 you can see a example image processing tree for a simple object recognition task. The black arrows show the transport of images, the black dotted arrows transport metainformation, here a vector of region of interests. The gray dotted arrows specify the control flow.

5   Methods

In this section the two most important points of the frameworks are described. First we present the method to change filter parameters at run-time. Afterwards we show a method to save computing time by evaluating only necessary subtrees.

5.1   Online-Parametrization

Each filter in the VIP framework can parametrized by xml. On start up of the vision framework each filter will be initialized. The initialized parameters were selected from a explicit indicated file or set on the default values. These parameters are now used in each processing step of the filter. If the filters task is to index e.g. all green ranges, the color values for green are stored in the parameters . This initialization of the filters is a comfortable way to adapt the system on the environmental conditions. The disadvantage of this procedure are the static parameters on runtime. If you would change the parameters you have to start the system once again. A solution would be running a few filters with different parametrization parallel and use only the best result. This is however not always possible from view of the computing time. The solution is a procedure, which can change certain parameters at run-time. In order not to have a own filter for e.g. each potential color, a filter can be extended by an CORBA interface. Now you can change using these interface the predefined parameter values. The changed values are used in the next call of the filter. So the next image result based on the new values. Thus parametrization of a filter consists of a static parametrization by xml and an online parametrization by CORBA. By binding with a CORBA interface now any client program (or also a other filter) can adapt the parameters exactly for its purposes. With this structure you can create general attention filters for attention tasks with nearly the same task. By steering from the client the filter is specified for the actual task. So you have a very task specific attention and you can save calculation power for different attention filters.

5.2   On-Demand Calculation

With the online changing of parameters it is good to adapt a filter to the actual needs. However the complete task changes e.g. instead of the object recognition for clawing a object to a driving robot recognize obstacles on its driving path. A completely different image preprocessing is necessary and the recognition methods for such objects need probably completely different inputs. Hence it is necessary to use different attention methods. This can not manage a parametrization of a filter any longer. As in the figure 2 you can see the information flow divides in the image processing into several subtrees. If every subtrees are processed by every frame, it is possible to calculate unnecessary data. The VIP framework was designed to manage this problem by calculation subtrees on demand. The pictures of the cam are triggered taken up but only processed if the result of later filters in the control flow are queried. So it is possible for the application program to steer only the image processing step which is needed at the moment. So you can define different image processing steps in different subtrees. It is not necessary to have ample calculation time for all subtrees for every frame in parallel.

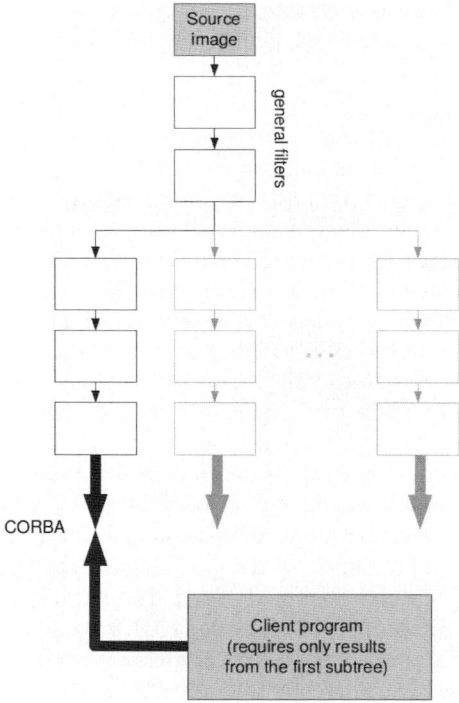

Fig. 2. The client probram requires only a result of one subtree. So the other subtrees are not calculated.

6   Example Application

In this section we demonstrate briefly three applications from our laboratory , which shows the advantages of this frameworks. On the one hand a rather static environment with a robot including a very slow computer and two other applications in the RoboCup scenario.

6.1   MirrorBot; Object Recognition

In this scenario a PeopleBot (robot from RWI) is located close to a table and has the main task to manipulate the objects on the table. The robot moves only very slowly (max translation $700\frac{mm}{s}$). Therefore the major task of the robot is the recognition of objects on the table as fast as possible. The robot gets its actual task by phonetic input. Now if the robot should look for a certain object and has to manipulate it, the results of the attention controll can be limited by online parametrizing the attention filter. If the robot receives the instruction to drive to the orange, the application program steers the image processing by CORBA, because an orange has certain characteristics with were stored before. In this scenario mainly colors and simple form characteristics are used. In figure 3(a) you can see all ROIs if the image processing system don't know any information of the actual task. In figure 3(b) only ROIs with green objects are shown. In this case the application program sends information to the attention filter to find only green objects with a certain size. So only few ROI are shown and later image processing steps have to work on a smaller part of the image.

(a)   (b)

Fig. 3. a) Result of an unspecific attention filter. Every object is framed by a ROI., b) Result of a specific attention filter for green. Only green objects are framed.

In Table 1 you can see the saving of the computing time for this example picture. The computing time for the pictures was measured directly on the robot. The robot works with a 800Mhz Pentium III processor and 126 MB RAM. For complex computations, like the neural object recognition [7] (second line in table), a Intel 2 dual Core 2,00 GHz is used. You already see the fact that the image processing is faster by the restriction of the objects. So some filters does not have to applied to the total picture. Likewise saving of time are to be seen during the object recognition, because only few objects must be classified.

With fewer ROIs you reduce computing time in nearly every step behind of the attention filter. In image processing and in the application programs time can be saved.

Table 1. Calculation time for the image processing for every frame and time for object recognition

	Calculation time 7 objects	Calculation time 2 objects
VIP	0.083 s	0.058 s
VIP + classification	0.083s + 0.013 s	0.058 s + 0.0018 s

6.2   Online Color Tracking on a RoboCup Robot

In many applications of robotics for simplification and for saving computing time the original picture is indexed . So the original colors are converted to only few pre-defined colors. In this index images you can find color blobs very easy and fast. This procedure presupposes the fact that the color calibration is done before system starts. The index color table was made by hand or semiautomatic and is initially once loaded. The problem here are changings in the lighting conditions. Because if the light changes e.g. by sunlight the index image is not very well indexed any more. To save this problem we developed a online color tracking system [8] , which works during the robots run. It adapts the indexed colors over the time. The procedure needs around 100ms to construct a whole new index table . This step is necessary only every few minutes, so the computation can be divided into several partial computations over the time. The online color tracking is included in a own filter. Owing the on damand implementation the application program can call the colortracking filter for example every 100ms . This task stores then new data and generates a new color chart without loading the system too much. With this procedure the attention filter can adapt to the lighting conditions steered by a client program. If the client detects changes in the light or bad indexed images so it can steer to update the colortable.

6.3   Processing Variability by CPU-Load

If every individual task of attention is in a own subtree of the imaging process, you can give the trees priorities and so you can create a priority-based application program. It is useful in several scenarios, for example in RoboCup. If there are many different objects to detect but not enough calculation time on the CPU free, the client can seer the calculation on priorities. In this example the robot has to recognize the ball very fast. For the self localization it needs the fieldlines and the goals. For ambitious behaviors it is good to recognize the other robots on the football field.

Thus you give a high priority (low number) to the attention control for balls and a little higher number for attention of lines and goals . Robotdetection gets a higher number than lines and goals. Now all ROIs are computed for balls and classified (if possible in each frame). Is still computing time on the processor free, additionally the filter subtree for lines and Goals is called. Likewise you proceed for the subtree of robot detection. If the computing time rises too much, first tasks with low priority are not any longer called. This leads then usually that tasks with low priorities are not computed for each frame. But free computing time is used for not so important tasks. In figure 4 you can see a example for prioritize image processing in RoboCup in our xml format. The

```
<section name="Filter_arbitrate" >
    <parameter name="Load" value= "0.85">
    <parameter name="BallDetection" value= "0">
    <parameter name="LineDetection" value= "1">
    <parameter name="GoalDetection" value= "1">
    <parameter name="RobotDetection" value= "2">
</section>
```

Fig. 4. Example for image processing priorities

priority 0 for BallDetection tells the application to demand the ball position for every frame. LineDetection and GoalDetection are calculate if there is free calculating time. The load of the last half second gives the decision how much filtertrees are calculated. Is the load below the value in the xml file (in our example 0.85) then a new subtree with a lower priority is called additionally. If the load is to high subtees with lower priorities are not called for the next frame.

7   Conclusion - Future Work

In this paper we presented a framework for flexible top-down visual attention. We showed that is not necessary to compute image operations you don't need at the time. So you reduce the CPU costs on runtime. With the CORBA interface to a filter it is very easy to change parameter on runtime, so you can keep your processing tree very simple for different attention jobs. The advantage of this framework is that the application client can control the control-flow of the image processing and can also change parameters in a filter.

In future work we want to develop a userfriendly and flexible interface to give the tasks priorities. And the system should know the mean processing time of a subtree of precalculate the CPU load. So the top of the defined load is never reached.

Acknowledgment

The work described in this paper was supported by DFG SPP 1125 within the project *Adaptivity and Learning in Teams of Cooperating Mobile Robots*. Miro, including its video image processing facilities and documentation, is available at http://smart.informatik.uni-ulm.de/MIRO/.

References

1. imlib3d. Available via, http://imlib3d.sourceforge.net/
2. Intel performance primitives (ipp). More information on, http://www.intel.com/software/products/perflib/
3. Lti-lib. Available via, http://ltilib.sourceforge.net/doc/homepage/index.shtml
4. Vxl. Available via, http://vxl.sourceforge.net/
5. Canny, J.F.: A computational approach to edge detection. In: IEEE Transactions on Pattern Analysis and Machine Intelligence, pp. 679–698 (1986)

6. Carter, M.K., Crennell, K.M., Golton, E., Maybury, R., Bartlett, A., Hammarling, S., Old-field, R.: The design and implementation of a portable image processing algorithms library in fortran and c. In: Proceedings of the 3rd IEE International Conference on Image Processing and its Applications, pp. 516–520 (1989)

7. Fay, R., Kaufmann, U., Schwenker, F., Palm, G.: Learning Object Recognition in a Neu-roBotic System. In: Gro, H.-M., Debes, K., Bhme, H.-J. (eds.) 3rd Workshop on SelfOrgani-zation of AdaptiVE Behavior SOAVE 2004, VDI, Dsseldorf, pp. 198–209 (2004)

8. Kaufmann, U., Reichle, R., Hoppe, C., Baer, P.: An Unsupervised Approach for Adaptive Color Segmentation. In: Proceedings of VISAPP 2007, Springer, Heidelberg (2007)

9. Tamura, H., Sakane, S., Tomita, F., Yokoya, N.: Design and implementation of spider-a trans-portable image processing package. Computer Vision, Graphics and Image Processing 23(3), 273–294 (1983)

10. Umbaugh, S.E.: Computer Vision and Image Processing: A Practical Approach Using CVIP-tools. Prentice-Hall, Englewood Cliffs (1998)

11. Utz, H., Kaufmann, U., Mayer, G., Kraetzschmar, G.K.: Vip – a framework-based approach to robot vision. Special Issue on Software Development and Integration in Robotics 3(1), 67–72 (2006)

12. Utz, H., Sablatng, S., Enderle, S., Kraetzschmar, G.K.: Miro – middleware for mobile robot applications. Special Issue on Object-Oriented Distributed Control Architectures 18(4), 493–497 (2002)

Visual Topological Mapping

Karel Košnar, Tomáš Krajník, and Libor Přeučil

The Gerstner Laboratory for Intelligent Decision Making and Control, Department of
Cybernetics, Faculty of Electrical Engineering, Czech Technical University in Prague
{kosnar,tkrajnik,preucil}@labe.felk.cvut.cz

Summary. We present an outdoor topological exploration system based on visual recognition.
Robot moves through a graph-like environment and creates a topological map, where edges represent paths and vertices their intersections. The algorithm can handle indistinguishable crossings
and close loops in the environment with the help of one marked place. The visual navigation system supplies path traversing and crossing detection abilities. Path traversing is purely reactive
and relies on color segmentation of an image taken by on-board camera. The crossing passage
algorithm reports azimuths of paths leading out of a crossing to the topological subsystem, which
decides what path to traverse next. Compass and odometry is then utilized to move the robot to
the beginning of picked path. The proposed system performance is tested in simulated and real
outdoor environment using a P3AT robotic platform.

Keywords: topological mapping, visual navigation, exploration.

1 Introduction

The problem of autonomous exploration and mapping of an unknown terrain remains
a fundamental problem in mobile robotics. The key question during exploration is to
determine where to move the robot in order to minimize the time needed to completely
explore the environment. For known, graph-like environments, the mapping task sets
up the problem of finding the shortest round trip through all edges of the graph. This
denotes in principle the well-known chinese postman problem.

This paper describes a technique to explore an unknown environment and buildup
a proper topological map of it [2, 3]. Topological maps [6],[5] mainly rely on graph
representation of spatial properties of the environment as vertices represent places and
edges denote paths between corresponding places. Therefore, edges represent procedural knowledge how to navigate from one place to another.

Environment is sensed by a visual recognition system similar to [1, 7] capable to
navigate paths, recognize and traverse crossings. It is assumed, that all paths are distinguishable from non-traversable terrain by color. Identified crossings correspond to
nodes, interconnecting paths refer to edges. Paths forming a crossing are distinguished
by an azimuth at which they lead out.

The paper is organized as follows: The introduction presents a brief overview of
visual based topological mapping. The next chapter presents topological mapping algorithms. The following division describes how the robot navigates paths and recognizes
crossings. After that, the experimental setups and results are described. The last section
concludes results and proposes next research directions.

H. Bruyninckx et al. (Eds.): European Robotics Symposium 2008, STAR 44, pp. 333–342, 2008.
springerlink.com

2 Topological Exploration

The exploration and mapping problem denotes the process of finding a spatially consistent map. A topological map generally represents spatial knowledge as a graph $G = (V, E)$, describing locations and objects of interest as vertices $v \in V$ and their spatial relations as edges $e \in E$. The edges can also reflect the procedural knowledge and control laws used to navigate between vertices.

Vertices - places of interest - are crossings of the roads. Crossings are places where robot can take a decision and change the way. Only one vertex is marked and therefore distinguishable from others. This vertex is called a "base". All other vertices are handled as indistinguishable. Vertices are detected by GeNav system(see sect. 3). The description of crossing detection procedures in GeNav is in the section 3.2.

Edges reflect roads and also store directions of outgoing roads signed by azimuth referred to magnetic field of the Earth. This information is used for crossing navigation. The robot uses the compass for determination of roads' azimuths. The exploration algorithm assumes that azimuths are determined with certain precision independently on the time. It is assumed that angle between any two edges leading from one vertex is not smaller than azimuth assessment precision.

The exploration and mapping algorithm presented here is based on multi-robot topological exploration algorithm described in details in [4]. Main difference is replacement of one robot with marker (base). This change allows to use only one robot.

Let us presume that a robot can determine its position status as being on a vertex. On the other hand, this robot position can not be distinguished from another, similar places at once except being on vertex marked as "base". It is assumed that if there is an edge between two vertices in a world model, the robot is able to move between these two vertices. This transition is executed by applying a single control strategy $c(e)$.

The control strategy $c(e)$ leads the robot along the edge e. As a first step, the robot uses compass data for turning to azimuth read from e. Subsequently, the robot follows the road using GeNav system path recognition (see 3.1).

It is also assumed that if the robot applies in certain vertex u control strategy $c(e), e = (u, v)$, it gets to the same vertex v at any time.

2.1 Algorithm

The algorithm consists of two phases, which are "exploration" and "vertices merging" procedures.

In the *exploration phase*, the robot moves through the environment and makes its own map $G_M = (V_M, E_M)$ of the world. As the robot cannot distinguish particular vertices from each other, it is also unable to close loop in exploration without visiting the "base" vertex (or interacting with some other robot). Moreover, every visited vertex must be handled as unvisited one until the robot proves the contrary.

The *vertices merging phase* starts whenever the robot detects the "base" vertex. This situation allows to close the loop and merge identical vertices. During the exploration phase, one place in the environment might be represented by more vertices in the map. This inconsistence is reduced in the "vertices merging" phase.

The algorithm works properly only if the robot is able to follow all detected edges. Existence of complementary edges is also necessary. Edge $\bar{e} \in E$ is complementary to edge $e \in E$ if and only if expression 1 holds.

$$\forall e \in E, e = (u,v) \quad \exists \bar{e} \in E, \bar{e} = (v,u) \tag{1}$$

Moreover, the robot knows complementary edge $\bar{e}$ after passing e. It means that the robot is able to backtrack its movement.

After the robot passes from vertex u to v, it also knows how to move from v to u even without passing this way back. In this implementation, this condition is fulfilled because if the robot knows azimuth from which it entered a vertex, it also knows the way back.

Exploration Phase

The robot moves through environment and stores vertices and edges into the map during this phase. At the beginning, the robot has no information about the environment. The robot starts to follow actual edge until a crossing is detected. This crossing is stored in the map as a first vertex. Nearest unexplored edge is used for further movement.

The exploration phase is based on graph depth-first search (DFS) algorithm. This algorithm is greedy because the robot follows the nearest vertex with unexplored edge. If there is more than one unexplored edge in actual vertex, it is chosen randomly with uniform probability. It is possible use also breadth-first search (BFS) algorithm, but the robot travels bigger distances with BFS.

When the robot arrives to a next vertex (crossing), the edge between this and previous vertex is added into the map. The complementary edge is known from entry azimuth and is also added into the map. If the robot visits the "base" vertex, vertices merging phase is executed.

Algorithm 1. exploration phase

---

follow edge $c(e)$;
if *detected crossing* **then**
    └ add new vertex v to the map V_M;

while *exists unexplored edge in the world* **do**
    if *all edges from u was explored* **then**
        find path to nearest node with unexplored edge;
        choose first edge e from path;
        use control strategy $c(e)$;
    else
        choose randomly unexplored edge e;
        store azimuth into $t(e)$; use control strategy $c(e)$ to move to v;
        use angle opposite to entry azimuth as $t(\bar{e})$;
        add vertex v into the map V_M;
        add edge $e = (u,v)$ to the map E_M;
        └ add edge $\bar{e} = (v,u)$ to the map E_M;

As the robot uses only greedy algorithm, exploration can take a long time. If the environment is tree-like with n crossings, exploration finishes in $2n$ steps. When cycles occur in the environment, the robot can get stuck in it for long time, especially if the cycle does not contain the "base". To ensure consistency of such environment map with greedy algorithm is also time consuming.

Therefore the *metrical heuristic function* is utilized. This heuristic function estimates metric position of the vertex from robot odometry. After the robot spends certain time in unexplored space, edges directing to the base are preferred. Edges are still chosen randomly but not with uniform probability. The Roulette-wheel selection is used. The parts for each edge are allocated according to its deviation from the direction to the base. Largest part of the wheel has edge with lowest deviation.

Random choice must be keeped because errors in computation of base position are affected by cumulative errors in odometry. Also roads may not be straight but can have different shapes.

Vertices Merging Phase

At first, the actual vertex recognized as base is merged with base vertex in the map. Next, the robot makes the map consistent.

Algorithm 2. vertices merging phase

$\text{merge}(v_{actual}, v_{base})$;
while $\exists u, v, w \in V_M :$ $(u,v),(u,w) \in E_M \wedge t(u,v) \approx t(u,w)$ **do**
    └ $\text{merge}(v,w)$;

By assumption, there may exist exactly one edge of each type leading from every vertex. Type of the edge is denoted $t(e)$ or $t(u,v)$. The same edge type means that the difference between azimuths is smaller than azimuth recognition precision. If two or more edges of the same type lead to different vertices, these vertices necessarily represent the same place in the world and therefore are merged. This is repeated recursively.

Algorithm 3. merge(u,v)

while $\exists x \in V_M : (v,x) \in E_M$ **do**
    │ **if** $(u,x) \notin E_M$ **then**
    │   └ add (u,x);
    └ remove (v,x);
while $\exists x \in V_M : (x,v) \in E_M$ **do**
    │ **if** $(x,u) \notin E_M$ **then**
    │   └ add (x,u);
    └ remove (x,v);
remove v;

Terminal Condition

The whole exploration procedure terminates whenever the environment is explored completely. It means that the robot has available a complete map of the environment at this time. By assumption, the map is complete if and only if no vertex of the map has unexplored edge. If local map acquired by the robot satisfies the condition of completeness, the robot stops its movement.

3 Visual Navigation

The GeNav (Gerstner Navigation) system was created for path and crossing recognition by calibrated camera aimed at a surface in front of the robot. The viewed area spans from 1 to 5 m in the direction of the robot movement and approximately 3 m to both sides. It is supposed, that color of the path is given by other method or sensor, or is known in advance. Path color can be also entered by an operator.

A hue-saturation-value (HSV) color space is utilized for path color specification, because a Cartesian product of HSV values color description offers greater invariance to the changing illumination than similar description in Red-Green-Blue color space. To prevent costly calculations of HSV description of every evaluated pixel during recognition procedures, a RGB lookup color table is first computed from the HSV color specification.

The system implements two behaviors: path traversing and crossing passage. In the path traversing mode, the algorithm attempts to keep the robot in the middle of recognized path while driving it forwards. It estimates the width of the recognized path and executes crossing recognition routines when this width changes rapidly. Once a crossing is recognized and approached, the robot switches to the crossing passage mode and sends crossing description to topological mapping module. The topological module either designates a pathway the robot should take when leaving the detected crossing or announces completion of exploration. The robot moves to designated exit path and reactivates the path traversing mode once the exit path is reached.

After start, GeNav checks, whether it can gain access to camera and the robot control board. If unsuccessful, it reads a predefined map and spawns a simulator. Then, it attempts to connect to the topological module. In case the topological module is not running, a random number generator for crossing turn decisions is initialized.

3.1 Path Traversing

In the first step of the algorithm, last row of acquired image is searched for pixels of path color and a mean value of their horizontal coordinate is computed. After that, an identification of path boundaries on this row is performed, i.e. a pixel sequence of other than path color is searched in both directions from the mean position. Path middle and width are then calculated out of the detected boundaries. If the width is greater than a predefined threshold, the algorithm proceeds to a higher row with search start position given by the current path middle coordinate. The search algorithm is completed when path width drops below this threshold.

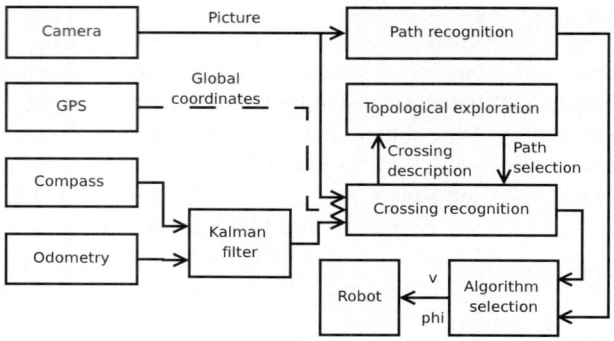

Fig. 1. Block scheme of GeNav system

Robot forward velocity v and turn speed φ vector is given by

$$\begin{pmatrix} v \\ \varphi \end{pmatrix} = \begin{pmatrix} \alpha(h-r) - \beta |\sum_{i=r}^{h}(\frac{w}{2} - m_i)| \\ \beta \sum_{i=r}^{h}(\frac{w}{2} - m_i) \end{pmatrix} \tag{2}$$

where h and w are the image height and width respectively, m_i is detected path middle of the ith row, r is the last processed row number and α, β are constants.

Because noise is usually present in the image, middle and width values are smoothed by second order linear adaptive filters.

Fig. 2. Detected path and crossing

3.2 Crossing Recognition

Unlike path recognition, employability and precision of this routine requires the camera to be calibrated. If the detected path width differs from the predicted one consecutively, crossing detection routines are activated. These search for continuous regions of path color on the periphery of the sensed image. Regions not connected by a path to the center of detected crossing are removed. Image coordinates of the remaining region centers are converted to the robot coordinate system (crossing is considered to be planar and collinear with robot undercarriage). The crossing description is then calculated out of these regions, detected crossing center and compass measurements. This description consists of a set of path bearings leading out of the crossing. Optionally, position of crossing center measured by odometry or GPS is added to the descriptor.

Finally, the image of crossing center is searched for a large blob of predefined color. If such blob is found, the crossing is designated as a "base".

The description is then delivered to the topological mapping module and the robot moves forward to the crossing center. A command with azimuth of output path is received and the robot turns to this direction. Afterwards, the robot moves forwards a short distance and path traversing routines are activated. For a short time, crossing detection routines are inhibited to prevent recognition of same crossing consecutively.

4   Experiments

4.1   Simulated World Experiments

Because real-world testing is a time costly process, system behavior has been first tested on a simulator. The robot behavior was simulated using MobileSim[1]. Synthetic camera images were automatically generated from a hand-drawn map of a part of Kinsky garden in Prague. In order to improve the realism of generated images, real-world textures were used and artificial noise was added.

Fig. 3. Generated view and textured map

The system was tested on four maps of various sizes (see figure 4). Ten test runs were performed for each map. Exploration time, number of failed exploration attempts and number of crossing passages were recorded (see table 1).

The topological exploration algorithm requires the system providing node information to be absolutely inerrant. Even that the GeNav system recognition success is approximately 98%, exploration success rate drops fast with the increasing number of passed crossings.

---

[1] http://robots.mobilerobots.com/MobileSim/

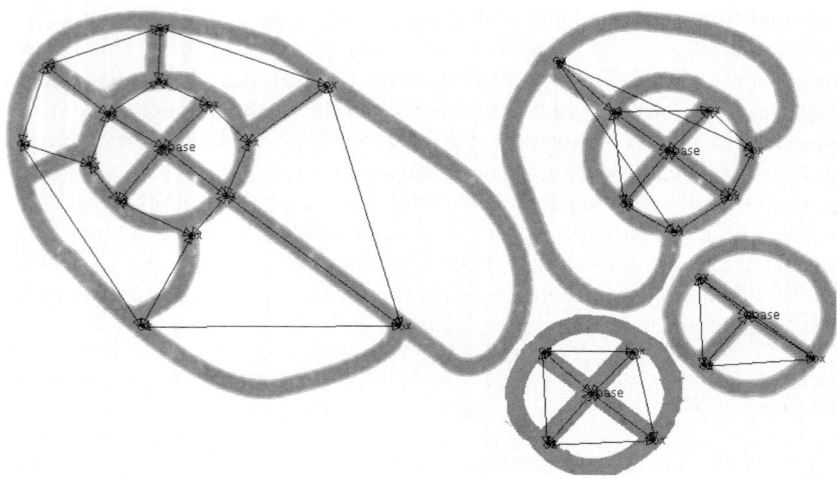

Fig. 4. Explored maps

Table 1. Simulation results

Map size (crossings)	Crossings traversed	Failures	Exploration Time (s)
Minimal (4)	11	0	365
Small (5)	14	0	418
Middle (8)	28	2	922
Large (14)	63.2	4	2283

Fig. 5. (a) Outdoor experiment map; (b) robotic platform

4.2   Real-World Experiments

Outdoor experiment was performed by Pioneer 3AT robotic platform with TCM2 compass. The robot was equipped with Fire i-400 camera providing 15 color images per second at 640x480 pixel resolution. The images were processed in real time by Intel Core 2 Duo notebook.

A rosarium at Kinsky garden[2] in Prague was chosen because of its narrow short paths and crossing abundance. In order to keep the mapped area reasonably small, the crossing descriptions send to the topological mapping system were reduced. Paths leading forward or to the right were not reported, resulting topological map was a cycle with four nodes.

5   Conclusion

The proposed system is capable to create topological maps of outdoor, graph like environments. Its main disadvantage relies in the fact, that visual recognition is purely reactive and therefore not capable to recognize larger crossings. The system has been successfully tested for small-scale environments. Exploration of larger maps may result in a failure, because topological mapping expects the vision system to be absolutely inerrant.

Our future research will be aimed towards increased robustness of the exploration system. The topological exploration algorithm will be improved to deal with occasional errors of the vision recognition. The visual recognition will be extended by building a local map of robot surrounding. Thus, navigation of crossings and paths larger than viewed area will be made possible and recognition reliability and precision will be raised. We also plan extending information exchange between both subsystems in order to use adaptive color segmentation. Methods allowing to distinguish passed crossings will be tested.

Acknowledgements

This work was supported by a research grant CTU0706113 of Czech Technical University in Prague and the Research program funded by the Ministry of Education of the Czech Republic No. MSM 684077038.

References

1. Bartel, A., Meyer, F., Sinke, C., Wiemann, T., Nchter, A., Lingemann, K., Hertzberg, J.: Real-time outdoor trail detection on a mobile robot. In: Proceedings of th 13th IASTED International Conference on Robotics, Applications and Telematics, pp. 477–482 (2007)
2. Bender, M.A., Slonim, D.K.: The power of team exploration: Two robots can learn unlabeled directed graphs. In: Proceedings of the 35th Symposium on Foundations of Computer Science (FOCS 1994), pp. 75–85 (1994)
3. Dudek, G., Jenkin, M., Milios, E., Wilkes, D.: Topological exploration with multiple robots. In: Proceedings of the 7th International Symposium on Robotics with Applications (ISORA 1998), Anchorage, Alaska (May 10-14, 1998)

---

[2] Kinsky gardens 50°4'53.579"N, 14°23'48.846"E.

4. Košnar, K., Přeučil, L., Štěpán, P.: Topological multi-robot exploration. In: Proceedings of the IEEE Systems, Man and Cybernetics Society United Kingdom & Republic of Ireland Capter 5th conference on Advances in Cybernetic System, pp. 137–141. IEEE - Systems, Man and Cybernetics Society, New York (2006)
5. Kuipers, B.J.: Modeling spatial knowledge. Cognitive Science 2, 129–153 (1978)
6. Kuipers, B.J.: The spatial semantic hierarchy. Artificial Intelligence 119, 191–233 (2000)
7. Thorpe, C., Hebert, M., Kanade, T., Shafer, S.: Vision and navigation for the carnegie-mellon navlab. IEEE Transactions on Pattern Analysis and Machine Intelligence 10(3), 362–373 (1988)

3D Mapping and Localization Using Leveled Map Accelerated ICP

Ondřej Jež

Dept. of Control and Instrumentation, Brno University of Technology, Kolejni 4, 612 00 Brno, Czech Republic
ondrej.jez@phd.feec.vutbr.cz

Summary. The ability of a robot to navigate itself in the environment is a crucial step towards its autonomy. In this article a method for simultaneous localization an mapping (SLAM) of mobile robots in six degrees of freedom (6DOF) is presented. As an input, the method is using 3D range data acquired from a continuously inclined laser rangefinder. The localization and mapping task is equal to the registration of multiple 3D images into a common frame of reference. For this purpose, an extended version of the Iterative Closest Point (ICP) algorithm is being used. In order to accelerate the time-demanding 6DOF image registration, the method is modified in the following way: first, a 3DOF registration is performed using leveled maps extracted from the 3D data, followed by a robust 6DOF registration. The proposed method compared to a single phase 6DOF registration gives promising results in structured environments.

Keywords: ICP, SLAM, Leveled Maps.

1   Introduction

The navigation is an essential step towards the autonomy of the robot. Only when the robot can localize itself in the environment and also sense its surroundings in an organized way, it can be used for different objectives. An autonomous mobile robot could fulfill different tasks which could range from utility of households, transportation of people and goods, exploration of unknown environments, automatic civil construction etc. The two aspects which make such applications of robots interesting are economical efficiency and also increased safety.

The main focus of this research is on the localization and map building in 3D using range measurements. These both task have to be performed simultaneously because the robot operates in an unknown environment: the position of the robot is determined for the currently updated map, and the map could not be updated if the position of the robot in the surrounding environment would not be determined. This method is in fact Simultaneous Localization and Mapping – and a SLAM algorithm is being used in this research. In this case we are acquiring three dimensional range data and the localization is being computed in six degrees of freedom: three translations and three rotations. Currently existing methods' main issue is the computational cost of the registration of different 3D range images into the common coordinate system. Therefore an important

H. Bruyninckx et al. (Eds.): European Robotics Symposium 2008, STAR 44, pp. 343–353, 2008.
springerlink.com

objective of this research was to accelerate the existing 6DOF SLAM methods. The SLAM method presented in this article is in fact an enhanced two stage ICP SLAM method.

At the early stages of this research it was also necessary to develop a device capable of sensing the environment. This article will therefore commence with the description of the existing 3D ranging technologies and the developed module for 3D range measurements. This will be followed by an insight into the theory of the leveled map accelerated 6DOF ICP SLAM algorithm and finally the achieved experimental results will be presented.

2   State of Range Measuring Systems for 6DOF SLAM

The localization and exploration tasks could only be solved if we acquire important information about the surrounding environment. The latest range sensors are able to provide 3D range measurements of the environment in such a precision and detail that the possibilities of using this information to navigate the robot are so far unused.

2.1   Time of Flight 2D and 3D Laser Rangefinders

The current mainstream in 3D mapping is the use of 2D laser rangefinders, which are installed in an actuating mechanism allowing the inclination of the sensor and thereafter 3D range data acquisition. The main characteristics of 2D laser rangefinders are their high precision of measurements over long distances (often 80m) and low minimum measured distance (down to 2cm).

There are more ways of inclining the ranging sensors in order to obtain the 3D data - different team's solutions will now be discussed. The most detailed analysis of the ways to use a 2D laser rangefinder for fast 3D scanning by rotating it around different axes was performed by Oliver Wulf from the University of Hannover [5]. The team from the ISE/RTS department has tested four methods of inclining the SICK laser rangefinder: pitching, rolling, and two different yawing methods. The analysis focuses mainly on the measurement of point distribution of various inclination principles. The analysis has shown that the density of measured points is always higher in areas where scanning beams are closer to the axis of inclination.

Thrun et al. and Zhao et al. are using multiple static mounted 2D laser rangefinders which are oriented in different axis and the planar data are merged into 3D model based on the current pose of the robot [4][8]. One scanner is oriented horizontally and another vertically, Zhao et al. use two additional scanners shifted by 45° from the previous two, to obtain a better model with less occlusions. Creating a 3D model requires movement of the robot and the errors which arise from the measurements of vehicle's odometry are also projected into the 3D model. The pose of the robot is also determined using a 3DOF SLAM on the data from the horizontal scanner and the odometry information; therefore the odometry error could be eliminated requiring a fast synchronization between the two scanners and slow movement of the robot.

3   3D Range Data Acquisition Platform

3.1   On Selection of the Rangefinder

In our research a planar laser rangefinder is being used for measuring the range data. The sensor SICK LMS 200 was chosen for the following paper characteristics: it is capable of data acquisition at scanning speeds up to 75Hz (plane scans per second), its angular range is 180 degrees and resolution at this speed is 1° or 0.5° in interpolated regime. The maximum distance range of the scanner is 80 meters and the systematic error at the distance of 8 meters is ±15mm. The RS-422 serial link at 500kbps was used as the communication subsystem.

3.2   3D Range Scanning System

In order to measure distances in three dimensions, the sensor has to be moved or inclined while scanning the planar ranges. In this research, the objective was to develop a robot-independent platform, which could be installed on different robot platforms available at our workplace. Therefore the inclining mechanism was selected so that a system for moving a robot when scanning is not required.

The orientation of the scanner was also an issue. The analysis performed by Wulf et al. concluded that the orientation of the scanner is very influential on where the highest density of scanned points would be [6]. In each application, this highest point density area should be directed to the area of interest, which in case of this research would be in front of the robot. This would imply the rolling inclination of the scanner (see Fig. 1 left). Though there were other criteria which were influencing the selection of the inclination method. A very important factor was the possible field of view; another aspect was the possibility of 3DOF SLAM use during the movement of the robot which requires horizontal alignment of the scanning plane with the ground. Taking all these factors into account, the "pitching" inclining method was selected: the principle of the pitching inclined mechanism is also presented in Fig. 1 (right).

For the actuating device powering the inclining mechanism, a DC motor was selected. The main operation regime is precise angular velocity regulation, meaning that the inclination is changing at a constant rate. The SICK LMS sensors have a very stable and precisely determined duration of one rotation of the mirror inside the sensor (13.32ms) which corresponds to double of the time it takes to measure 180 range values. If we obtain the information about one single pitch (inclination) angle position of the scanner at a given time and at specific rolling angle, the pitch angle ϑ_i of any scanned point can be determined given the constant inclining speed.

In order to ensure correct operation of the hardware, it was necessary to equip the DC motor with a precise incremental encoder and to ensure that the motor itself is powerful enough to overcome the disturbance of gravity and other influences (friction etc.). The resulting 3D image in case of continuous inclination is not in the regular matrix form since the measured points are positioned as if the scanned lines were tilted. Therefore, data processing is more complicated, requiring computation of the additional parameter: the inclination of each scanned point.

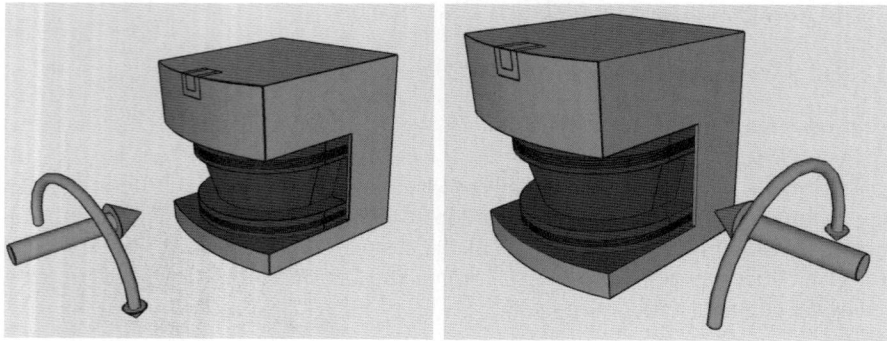

Fig. 1. Two considered principles of inclining the sensor: the "rolling" inclination around the median scanning angle (left) and the selected "pitching" inclination around the minimum and maximum scanning angle (right)

3.3 Available Robotic Platforms

The 3D range scanning system was accustomed mainly for one experimental robot: Universal Telepresence and Autonomous Robot (UTAR). This platform was developed especially for experimental objectives and is based on a skid-steered wheeled platform [7]. The robot can operate in both interior and exterior environment, allowing movement in light terrain (the maximum height of an obstacle is 10cm). Its maximum velocity is 2km/h; the drivers are digitally controlled, allowing readings from incremental encoders connected to both motors. It is equipped with an industrial PC and the communication is performed by a WiFi MIMO module. Another platform which is being used for the 3D scanning is the HERMES omni directional platform based robot. The mechanism of this platform is using a servo motor for inclining the sensor. This platform is also suitable for experiments since it allows movement of the robot in three degrees of freedom. Both robots are shown in Fig. 2.

4 On Leveled Map Accelerated SLAM Method

A block diagram of the overall navigation method is shown in Fig 3. In this diagram one can differentiate three main parts of the navigation system. The first part is naturally the sensor subsystem, using the above described scanner and inclination module. Then in the second part (not highlighted in the diagram), the 3D data are preprocessed and reduced in order to be matched and to expand the model of the environment in the final part – the SLAM core. The output from the SLAM core is the localization of the robot in six degrees of freedom (roll, pitch, yaw angles and x, y, z translations) though the created model of the environment could also be perceived as an output.

The measurement of the data was already discussed in the previous chapters. The preprocessing is mainly serving the purpose of computing the inclination angles and spherical and Cartesian coordinates of each scanned point. Also, the aligning of the

Fig. 2. Robots equipped with the 3D ranging devices: UTAR with the continuous inclining platform *(left)* and HERMES with the servo controlled inclining platform *(right)*

measured data with the robot's construction is done since there are problems in syn-
chronizing the range data readings with the absolute inclination values.

Then data reduction takes place, removing the robot and the inclining mechanism
from the 3D image since these would introduce additional error in the registration (the
robot and this mechanism is always in the same position in the data).

In this method the model of the system composes of two sub-models: the leveled map
model and the 3D environment model. This corresponds to the fact that the SLAM core
composes of two phases: in the first phase, leveled maps are extracted from the 3D data
and these are registered in a 3DOF ICP based matching algorithm. Then the 3D data
are transformed according to these results in 3DOF and passed to the robust 6DOF ICP
based matching algorithm. Since the ICP is used in both phases of the SLAM algorithm
it will be discussed in the next subchapter, followed by a detailed description of both
phases of the navigation algorithm.

4.1 Iterative Closest Point Algorithm

The Iterative Closest Point (ICP) algorithm is an iterative aligning algorithm first proposed in 1992 by P. Besl and N.D. McKay [1]. A detailed application of this algorithm
in 6DOF SLAM can be found in publications by Fraunhofer AIS institute [2]. The ICP
algorithm implemented in this project is very similar to the ICP presented by the Fraunhofer AIS. The Iterative Closest Point algorithm registers a data image – a set of points
D - into the model image – the set of points M. This step is based on finding the optimal
rotation R and translation t corresponding to the displacement and rotation of the robot
for which the matching of the new scan into the global frame of reference with previous
scans is most consistent. In ICP, this is expressed as minimizing of an energy function
as stated in equation (1).

$$E = f(R,t,M,D) = \sum_{i=1}^{m} \sum_{j=1}^{n} w_{i,j} \left\| m_i - (Rd_j + t) \right\|^2 \qquad (1)$$

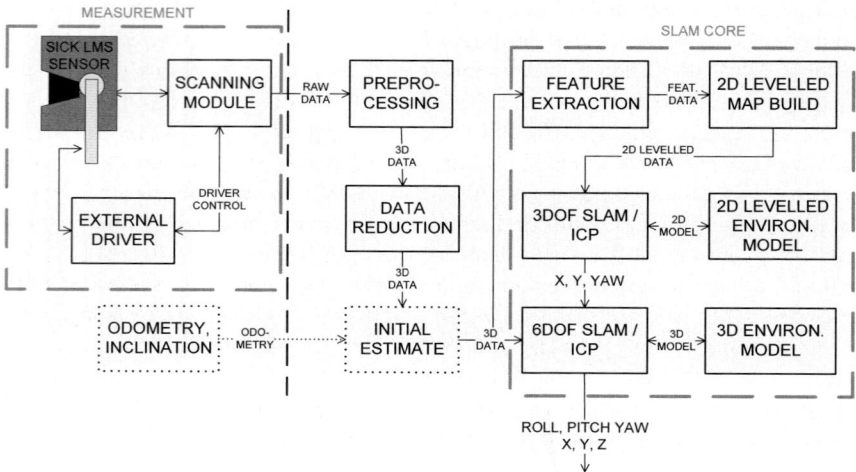

Fig. 3. The block diagram of the navigation method: 6DOF navigation system using 3D range data

$$M = \{m_1, ..., m_i\}, D = \{d_1, ..., d_j\} \tag{2}$$

$$R = VU^T, t = c_m - Rc_d \tag{3}$$

$$H = U\Lambda V^T = \sum_{i=1}^{N} (d_i - c_d) \cdot (m_i - c_m) \tag{4}$$

In the energy function, weight $w_{i,j}$ is 1 if the point d_i in D describes the same point as m_i in M (the points are corresponding), otherwise it is 0. The optimum rotation R is found as in (3), the two matrices U and V are obtained using singular value decomposition of matrix H as in (4). The matrix H is calculated by adding the multiplied vectors of corresponding (closest) data and model points d_i and m_i which are centered using data and model centroids c_d, c_m. Final translation t is derived from the centroids and the rotation as in (3). In each iterative step, the algorithm selects closest points as corresponding and calculates the transformation (R,t) according to the minimization of $E(R,t)$.

There are also other methods for minimizing the E function, easy to implement is a quaternion based method. Both mentioned methods are applied and evaluated by Fraunhofer AIS, in later work SVD is preferred due to its robustness and easy implementation [2]. The most computation demanding task of the ICP method is the determination of the closest points. There are methods to speed up the search of corresponding points, commonly used in 3D graphics programming. Majority of them is based on the structuring of point clouds using a tree structure: octrees, box decomposition trees and kd-trees [2][3]. The kd-trees are a generalization of binary trees where every nod represents a partition of a point set to the two successor nodes.

In the ICP algorithm applied in this research, kd-trees are built for the model using a sliding midpoint splitting rule proposed by D. Mount [9]. This is followed by the building of the H matrix using closest point queries to find matching model points for all data points. This was done in two variants: the first variant does not keep track whether the model point was already used and therefore one model point is possibly used for more data closest point matches. The validity of the match was only limited by a largest distance limit of the two points (typically set to 40cm). The second variant keeps track of the already used model points and restricts the reuse of them unless the distance of the current queried point from the model is lower that for the previous match.

After all data points are queried, the updated H matrix is decomposed using singular value decomposition. The rotation is obtained from the resulting matrixes and it is tested whether it is close to ones matrix. If so, the optimization does not alter the solution any more and the ICP is finished. In case the reuse of model points was restricted, the centroids are recalculated only using the matched points. The translation t is determined from the rotation and the centroids, as in (3).

4.2 First Phase: 3DOF Match Using Extracted Leveled Map

The main objective of the research was to implement and accelerate the 6DOF ICP SLAM method. The main focus was not given to the ICP algorithm itself and the search of the closest points, although the algorithm was modified so that it works with realistically overlapping data sets. The inevitable problem in 6DOF ICP SLAM is the number of points for which the closest point search has to be performed in many iterations. This is due to the size of the 3D data sets. Therefore the focus was on how to perform the registration initially in fewer dimensions and thus fewer degrees of freedom, decoupling the registration in certain degrees of freedom from the overall robust but slow 6DOF SLAM. Ideally, this 3DOF registration would be invariant to the remaining 3DOF pose of the robot.

In this solution, one aspect of the physical environment is being used to decouple the 3DOF registration from the 6DOF SLAM: it is the gravity. It is quite easy to measure the robot's pose in two degrees of freedom: the pitch and yaw inclination. Since we can measure these angles, the 3D data can be pre-aligned so that the 2D data extraction for the 3DOF SLAM can be independent on the pitch and roll inclination of the robot. This 2D extracted data set is called leveled map since it is aligned with the horizontal plain. In other words the leveled map created from the 3D data is invariant on the pitch and yaw inclination of the robot since we can measure these angles and align the 3D data in these two degrees of freedom before the extraction. Then we have to ensure that the 2D leveled map is also invariant on the remaining degree of freedom – the z ("upward") translation. This implies the type of object we will extract into the leveled map: vertical objects. We assume that the environment has such characteristics that in most times the robot will be able to see vertical objects in all heights (that is z translations), therefore the extracted leveled map will be independent on the robot's pose's z translation. This is actually true for most single-floor indoor environments and also for some outdoor environments. In the first phase of registration, the three degrees of freedom which will be registered are the horizontal translations x and y and the yaw angle ψ.

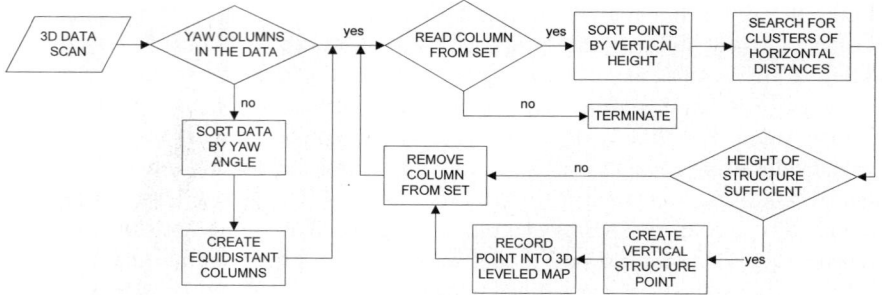

Fig. 4. The block diagram of the leveled map extraction algorithm

The leveled map extraction is done in the following algorithmic fashion:

1. First it is necessary to create vertical "columns" in the data. This could be either done by adapting the scanning platform (aligning the scanning planes with the vertical axis – not in case of this research) or by sorting the data by the yaw angle and extracting equidistant columns (in this implementation using Combsort algorithm).
2. Sorting these columns by the vertical height of points.
3. Search for clusters of points with similar horizontal distance (within a specified tolerance) in the columns, ensuring that these clusters contain points whose horizontal distance deviation is smaller than a specified criteria and that the cluster is higher than a specified minimum height and that it is not interrupted by more than a specified maximum height gap.
4. Each such cluster satisfying both conditions is considered to be a vertical structure and is recorded into the vertical map (computing the coordinates as a mean of the cluster's points' coordinates).

The algorithm of the vertical structure extraction is shown in Fig. 4. When the leveled 2D map is constructed, it is passed to the 3DOF ICP algorithm. A kd-tree is built for the leveled model set and then closest points to the leveled data points are queried, updating the matrix H and finally obtaining the rotation R using the SVD algorithm. This matching can use both variants, so we can choose if model points are repetitively matched to more data points. The result from this ICP registration is the rotation and translation in 3DOF (yaw angle – rotation around z axis, x and y translation).

4.3   Second Phase: 6DOF Match

The obtained estimate of the yaw rotation is applied to the 3D data set and this is then used in the 6DOF ICP match. Since the ICP is working with centered data sets, the application of the translation is not necessary at this point. The ICP algorithm as described earlier is applied on the 3D data, iteratively calculating the optimal rotation R, finally resulting in the full 6DOF match of the data. Then the data can be used to expand the overall model of the environment.

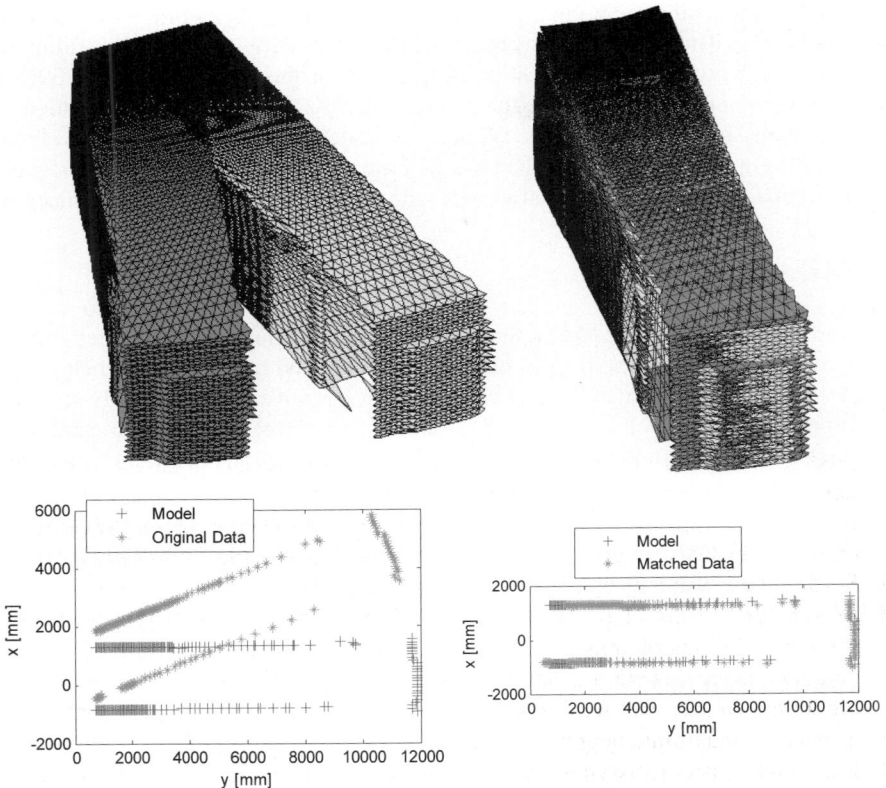

Fig. 5. Scans of a hallway: unmatched *(left)* and the matched solution *(right)*, 3D images *(upper)* and leveled maps *(bottom)*

5   Evaluation of the Method

In order to evaluate the potential of the accelerated 6DOF SLAM method, three different types of datasets were used: in the first dataset, the scans are taken in a single floor orthogonally structured hallway. The second dataset was a laboratory room, though this time it contained many scattered objects of different shapes. The last set contained scans of multi-storey indoor environment with scattered obstacles of different shapes and size. The performance of the method was compared to a non-accelerated version of the 6DOF SLAM, comparing both ICP versions with enforced or not enforced single matching of model points. In Fig. 5, the original and matched solutions are shown in both 3D visualization and Leveled maps.

Since the unique use of model points in the closest point queries proved to output more accurate results, the results for this setting will be presented (although this setting increases computing time). The computing effort to extract and match the leveled maps was very small compared to the time required to match the 3D scans (it was always under 3% of the total computing time – from 100 to 500ms depending on the data size and structuring).The overall improvement of the registration time was the following: 55% savings (in total time / closest point queries) for the hallway data set, 39% for the room with objects and 39% for the complex multi-storey scan. For illustration, the hallway scans shown in Fig. 5 took 12260ms to match when leveled maps accelerated the match compared to 27438ms when in single-phase ICP (the data and the model each consisted of approx. 30000 points after data reduction on Pentium M 1.8MHz machine).

6 Discussion

The implemented 6DOF SLAM method proved to be matching the 3D images correctly and the improvement of the overall 6DOF SLAM method performance when accelerated by leveled maps' match is most significant in case of indoor vertically structured environments. In all tested conditions, the robot did not travel in vertical direction therefore such sensitivity could not be evaluated. The rate with which the first phase – using leveled maps - accelerates the registration is higher for typical building environments while the second phase 6DOF registration algorithm itself is robust and suitable even for much more complicated environments; therefore robustness of the algorithm is preserved. Further evaluation and testing in different environments is subject to future research.

Acknowledgement

I would like to thank to Assoc. Prof. Luděk Žalud Ph.D. and Prof. František Šolc CSc. for their support during this research. This project was also supported by Intelligent Systems in Automation Grant MSM0021630529 and also by the Ministry of Education of the Czech Republic under Project 1M0567.

References

1. Besl, P.J., McKay, N.D.: A method for registration of 3-d shapes. Proc. of IEEE Transactions on Pattern Analalysis and Machine Intelligence 14(2), 239–256 (1992)
2. Nuchter, A., Lingemann, K., Hertzberg, J., Surmann, H.: 6D SLAM with Approximate Data Association. In: Proceedings of the 12th International Conference on Advanced Robotics (ICAR 2005), pp. 242–249 (2005), ISBN 0-7803-9178-0
3. Pulli, K.: Multiview Registration for Large Data Sets. In: Proc. of the 2nd International Conference on 3D Digital Imaging and Modeling (3DIM 1999), Ottawa, pp. 160–168 (1999)
4. Thrun, S., Hahnel, D., Burgard, W.: Learning Compact 3D Models of Indoor and Outdoor Environments with a Mobile Robot. In: Proceedings of the fourth European workshop on advanced mobile robots (EUROBOT 2001), Lund, Sweden (2001)
5. Wulf, O., Brenneke, C., Wagner, B.: Colored 2D Maps for Robot Navigation with 3D Sensor Data. In: Proceedings of the IEEE/RSJ International Conference on Intelligent Robots and Systems (IROS), Sendai, Japan (2004)
6. Wulf, O., Wagner, B.: Fast 3D-Scanning Methods for Laser Measurement Systems. In: Proceedings of the International Conference on Control Systems and Computer Science (CSCS14), Bucharest, Romania (2003)

7. Zalud, L., Kopecny, L., Neuzil, T.: Laser Proximity Scanner Correlation Based Method for Cooperative Localization and Map Building. In: Proc. of the 7th International Workshop on Advanced Motion Control, Maribor, Slovenia, pp. 480–486 (2002)
8. Zhao, H., Shibasaki, R.: Reconstructing Textured CAD Model of Urban Environment Using Vehicle-Borne Laser Range Scanners and Line Cameras. In: Schiele, B., Sagerer, G. (eds.) ICVS 2001. LNCS, vol. 2095, pp. 284–295. Springer, Heidelberg (2001)
9. Mount, D.M.: Aproximate Nearest Neighbour Library (22.10.2009) (2009), http://www.cs.umd.edu/~mount/ANN/

Author Index

Springer Tracts in Advanced Robotics

Edited by B. Siciliano, O. Khatib and F. Groen

Vol. 22: Christensen, H.I. (Ed.)
European Robotics Symposium 2006
209 p. 2006 [978-3-540-32688-5]

Vol. 21: Ang Jr., H.; Khatib, O. (Eds.)
Experimental Robotics IX – The 9th International
Symposium on Experimental Robotics
618 p. 2006 [978-3-540-28816-9]

Vol. 20: Xu, Y.; Ou, Y.
Control of Single Wheel Robots
188 p. 2005 [978-3-540-28184-9]

Vol. 19: Lefebvre, T.; Bruyninckx, H.;
De Schutter, J. Nonlinear Kalman Filtering
for Force-Controlled Robot Tasks
280 p. 2005 [978-3-540-28023-1]

Vol. 18: Barbagli, F.; Prattichizzo, D.;
Salisbury, K. (Eds.)
Multi-point Interaction with Real
and Virtual Objects
281 p. 2005 [978-3-540-26036-3]

Vol. 17: Erdmann, M.; Hsu, D.; Overmars, M.;
van der Stappen, F.A (Eds.)
Algorithmic Foundations of Robotics VI
472 p. 2005 [978-3-540-25728-8]

Vol. 16: Cuesta, F.; Ollero, A.
Intelligent Mobile Robot Navigation
224 p. 2005 [978-3-540-23956-7]

Vol. 15: Dario, P.; Chatila R. (Eds.)
Robotics Research – The Eleventh
International Symposium
595 p. 2005 [978-3-540-23214-8]

Vol. 14: Prassler, E.; Lawitzky, G.; Stopp, A.;
Grunwald, G.; Hägele, M.; Dillmann, R.;
Iossifidis. I. (Eds.)
Advances in Human-Robot Interaction
414 p. 2005 [978-3-540-23211-7]

Vol. 13: Chung, W.
Nonholonomic Manipulators
115 p. 2004 [978-3-540-22108-1]

Vol. 12: Iagnemma K.; Dubowsky, S.
Mobile Robots in Rough Terrain –
Estimation, Motion Planning, and Control
with Application to Planetary Rovers
123 p. 2004 [978-3-540-21968-2]

Vol. 11: Kim, J.-H.; Kim, D.-H.; Kim, Y.-J.;
Seow, K.-T.
Soccer Robotics
353 p. 2004 [978-3-540-21859-3]

Vol. 10: Siciliano, B.; De Luca, A.; Melchiorri, C.;
Casalino, G. (Eds.)
Advances in Control of Articulated and
Mobile Robots
259 p. 2004 [978-3-540-20783-2]

Vol. 9: Yamane, K.
Simulating and Generating Motions
of Human Figures
176 p. 2004 [978-3-540-20317-9]

Vol. 8: Baeten, J.; De Schutter, J.
Integrated Visual Servoing and Force
Control – The Task Frame Approach
198 p. 2004 [978-3-540-40475-0]

Vol. 7: Boissonnat, J.-D.; Burdick, J.;
Goldberg, K.; Hutchinson, S. (Eds.)
Algorithmic Foundations of Robotics V
577 p. 2004 [978-3-540-40476-7]

Vol. 6: Jarvis, R.A.; Zelinsky, A. (Eds.)
Robotics Research – The Tenth
International Symposium
580 p. 2003 [978-3-540-00550-6]

Vol. 5: Siciliano, B.; Dario, P. (Eds.)
Experimental Robotics VIII – Proceedings
of the 8th International Symposium ISER02
685 p. 2003 [978-3-540-00305-2]

Vol. 4: Bicchi, A.; Christensen, H.I.;
Prattichizzo, D. (Eds.)
Control Problems in Robotics
296 p. 2003 [978-3-540-00251-2]

Vol. 3: Natale, C.
Interaction Control of Robot Manipulators –
Six-degrees-of-freedom Tasks
120 p. 2003 [978-3-540-00159-1]

Vol. 2: Antonelli, G.
Underwater Robots – Motion and Force Control
of Vehicle-Manipulator Systems
268 p. 2006 [978-3-540-31752-4]

Vol. 1: Caccavale, F.; Villani, L. (Eds.)
Fault Diagnosis and Fault Tolerance for
Mechatronic Systems – Recent Advances
191 p. 2003 [978-3-540-44159-5]

Printing: Krips bv, Meppel, The Netherlands
Binding: Stürtz, Würzburg, Germany